ATOMIZATION AND SPRAYS

Combustion: An International Series
Norman Chigier, *Editor*

Lefebvre, Atomization and Sprays
Lefebvre, Gas Turbine Combustion

Forthcoming titles
Chigier, Stevenson, and Hirleman, Flow Velocity and Particle Size Measurement

ATOMIZATION AND SPRAYS

Arthur H. Lefebvre
Purdue University
West Lafayette, Indiana

⬤HEMISPHERE PUBLISHING CORPORATION

A member of the Taylor & Francis Group

New York   Washington   Philadelphia   London

ATOMIZATION AND SPRAYS

1 2 3 4 5 6 7 8 9 0   B C B C   8 9 8 7 6 5 4 3 2 1 0 9 8

This book was set in Times Roman by Edwards Brothers, Inc. The editors were Sandra Tamburrino and Mary Prescott.
Cover design by Sharon Martin DePass.
BookCrafters, Inc. was printer and binder.

Library of Congress Cataloging-in-Publication Data

Lefebvre, Arthur Henry, date
  Atomization and sprays.

  (Combustion)
  Bibliography: p.
  Includes index.
  1. Spraying. 2. Atomization. I. Title. II. Series:
Combustion (Hemisphere Publishing Corporation)
TP156.S6L44 1989   660.2'961   88-30101
ISBN 0-89116-603-3
ISSN 1040-2756

CONTENTS

PREFACE

The transformation of bulk liquid into sprays and other physical dispersions of small particles in a gaseous atmosphere is of importance in several industrial processes. These include: combustion—spray combustion in furnaces, gas turbines, diesel engines, and rockets; process industries—spray drying, evaporative cooling, powdered metallurgy, and spray painting; agriculture—crop spraying; and many other applications in medicine and meteorology. Numerous spray devices have been developed, and they are generally designated as atomizers or nozzles.

As is evident from the above applications, the subject of atomization is wide ranging and important. During the past decade there has been a tremendous expansion of interest in the science and technology of atomization, which has now developed into a major international and interdisciplinary field of research. This growth of interest has been accompanied by large strides in the areas of laser diagnostics for spray analysis and in a proliferation of mathematical models for spray combustion processes. It is becoming increasingly important for engineers to acquire a better understanding of the basic atomization process and to be fully conversant with the capabilities and limitations of all the relevant atomization devices. In particular, it is important to know which type of atomizer is best suited for any given application and how the performance of any given atomizer is affected by variations in liquid properties and operating conditions.

This book owes its inception to a highly successful short course on atomization and sprays held at Carnegie Mellon University in April 1986 under the direction of Professor Norman Chigier. As an invited lecturer to this course, my task was by no means easy because most of the relevant information on atomization is dispersed throughout a wide variety of journal articles and conference proceedings. A fairly thorough survey of this literature culminated in the preparation of extensive course notes. The enthusiastic response accorded to this course encouraged me to expand these notes into this book, which will serve many purposes, including those of text, design manual, and research reference in the areas of atomization and sprays.

The book begins with a general review of atomizer types and their applications (Chapter 1). This chapter also includes a glossary of terms in widespread use throughout the atomization literature. Chapter 2 provides a detailed introduction to the various mechanisms of liquid particle breakup and to the manner in which a liquid jet or sheet emerging from an atomizer is broken down into drops.

Owing to the heterogeneous nature of the atomization process, most practical atomizers generate drops in the size range from a few micrometers up to around 500 μm. Thus, in addition to mean drop size, which may be satisfactory for many engineering purposes, another parameter of importance in the definition of a spray is the distribution of drop sizes it contains. The various mathematical and empirical relationships that are used to characterize the distribution of drop sizes in a spray are described in Chapter 3.

In Chapter 4 the performance requirements and basic design features of the main types of atomizers in industrial and laboratory use are described. Primary emphasis is placed on the atomizers employed in industrial cleaning, spray cooling, and spray drying, which, along with liquid fuel-fired combustion, are their most important applications.

Chapter 5 is devoted primarily to the internal flow characteristics of plain-orifice and pressure-swirl atomizers, but consideration is also given to the complex flow situations that arise on the surface of a rotating cup or disk. These flow characteristics are important because they govern the quality of atomization and the distribution of drop sizes in the spray.

Atomization quality is usually described in terms of a mean drop size. Because the physical processes involved in atomization are not well understood, empirical equations have been developed for expressing the mean drop size in a spray in terms of liquid properties, gas properties, flow conditions, and atomizer dimensions. The equations selected for inclusion in Chapter 6 are considered to be the best available for the types of atomizers described in Chapter 4.

The function of an atomizer is not merely to disintegrate a bulk liquid into small drops, but also to discharge these drops into the surrounding gas in the form of a symmetrical, uniform spray. The spray characteristics of most practical importance are discussed in Chapter 7. They include cone angle, penetration, radial liquid distribution, and circumferential liquid distribution.

Although evaporation processes are not intrinsic to the subject of atomization and sprays, it cannot be overlooked that in many applications the primary purpose of atomization is to increase the surface area of the liquid and thereby enhance its rate of evaporation. In Chapter 8 attention is focused on the evaporation of fuel drops over wide ranges of ambient gas pressures and temperatures. Consideration is given to both steady-state and unsteady-state evaporation. The concept of an effective evaporation constant is introduced, which is shown to greatly facilitate the calculation of evaporation rates and drop lifetimes for liquid hydrocarbon fuels.

The spray patterns produced by most practical atomizers are so complex that fairly precise measurements of drop size distributions can be obtained only if accurate and reliable instrumentation and data reduction procedures are combined with a sound appreciation of their useful limits of application. In Chapter 9 the various methods employed in drop size measurement are reviewed. Primary emphasis is placed on optical methods that have the important advantage of allowing size measurements to be made without the insertion of a physical probe into the spray. For ensemble measurements the light diffraction method has much to commend it and is now in widespread use as a general purpose tool for spray analysis. Of the remaining methods discussed, the advanced optical techniques have the capability of measuring drop velocity and number density as well as size distribution.

Much of the material covered in this book is based on knowledge acquired during my work on atomizer design and performance over the past thirty years. However, the reader will observe that I have not hesitated in drawing on the considerable practical experience of my industrial colleagues, notably Ted Koblish of Fuel Systems TEXTRON, Hal Simmons of the Parker Hannifin Corporation, and Roger Tate of Delavan Incorporated. I am also deeply indebted to my graduate students in the School of Mechanical Engineering at Cranfield and the Gas Turbine Combustion Laboratory at Purdue. They have made significant contributions to this book through their research, and their names appear throughout the text and in the lists of references.

Professor Norman Chigier has been an enthusiastic supporter in the writing of this book. Other friends and colleagues have kindly used their expert knowledge in reviewing and commenting on individual chapters, especially Chapter 9, which covers an area that in recent years has become the subject of fairly intense research and development. They include Dr. Will Bachalo of Aerometrics Inc., Dr. Lee Dodge of Southwest Research Institute, Dr. Patricia Meyer of Insitec, and Professor Arthur Sterling of Louisiana State University. In the task of proofreading, I have been ably assisted by Professor Norman Chigier, Professor Ju Shan Chin, and my graduate student Jeff Whitlow; their help is hereby gratefully acknowledged.

I am much indebted to Betty Gick and Angie Myers for their skillful typing of the manuscript and to Mark Bass for the high quality artwork he provided for this book. Finally, I would like to thank my wife, Sally, for her encouragement and support during my undertaking of this time-consuming but enjoyable task.

Arthur H. Lefebvre

GENERAL CONSIDERATIONS

INTRODUCTION

The transformation of bulk liquid into sprays and other physical dispersions of small particles in a gaseous atmosphere is of importance in several industrial processes and has many other applications in agriculture, meteorology, and medicine. Numerous spray devices have been developed, and they are generally designated as atomizers or nozzles. The process of atomization is one in which a liquid jet or sheet is disintegrated by the kinetic energy of the liquid itself, or by exposure to high-velocity air or gas, or as a result of mechanical energy applied externally through a rotating or vibrating device. Because of the random nature of the atomization process the resultant spray is usually characterized by a wide spectrum of drop sizes.

Natural sprays include waterfall mists, rains, and ocean sprays. In the home, sprays are produced by shower heads, garden hoses, and hair sprays. They are commonly used in applying agricultural chemicals to crops, paint spraying, spray drying of wet solids, food processing, cooling of nuclear cores, gas-liquid mass transfer applications, dispersing liquid fuels for combustion, and many other applications.

Combustion of liquid fuels in diesel engines, spark ignition engines, gas turbines, rocket engines, and industrial furnaces is dependent on effective atomization to increase the specific surface area of the fuel and thereby achieve high rates of mixing and evaporation. In most combustion systems, reduction in mean fuel drop size leads to higher volumetric heat release rates, easier lightup, a wider burning range, and lower exhaust concentrations of pollutant emissions [1–3]. In other applications, however, such as crop spraying, small droplets must be avoided

1

because their settling velocity is low and, under certain meteorological conditions, they can drift too far downwind. Drop sizes are also important in spray drying and must be closely controlled to achieve the desired rates of heat and mass transfer.

During the past decade there has been a tremendous expansion of interest in the science and technology of atomization, which has now developed into a major international and interdisciplinary field of research. This growth of interest has been accompanied by large strides in the area of laser diagnostics for spray analysis and by a proliferation of mathematical models for spray combustion processes. It is becoming increasingly important for engineers to acquire a better understanding of the basic atomization process and to be fully conversant with the capabilities and limitations of all the relevant atomization devices. In particular, it is important to know which type of atomizer is best suited for any given application and how the performance of any given atomizer is affected by variations in liquid properties and operating conditions.

ATOMIZATION

Sprays may be produced in various ways. There are several basic processes associated with all methods of atomization, such as the hydraulics of the flow within the atomizer, which governs the turbulence properties of the emerging liquid stream. The development of the jet or sheet and the growth of small disturbances, which eventually lead to disintegration into ligaments and then drops, are also of primary importance in determining the shape and penetration of the resulting spray as well as its detailed characteristics of number density, drop velocity, and drop size distributions as functions of time and space. All these characteristics are markedly affected by the internal geometry of the atomizer, the properties of the gaseous medium into which the liquid stream is discharged, and the physical properties of the liquid itself. Perhaps the simplest situation is the disintegration of a liquid jet issuing from a circular orifice, where the main velocity component lies in the axial direction and the jet is in laminar flow. In his classic study [4], Lord Rayleigh postulated the growth of small disturbances that eventually lead to breakup of the jet into drops having a diameter nearly twice that of the jet. A fully turbulent jet can break up without the application of any external force. Once the radial components of velocity are no longer confined by the orifice walls, they are restrained only by surface tension, and the jet disintegrates when the surface tension forces are overcome. The role of viscosity is to inhibit the growth of instabilities and generally delay the onset of disintegration. This causes atomization to occur farther downstream in regions of lower relative velocity; consequently, drop sizes are larger. In most cases, turbulence in the liquid, cavitation in the nozzle, and aerodynamic interaction with the surrounding air, which increases with air density, all contribute to atomization.

Many applications call for a conical or flat spray pattern to achieve the desired dispersion of drops for liquid-gas mixing. Conical sheets may be produced by pressure-swirl nozzles in which a circular discharge orifice is preceded by a chamber in which tangential holes or slots are used to impart a swirling motion to the liquid as it leaves the nozzle. Flat sheets are generally produced either by forcing the liquid through a narrow annulus, as in fan spray nozzles, or by feeding it to the center of a rotating disk or cup. To expand the sheet against the contracting force of surface tension, a minimum sheet velocity is required and is produced by pressure in pressure-swirl and fan spray nozzles and by centrifugal force in rotary atomizers. Regardless of how the sheet is formed, its initial hydrodynamic instabilities are augmented by aerodynamic disturbances, so as the sheet expands away from the nozzle and its thickness declines, perforations are formed that expand toward each other and coalesce to form threads and ligaments. As these ligaments vary widely in diameter, when they collapse the drops formed also vary widely in diameter. Some of the larger drops created by this process disintegrate further into smaller droplets. Eventually a range of drop sizes is produced whose average diameter depends mainly on the initial thickness of the liquid sheet, its velocity relative to the surrounding gas, and the liquid properties of viscosity and surface tension.

A liquid sheet moving at high velocity can also disintegrate in the absence of perforations by a mechanism known as *wavy-sheet* disintegration, whereby the crests of the waves created by aerodynamic interaction with the surrounding gas are torn away in patches. Finally, at very high liquid velocities, corresponding to high injection pressures, sheet disintegration occurs close to the nozzle exit. However, although several modes of sheet disintegration have been identified, in all cases the final atomization process is one in which ligaments break up into drops according to the Rayleigh mechanism.

With prefilming airblast atomizers, a high relative velocity is achieved by exposing a slow-moving sheet of liquid to high-velocity air. Photographic evidence suggests that for low-viscosity liquids the basic mechanisms involved are essentially the same as those observed in pressure atomization, namely the production of drops from ligaments created by perforated-sheet and/or wavy-sheet disintegration.

A typical spray includes a wide range of drop sizes. Some knowledge of drop size distribution is helpful in evaluating process applications in sprays, especially in calculations of heat or mass transfer between the dispersed liquid and the surrounding gas. Unfortunately, no complete theory has yet been developed to describe the hydrodynamic and aerodynamic processes involved when jet and sheet disintegration occurs under normal atomizing conditions, so that only empirical correlations are available for predicting mean drop sizes and drop size distributions. Comparison of the distribution parameters in common use reveals that all of them have deficiencies of one kind or another. In one the maximum drop diameter is unlimited; in others the minimum possible diameter is zero or even negative. So far, no single parameter has emerged that has clear advantages over the others. For any given application the best distribution function is one that is easy to manipulate and provides the best fit to the experimental data.

The difficulties in specifying drop size distributions in sprays have led to widespread use of various mean or median diameters. A median droplet diameter divides the spray into two equal parts by number, length, surface area, or volume [5]. Median diameters may be determined from cumulative distribution curves of the types shown in Fig. 3.6. In a typical spray the value of the median diameter, expressed in micrometers, will vary by a factor of about four depending on the median diameter selected for use. It is important therefore to decide which measure is the most suitable for a particular application. Some diameters are easier to visualize and comprehend, while others may appear in prediction equations that have been derived from theory or experiment. Some drop size measurement techniques yield a result in terms of one particular median diameter. In some cases a given median diameter is selected to emphasize some important characteristic, such as the total surface area in the spray. For liquid fuel-fired combustion systems and other applications involving heat and mass transfer to liquid drops, the Sauter mean diameter, which represents the ratio of the volume to the surface area of the spray, is often preferred. The mass median diameter, which is about 15 to 25% larger than the Sauter mean diameter, is also widely used. As Tate [5] has pointed out, the ratio of these two diameters is a measure of the spread of drop sizes in the spray.

ATOMIZERS

Sprays may be produced in various ways. Essentially, all that is needed is a high relative velocity between the liquid to be atomized and the surrounding air or gas. Some atomizers accomplish this by discharging the liquid at high velocity into a relatively slow-moving stream of air or gas. Notable examples include the various forms of pressure atomizers and also rotary atomizers, which eject the liquid at high velocity from the periphery of a rotating cup or disk. An alternative approach is to expose the relatively slow-moving liquid to a high-velocity airstream. The latter method is generally known as *twin-fluid, air-assist,* or *airblast* atomization.

Pressure Atomizers

When a liquid is discharged through a small aperture under high applied pressure, the pressure energy is converted into kinetic energy (velocity). For a typical hydrocarbon fuel, in the absence of frictional losses a nozzle pressure drop of 138 kPa (20 psi) produces an exit velocity of 18.6 m/s. As velocity increases as the square root of the pressure, at 689 kPa (100 psi) a velocity of 41.5 m/s is obtained, while 5.5 MPa (800 psi) produces 117 m/s.

Plain orifice. A simple circular orifice is used to inject a round jet of liquid into the surrounding air. Finest atomization is achieved with small orifices but, in practice, the difficulty of keeping liquids free from foreign particles usually limits the minimum orifice size to around 0.3 mm. Combustion applications for plain-orifice atomizers include turbojet afterburners, ramjets, diesel engines, and rocket engines.

Pressure-swirl (simplex). A circular outlet orifice is preceded by a swirl chamber into which liquid flows through a number of tangential holes or slots. The swirling liquid creates a core of air or gas that extends from the discharge orifice to the rear of the swirl chamber. The liquid emerges from the discharge orifice as an annular sheet, which spreads radially outward to form a hollow conical spray. Included spray angles range from 30° to almost 180°, depending on the application. Atomization performance is generally good. Finest atomization occurs at high delivery pressures and wide spray angles.

For some applications a spray in the form of a solid cone is preferred. This can be achieved by using an axial jet or some other device to inject droplets into the center of the hollow conical spray pattern produced by the swirl chamber. These two modes of injection create a bimodal distribution of drop sizes, droplets at the center of the spray being generally larger than those near the edge.

Square spray. This is essentially a solid-cone nozzle, but the outlet orifice is specially shaped to distort the conical spray into a pattern that is roughly in the form of a square. Atomization quality is not as high as with conventional hollow-cone nozzles but, when used in multiple-nozzle combinations, a fairly uniform coverage of large areas can be achieved.

Duplex. A drawback of all types of pressure nozzles is that the liquid flow rate is proportional to the square root of the injection pressure differential. In practice, this limits the flow range of simplex nozzles to about 10:1. The duplex nozzle overcomes this limitation by feeding the swirl chamber through two sets of distributor slots, each having its own separate liquid supply. One set of slots is much smaller in cross-sectional area than the other. The small slots are termed *primary* and the large slots *secondary*. At low flow rates all the liquid to be atomized flows into the swirl chamber through the primary slots. As the flow rate increases, so does the injection pressure. At some predetermined pressure level a valve opens and admits liquid into the swirl chamber through the secondary slots.

Duplex nozzles allow good atomization to be achieved over a range of liquid flow rates of about 40:1 without the need to resort to excessively high delivery pressures. However, near the point where the secondary liquid is first admitted into the swirl chamber, there is a small range of flow rates over which atomization quality is poor. Moreover, the spray cone angle changes with flow rate, being widest at the lowest flow rate and becoming narrower as the flow rate is increased.

Dual orifice. This is similar to the duplex nozzle except that two separate swirl chambers are provided, one for the primary flow and the other for the secondary flow. The two swirl chambers are housed concentrically within a single nozzle body to form a "nozzle within a nozzle." At low flow rates all the liquid passes through the inner primary nozzle. At high flow rates liquid continues to flow through the primary nozzle but most of the liquid is passed through the outer secondary nozzle, which is designed for a much larger flow rate. As with the duplex nozzle, there is a transition phase, just after the pressurizing valve opens, when the secondary spray draws its energy for atomization from the primary spray, so the overall atomization quality is relatively poor.

Dual-orifice nozzles offer more flexibility than duplex nozzles. For example, if desired it can be arranged for the primary and secondary sprays to merge just downstream of the nozzle to form a single spray. Alternatively, the primary and secondary nozzles can be designed to produce different spray angles, the former being optimized for low flow rates and the latter optimized for high flow rates.

Spill return. This is essentially a simplex nozzle, but with a return flow line at the rear or side of the swirl chamber and a valve to control the quantity of liquid removed from the swirl chamber and returned to supply. Very high turndown ratios are attainable with this design. Atomization quality is always good because the supply pressure is held constant at a high value, reductions in flow rate being accommodated by adjusting the valve in the spill return line. This construction provides a hollow-cone spray pattern, with some increase in spray angle as the flow is reduced.

Fan spray. Several different concepts are used to produce flat or fan-shaped sprays. The most popular type of nozzle is one in which the orifice is formed by the intersection of a V groove with a hemispheric cavity communicating with a cylindrical liquid inlet [5]. It produces a liquid sheet parallel to the major axis of the orifice, which disintegrates into a narrow elliptical spray.

An alternative method of producing a fan spray is by discharging the liquid through a plain circular hole onto a curved deflector plate. The deflector method produces a somewhat coarser spray pattern. Wide spray angles and high flow rates are attainable with this type of nozzle. Because the nozzle flow passages are relatively large, the problem of plugging is minimized.

A fan spray can also be produced by the collision of impinging jets. If two liquid jets are arranged to collide outside the nozzle, a flat liquid sheet is formed that is perpendicular to the plane of the jets. The atomization performance of this type of injector is relatively poor, and high stream velocities are necessary to approach the spray quality obtainable with other types of pressure nozzles. Extreme care must be taken to ensure that the jets are properly aligned. The main advantage of this method of atomization is the isolation of different liquids until they collide outside the nozzle.

Rotary Atomizers

One widely used type of rotary atomizer comprises a high-speed rotating disk with means for introducing liquid at its center. The liquid flows radially outward across the disk and is discharged at high velocity from its periphery. The disk may be smooth and flat or may have vanes or slots to guide the liquid to the periphery. At low flow rates, droplets form near the edge of the disk. At high flow rates, ligaments or sheets are generated at the edge and disintegrate into droplets. Small disks operating at high rotational speeds and low flow rates are capable of producing sprays in which drop sizes are fairly uniform. A 360° spray pattern is developed by rotating disks, which are usually installed in a cylindrical or conical chamber where an umbrellalike spray is created by downward gas currents [5].

Some rotary atomizers employ a cup instead of a disk. The cup is usually smaller in diameter and is shaped like an elongated bowl. In some designs the edge of the cup is serrated to encourage a more uniform drop size distribution in the spray. A flow of air around the periphery is sometimes used to shape the spray and to assist in transporting the droplets away from the atomizer. In contrast to pressure nozzles, rotary atomizers allow independent variation of flow rate and disk speed, thereby providing more flexibility in operation.

Air-Assist Atomizers

In this type of nozzle the liquid is exposed to a stream of air or steam flowing at high velocity. In the *internal-mixing* configuration, gas and liquid mix within the nozzle before discharging through the outlet orifice. The liquid is sometimes supplied through tangential slots to encourage a conical discharge pattern. However, the maximum spray angle is limited to about 60°. The device tends to be energy inefficient, but it can produce a finer spray than simple pressure nozzles.

As its name suggests, in the *external-mixing* form of air-assist nozzle the high-velocity gas or steam impinges on the liquid at or outside the liquid discharge orifice. Its advantage over the internal-mixing type is that problems of back pressures are avoided because there is no internal communication between gas and liquid. However, it is less efficient than the internal-mixing concept, and higher gas flow rates are needed to achieve the same degree of atomization. Both types of nozzles can atomize high-viscosity liquids effectively.

Airblast Atomizers

These devices function in a very similar manner to air-assist nozzles, and both types fall in the general category of twin-fluid atomizers. The main difference between air-assist and airblast atomizers is that the former use relatively small quantities of air or steam flowing at very high velocities (usually sonic), whereas the latter employ large amounts of air flowing at much lower velocities (<100 m/s). Airblast nozzles are thus ideally suited for atomizing liquid fuels in continuous-flow combustion systems, such as gas turbines, where air velocities of this magnitude are usually readily available. The most common form of airblast atomizer is one in which the liquid is first spread into a thin conical sheet and then exposed to high-velocity airstreams on both sides of the sheet. The atomization performance of this *prefilming* type of airblast nozzle is superior to that of the alternative *plain-jet* airblast nozzle, in which the liquid is injected into the airstream in the form of one or more discrete jets.

Other Types

Most practical atomizers are of the pressure, rotary, or twin-fluid type. However, many other forms of atomizers have been developed that are useful in special applications.

Electrostatic. A liquid jet or film is exposed to an intense electrical pressure that tends to expand its area. This expansion is opposed by surface tension forces. If the electrical pressure predominates, droplets are formed. Droplet size is a function of the electrical pressure, the liquid flow rate, and the physical and electrical properties of the liquid. The low liquid flow rates associated with electrostatic atomizers have tended to limit their practical applications to electrostatic painting and nonimpact printing.

Ultrasonic. The liquid to be atomized is fed through or over a transducer and horn, which vibrates at ultrasonic frequencies to produce the short wavelengths necessary for fine atomization. The system requires a high-frequency electrical input, two piezoelectric transducers, and a stepped horn. The concept is well suited for applications that require very fine atomization and a low spray velocity. At present, an important application of ultrasonic atomizers (nebulizers) is for medical inhalation therapy, where very fine sprays and the absence of gas to effect atomization are important attributes.

Sonic (whistle). Gas is accelerated within the device to sonic velocity and impinges on a plate or annular cavity (resonation chamber). The sound waves produced are reflected into the path of the incoming liquid [6]. The frequency of the sound waves is around 20 kHz, and this serves to disintegrate the liquid into small droplets ranging downward in size from 50 μm. The sonic and pneumatic effects are difficult to isolate from each other. Efforts have been made to design nozzles that operate above the audible frequency limit to reduce the nuisance of noise [7]. However, in some applications the attendant sound field may benefit the process (e.g., combustion) for which the resultant spray is required.

Windmill. Many aerial applications of pesticides require a narrow spectrum of drop sizes. Conventional rotary disk atomizers can provide such a spectrum but only when operating in the ligament mode at low flow rates. By making radial cuts at the periphery of a disk and twisting the tips of the segments, the disk can be converted into a windmill that will rotate rapidly when inserted into an airflow at aircraft flight speed. According to Spillmann and Sanderson [8], the disk windmill constitutes an ideal rotary atomizer for the aerial application of pesticides. It provides a narrow spectrum of drop sizes in the range most suitable for herbicides, at relatively high flow rates.

Vibrating capillary. This type of droplet generator was first used to study the collision and coalescence of small water droplets. It consists of a hypodermic needle vibrating at its resonant frequency and can produce uniform streams of drops down to 30 μm in diameter. The size and frequency with which the droplets can be produced depend on the flow rate of the liquid through the needle, the needle diameter, the resonant frequency, and the amplitude of oscillation of the needle tip.

Flashing liquid jets. An orifice downstream of which a high-pressure liquid flash vaporizes to shatter the liquid into small droplets can produce a fairly regular spray pattern. Flashing dissolved gas systems have been studied by Brown and York [9], Sher and Elata [10], and Marek and Cooper [11]. Solomon et al. [12] have shown that flashing even small quantities of dissolved gas (mole fractions less than 15%) can effect a significant improvement in atomization. However, these beneficial effects cannot be realized unless they are promoted by fitting an expansion chamber just upstream of the discharge orifice. The need for this expansion chamber stems from the low bubble growth rate for dissolved gas systems. This low bubble growth rate appears to pose a fundamental limitation to the practical application of flashing injection by means of dissolved gas systems.

Effervescent atomization. This method of atomization overcomes the basic problems associated with flashing dissolved gas systems. No attempt is made to dissolve any air or gas in the liquid. Instead, the gas is injected at low velocity into the flowing liquid stream at some point upstream of the discharge orifice. The pressure differential between the atomizing gas and the liquid into which it is injected is only a few centimeters of water and is only what is needed to prevent the liquid from flowing back up the gas line. Studies by Lefebvre et al. [13] have shown that good atomization can be achieved at much lower liquid injection pressures than the values normally associated with pressure atomization.

Schematic diagrams illustrating the principal design features of the most important of the atomizers described above are shown in Fig. 1.1. The relative merits of these and other atomizers are listed in Table 1.1.

FACTORS INFLUENCING ATOMIZATION

The performance of any given type of atomizer depends on its size and geometry and on the physical properties of the dispered phase (i.e., the liquid being atomized) and the continuous phase (i.e., the gaseous medium into which the droplets are discharged).

For plain-orifice pressure nozzles and plain-jet airblast atomizers, the dimen-

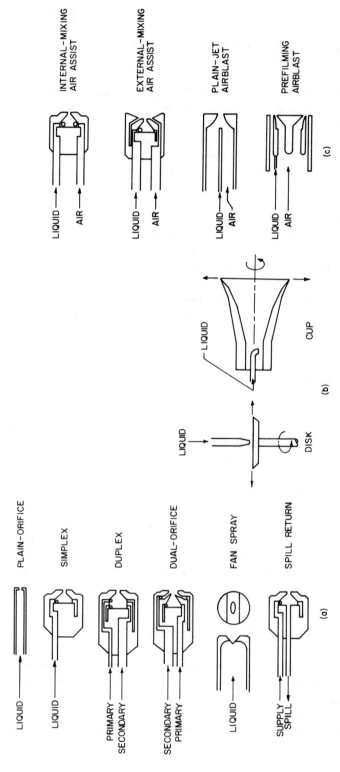

Figure 1.1 (a) Pressure atomizers. (b) Rotary atomizers. (c) Twin-fluid atomizers.

sion most important for atomization is the diameter of the final discharge orifice. For pressure-swirl, rotary, and prefilming airblast atomizers, the critical dimension is the thickness of the liquid sheet as it leaves the atomizer. Theory predicts, and experiment confirms, that mean drop size is roughly proportional to the square root of the liquid jet diameter or sheet thickness. Thus, provided the other key parameters that affect atomization are maintained constant, an increase in atomizer scale (size) will impair atomization.

Liquid Properties

The flow and spray characteristics of most atomizers are strongly influenced by the liquid properties of density, viscosity, and surface tension. In theory, the mass flow rate through a pressure nozzle varies with the square root of liquid density. However, as Tate [5] has pointed out, in practice it is seldom possible to change the density without affecting some other liquid property, so this relationship must be interpreted cautiously. The significance of density for atomization performance is diminished by the fact that most liquids exhibit only minor differences in this property. Moreover, the modest amount of available data on the effect of liquid density on mean drop size suggests that its influence is quite small.

One way of defining a spray is in terms of the increase in liquid surface area resulting from atomization. The surface area before breakup is simply that of the liquid cylinder as it emerges from the nozzle. After atomization, the area is the sum of the surface areas of all the individual droplets. This multiplication factor provides a direct indication of the level of atomization achieved and is useful in applications that emphasize surface phenomena such as evaporation and absorption. Surface tension is important in atomization because it represents the force that resists the formation of new surface area. The minimum energy required for atomization is equal to the surface tension multiplied by the increase in liquid surface area. Whenever atomization occurs under conditions where surface tension forces are important, the Weber number, which is the ratio of the inertial force to the surface tension force, is a useful dimensionless parameter for correlating drop size data. Commonly encountered surface tensions range from 0.073 kg/s^2 for water to 0.027 kg/s^2 for petroleum products. For most pure liquids in contact with air, the surface tension decreases with an increase in temperature and is independent of the age of the surface [14].

In many respects, viscosity is the most important liquid property. Although in an absolute sense its influence on atomization is no greater than that of surface tension, its importance stems from the fact that it affects not only the drop size distributions in the spray but also the nozzle flow rate and spray pattern. An increase in viscosity lowers the Reynolds number and also hinders the development of any natural instabilities in the jet or sheet. The combined effect is to delay disintegration and increase the size of the drops in the spray.

The effect of viscosity on flow within the nozzle is complex. In hollow-cone nozzles a modest increase in viscosity can actually increase the flow rate. It does this by thickening the liquid film in the discharge orifice, thereby raising the ef-

Table 1.1 Relative merits of various types of atomizers

Type	Description	Advantages	Drawbacks	Applications
Pressure atomizer	Plain orifice	1. Simple, cheap 2. Rugged	1. Narrow spray angle 2. Solid spray cone	Diesel engines, jet engine afterburners, ramjets
	Simplex	1. Simple, cheap 2. Wide spray angle (up to 180°)	1. Needs high supply pressures 2. Cone angle varies with pressure differential and ambient gas density	Gas turbines and industrial furnaces
	Duplex	Same as simplex, plus good atomization over a very wide range of liquid flow rates	Spray angle narrows as liquid flow rate is increased	Gas turbine combustors
	Dual orifice	1. Good atomization 2. Turndown ratio as high as 50:1 3. Relatively constant spray angle	1. Atomization poor in transition range 2. Complexity in design 3. Susceptibility of small passages to blockage	Wide range of aircraft and industrial gas turbines
	Spill return	1. Simple construction 2. Good atomization over entire flow range 3. Very large turndown ratio 4. Large holes and flow passages obviate risk of blockage	1. Spray angle varies with flow rates 2. Power requirements higher than with other pressure nozzles except at maximum discharge	Various types of combustor Has good potential for slurries and fuels of low thermal stability
	Fan spray	1. Good atomization 2. Narrow elliptical pattern sometimes advantageous	Needs high supply pressures	High-pressure coating operations Annular combustors
Rotary	Spinning disk	1. Nearly uniform atomization possible with small disks rotating at high speeds 2. Independent control of atomization quality and flow rate	Produces a 360° spray pattern	Spray drying Crop spraying
	Rotary cup	Capable of handling slurries	May require air blast around periphery	Spray drying Spray cooling

	Advantages	Disadvantages	Applications
Air assist — Internal mixing	1. Good atomization 2. Large passages prevent clogging 3. Can atomize high-viscosity liquids	1. Liquid can back up in air line 2. Requires auxiliary metering device 3. Needs external source of high-pressure air or steam	Industrial furnaces Industrial gas turbines
External mixing	Same as internal mixing, plus construction prevents backing up of liquid into the air line	1. Needs external source of air or steam 2. Does not permit high liquid/air ratios	Same as internal mixing
Airblast — Plain jet	1. Good atomization 2. Simple, cheap	1. Narrow spray angle 2. Atomizing performance inferior to prefilming airblast	Industrial gas turbines
Prefilming	1. Good atomization especially at high ambient air pressures 2. Wide spray angle	Atomization poor at low air velocities	Wide range of industrial and aircraft gas turbines
Ultrasonic	1. Very fine atomization 2. Low spray velocity	Cannot handle high flow rates	Medical sprays Humidification Spray drying Acid etching Combustion
Electrostatic	Very fine atomization	Cannot handle high flow rates	Paint spraying Printing

fective flow area. At high viscosities, however, the flow rate usually diminishes with increasing viscosity. With pressure-swirl nozzles an increase in viscosity generally produces a narrower spray angle. At very high viscosities the normal conical spray may collapse into a straight stream of relatively large ligaments and drops. An increase in liquid viscosity invariably has an adverse effect on atomization quality, because when viscous losses are large, less energy is available for atomization and a coarser spray results. In airblast atomizers, liquid velocities are usually much lower than in pressure nozzles. In consequence, the drop sizes produced by airblast nozzles tend to be less sensitive to variations in liquid viscosity.

Table 1.2 lists the relevant physical properties of some of the liquids used in spray applications. The viscosity of these liquids ranges from 0.001 kg/m s for water to 0.5 kg/m s for heavy fuel oil. The viscosity of liquids generally decreases with an increase in temperature. It is customary to heat up many of the heavier fuel oils, partly to reduce pumping power requirements but also to improve atomization.

Some fluids, for example, slurries of liquids and solid powders, are characterized by a nonlinear relationship between shear stress and shear strain rate. For such liquids, which are called non-Newtonian, it is necessary to specify the shear rate with the viscosity. The apparent reduction in viscosity with increasing shear rate highlights the need to minimize pressure losses in the supply lines and nozzles. This reduction is also desirable because the viscosities of non-Newtonian fluids have less effect on atomization if a high shear rate is produced in the liquid film formed by the nozzle [14]. Very little secondary atomization will occur once the drops are formed, due to the increase in apparent viscosity at the lower shear rate.

The influence of temperature on viscosity and surface tension is illustrated for some widely used hydrocarbon fuels in Figs. 8.31 and 1.2, respectively.

Ambient Conditions

The ambient gas into which sprays are injected can vary widely in pressure and temperature. This is especially true of liquid fuel-fired combustion systems. In diesel engines, critical and supercritical pressure and temperature conditions are encountered. In gas turbine combustors, fuel sprays are injected into highly turbulent, swirling recirculating streams of reacting gases. In industrial furnaces, the fuel is sprayed into high-temperature flames of recirculating combustion products. With pressure-swirl atomizers the spray angle decreases markedly with increase in ambient gas density until a minimum angle is reached beyond which any further increase in ambient density has no effect on spray angle. Ambient gas density also has a strong influence on the mean drop sizes produced by pressure-swirl atomizers. If the ambient pressure is raised continuously above the normal atmospheric value, the mean drop size increases initially until a maximum value is reached and then slowly declines. The reasons for this unusual relationship between ambient pressure and mean drop size are discussed in Chapter 6.

The spray patterns generated by pressure-swirl atomizers are also affected by

Table 1.2 Properties of liquids

Liquid	Temperature (K)	Viscosity (kg/m s)	Density (kg/m³)	Surface tension (kg/s²)
Acetone	273	0.000400		0.0261
	293	0.00032	792	0.0237
	300	0.000300		
	303	0.000295		
	313	0.00028		0.02116
Ammonia	284			0.0234
	300	0.00013		
	307			0.0181
Aniline	283	0.0065		0.0441
	288	0.0053	1000	0.040
	323	0.00185		0.0394
	373	0.00085		
Benzene	273	0.00091	899	
	283	0.00076		0.0302
	293	0.00065	880	0.0290
	303	0.00056		0.0276
	323	0.00044		
	353	0.00039		
Butane	300	0.00016		0.0116
Carbon tetrachloride	273	0.00133		
	293	0.00097	799	0.0270
	323	0.00065		
	373	0.00038		0.0176
Castor oil	283	2.42		
	288		969	
	293	0.986		
	303	0.451		
	313	0.231		
	373	0.0169		
Chloroform	273	0.0007		
	293	0.00058	1489	0.02714
	303	0.00051		
Cottonseed oil	289		926	
	293	0.0070		
Creosote	313	0.0070		
n-Decane	293	0.00092		
	300			0.0233
Ethane	300	0.000035		0.0007
Ethyl alcohol	273	0.00177		0.02405
	293	0.0012	791	0.02275
	303	0.0010		0.02189
	323	0.0007		
	343	0.00050		
Ethylene glycol	293	0.01990		
	313	0.00913		
	333	0.00495		
	353	0.00302		
	373	0.00199		
Fuel oil (light)	288	0.172	930	0.0250
	313	0.047	916	0.0230
	356	0.0083	880	0.0210

Table 1.2 Properties of liquids (*Continued*)

Liquid	Temperature (K)	Viscosity (kg/m s)	Density (kg/m³)	Surface tension (kg/s²)
Fuel oil (medium)	313	0.215	936	0.0230
	378	0.0134	897	0.0200
Fuel oil (heavy)	313	0.567	970	0.0230
	366	0.037	920	0.0210
	400	0.015	900	0.0200
Gas oil	288	0.0060	850	0.0240
	313	0.0033	863	0.0230
Glycerin	273	12.1	1260	0.0630
	288	2.33		
	293	0.622		0.0630
Heptane	273	0.00052		
	300	0.00038		0.0194
	313	0.00034		
	343	0.00026		0.0194
Hexane	273	0.00040		
	293	0.00033		0.0184
	300	0.00029		0.0176
	323	0.00025		
Hydrazine	274	0.00129		
	293	0.00097		
	298			0.0915
Kerosine	293	0.0016	800	0.0260
Linseed oil	288		942.0	
	303	0.0331		
	363	0.0071		
Machine oil (light)	288.6	0.114		
	310.8	0.0342		
	373	0.0049		
Machine oil (heavy)	288.6	0.661		
	310.8	0.127		
Mercury	273	0.0017	13600	
	293	0.00153	13550	0.480
	313	0.00045		
Methyl alcohol	273	0.00082	810	0.0245
	293	0.00060		0.0226
	300	0.00053		0.0221
	323	0.00040		
Naphthalene	353	0.00097		
	373	0.00078		
	400			0.0288
Nonane	300			0.0223
n-Octane	293	0.00054		0.0218
	300	0.0005		0.0210
Olive oil	283	0.138		
	288		918	
	291			0.0331
	293	0.0840		
	343	0.0124		
Pentane	273	0.00029		
	300	0.00022		0.0153
Propane	300	0.000098		0.0064

Table 1.2 Properties of liquids (*Continued*)

Liquid	Temperature (K)	Viscosity (kg/m s)	Density (kg/m³)	Surface tension (kg/s²)
Toluene	273	0.00077		0.0277
	293	0.00059		0.0285
	303	0.00053		0.0274
	343	0.00035		
Turpentine	273	0.00225	870	
	283	0.00178		0.0270
	303	0.00127		
	343	0.000728		
Water	291			0.073
	300	0.00085		0.0717

the liquid injection pressure differential ΔP_L. The ejector action of the high-velocity spray generates air currents, which cause the spray angle to contract. This effect is aggravated by the increase in spray velocity that accompanies an increase in ΔP_L. Thus, although increasing ΔP_L has no effect on the spray angle immediately downstream of the nozzle, it causes appreciable contraction of the spray pattern farther downstream.

With plain-orifice atomizers an increase in ambient gas density leads to a wider spray angle. This is because the increase in aerodynamic drag on the drop-

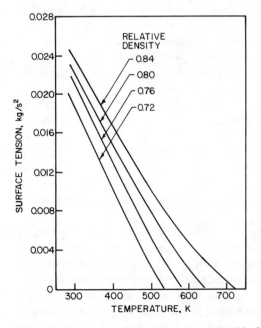

Figure 1.2 Surface tension-temperature relationship for hydrocarbon fuels of varying relative densities.

lets, created by an increase in gas density, tends to produce a greater deceleration in the axial direction than in the radial direction.

The spray patterns produced by airblast atomizers tend to be fairly insensitive to variations in ambient gas density. All but the largest drops in the spray tend to follow the streamlines of the airflow pattern generated at the nozzle exit by the various swirlers and shaped passages within the nozzle. This airflow pattern generally remains fixed and independent of air density, apart from second-order Reynolds number and Mach number effects. However, if the "natural" cone angle of the spray, i.e., the spray angle with no air flowing, is markedly different from that of the air, then the change in aerodynamic drag forces produced by a change in air density will affect the resulting spray pattern. In general, an increase in air density will cause the spray pattern to adhere more closely to the streamlines of the atomizing air.

SPRAY CHARACTERISTICS

The importance of drop size for the many combustion and industrial processes that utilize liquid sprays is attested to by the proliferation of techniques available for drop size measurement. No single technique is completely satisfactory but each technique has its own advantages and drawbacks, depending on the application. Direct methods include those in which individual drops are collected on slides for subsequent measurement and counting or in which droplets are frozen and sized as solid particles. With the impaction method the drops are sorted on the basis of inertial differences. Depending on its size, a droplet may impact or fail to impact on a solid surface or may follow a different trajectory. This allows all the drops in a spray to be sorted into different size categories.

High-speed photography is often used to provide instantaneous images of the drops in a spray, which are recorded for subsequent counting or analysis. The method has the important advantage of being nonintrusive, but it cannot give the temporal distribution of drop sizes as produced by the atomizer. This is because the rate of deceleration of the drops downstream of the atomizer is higher for the smaller droplets, so photographs taken in this region indicate a preponderance of small drops. The spatial distribution of drop sizes given by instantaneous photography can be converted into the true temporal distribution by multiplying the number of droplets in any given size range by the average velocity for that size range.

In recent years considerable advances have been made in the development of laser diagnostic techniques for measuring particle size and velocity in sprays. High-speed pulsed microphotography, cinematography, and holography are being used to study drop size distributions and spray structure. Much of the tedium normally involved in detailed studies of drop size distributions in various regions of the spray can now be alleviated using automatic image analysis, in which droplet images are enlarged, counted, and sorted by electronic scanning devices and analyzed by microprocessors [15].

A most convenient and effective method for assessing and comparing sprays is Fraunhofer diffraction particle sizing, which uses a line-of-sight measurement through the spray. Various models of this instrument are commercially available for applications to both continuous and intermittent sprays. Recent developments include new computer software for providing a rapid printout of size distributions for on-line analysis.

Laser Doppler anemometry and interferometry, interfaced with signal processing and high-speed data acquisition systems, are now being used in simultaneous measurements of drop size, drop velocity, and local drop drag coefficients.

Summaries of the methods used in spray analysis prior to 1970 have been prepared by Putnam et al. [16] and Tate and Olsen [17]. These and more recent developments are described in some detail in Chapter 9.

Table 1.3 Spray applications

Production or processing
  Spray drying (dairy products, coffee and tea, starch pharmaceuticals, soaps and detergents, pigments, etc.)
  Spray cooling
  Spray reactions (absorption, roasting, etc.)
  Atomized suspension technique (effluents, waste liquors, etc.)
  Powdered metals
Treatment
  Evaporation and aeration
  Cooling (spray ponds, towers, reactors, etc.)
  Humidification and misting
  Air and gas washing and scrubbing
  Industrial washing and cleaning
Coating
  Surface treatment
  Spray painting (pneumatic, airless, and electrostatic)
  Flame spraying
  Insulation, fibers, and undercoating materials
  Multicomponent resins (urethanes, epoxies, polyesters, etc.)
  Particle coating and encapsulation
Combustion
  Oil burners (furnaces and heaters, industrial and marine boilers)
  Diesel fuel injection
  Gas turbines (aircraft, marine, automotive, etc.)
  Rocket fuel injection
Miscellaneous
  Medicinal sprays
  Dispersion of chemical agents
  Agricultural spraying (insecticides, herbicides, fertilizer solutions, etc.)
  Foam and fog suppression
  Printing
  Acid etching

Source: Tate [5].

APPLICATIONS

A compilation by Tate [5] of some of the most important applications is contained in Table 1.3. As the cost, complexity, atomizing performance, and energy consumption vary widely between different types of atomizers, it is important to select the best atomizer for any given application. The following factors enter into the proper selection: properties of the liquid to be atomized, e.g., density, viscosity, surface tension, and temperature; ambient gas properties, such as pressure, temperature, and flow pattern; particle sizes and percent solids in suspensions, slurries, and pastes; maximum flow rate; range of flow rates (turndown ratio); required mean drop size and drop size distribution; liquid or gas pressures available for nozzles, or power required for rotary atomizers; conditions that may contribute to wear and corrosion; size and shape of vessel, enclosure, or combustor containing spray; economics of spray operation taking into account initial cost, operating expenses, and depreciation; and safety considerations [5].

GLOSSARY

Some of the terms frequently used in descriptions of atomizers and sprays are defined below. These definitions are necessarily brief, and no attempt has been made to include all qualifying considerations.

Air-assist nozzle. Nozzle in which high-velocity air or steam is used to enhance pressure atomization at low liquid flow rates.

Airblast atomizer. Atomizer in which a liquid jet or sheet is exposed to air flowing at high velocity.

Air core. Cylindrical void space within the rotating liquid in a simplex swirl chamber.

Arithmetic mean diameter. Linear mean diameter of drops in spray.

Atomization. Process whereby a volume of liquid is disintegrated into a multiplicity of small drops.

Beam steering. Refraction of a laser beam due to density gradients in the continuous phase.

Breakup length. Length of continuous portion of jet measured from nozzle exit to point where breakup occurs.

Cavitation. Formation of bubbles by gas or vapor released in flow regions of low static pressure; affects discharge coefficient and jet breakup.

Cavitation number. Ratio of pressure differential to downstream pressure; indicator of propensity for cavitation.

Combined spray. Spray produced when both stages flow simultaneously in a dual-orifice or piloted-airblast nozzle.

Continuous phase. Medium, usually gaseous, in which atomization occurs.

Critical flow rate. Liquid flow rate corresponding to the transition from one mode of atomization to another.

Critical Weber number. Value of Weber number above which a single drop will split into two or more drops.

Cumulative distribution. Plot of percentage by number, surface area, or volume of drops whose diameter is less than a given drop diameter.

Discharge coefficient. Ratio of actual flow rate to theoretical flow rate.

Discharge orifice. Final orifice through which liquid is discharged into the ambient gas.

Dispersed phase. Liquid to be atomized.

Dispersion. Ratio of the volume of a spray to the volume of the liquid contained within it.

Drooling. Sluggish dripping of liquid from a nozzle while spraying, usually caused by impingement of the spray on some surface other than the orifice from which the liquid is discharging [18].

Drop coalescence. Collision of two drops to form a single drop.

Droplet size. Diameter of a spherical droplet, usually expressed in micrometers.

Droplet uniformity index. Indication of the range of drop sizes in a spray relative to the median diameter.

Drop saturation. Droplet population exceeding the capability of the sizing instrument or method.

Dual-orifice atomizer. Atomizer consisting of two simplex nozzles fitted concentrically one inside the other.

Duplex nozzle. Nozzle featuring a swirl chamber with two sets of tangential swirl ports, one set being the primary ports for low flows and the other the larger secondary ports for handling high flow rates.

Effective evaporation constant. Value of evaporation constant that includes heat-up period and convective effects.

Electrostatic atomizer. Atomizer in which electrical pressure is used to overcome surface tension forces and achieve atomization.

Equivalent spray angle. Angle formed by drawing two straight lines from the nozzle discharge orifice through the center of the liquid mass in the left and right lobes of the spray.

Evaporation constant. Indication of the rate of change of drop surface area during steady-state evaporation.

External mixing nozzle. Air-assist atomizer in which high-velocity gas impinges on a liquid at or outside the final orifice.

Extinction. Percentage of light removed from original direction; indicator of the extent to which measurements of mean drop size are affected by multiple scattering. Also termed *obscuration.*

Fan spray. Spray in the shape of a sector of a circle of about 75° angle; elliptical in cross section.

Film thickness. Thickness of annular liquid sheet as it discharges from the atomizer.

Flat spray. Same as fan spray.

Flow number. Effective flow area of a nozzle, usually expressed as the ratio of mass or volumetric flow rate to the square root of injection pressure differential.

Flow rate. Amount of liquid discharged during a given period of time; normally identified with all factors that affect flow rate, such as pressure differential and liquid density.

Frequency distribution curve. Plot of liquid volume per size class.

Heat transfer number. Indicator of rate of evaporation due to heat transfer to droplet from surrounding gas.

Heat-up period. Initial phase of droplet evaporation prior to attainment of steady-state conditions.

Hollow-cone spray. Spray in which most of the droplets are concentrated at the outer edge of a conical spray pattern.

Impingement. Collision of two round liquid jets or collision of a jet of liquid with a stationary deflector.

Impinging jet atomizer. Atomizer in which two liquid jets collide outside the nozzle to produce a liquid sheet perpendicular to the plane of the jets.

Internal mixing nozzle. Air-assist atomizer in which gas and liquid mix within the nozzle before discharging through the outlet orifice.

Mass transfer number. Indicator of rate of evaporation due to mass transfer.

Mass (volume) median diameter. Diameter of a drop below or above which 50% of the total mass (volume) of drops lies.

Mean drop size. A given spray is replaced by a fictitious one in which all the drops have the same diameter while retaining certain characteristics of the original spray.

Monodisperse spray. Spray containing drops of uniform size.

Multiple scattering. When spray number density is high some drops obscure part of signal generated by others, leading to biased diffraction patterns.

Normal distribution. Distribution of drop sizes based on the random occurrence of a given drop size.

Obscuration. Percentage of light removed from original direction; indicator of the extent to which measurements of mean drop size are affected by multiple scattering. Also known as *extinction*.

Ohnesorge number. Dimensionless group obtained by dividing the square root of Weber number by Reynolds number, which eliminates velocity from both; indicator of jet or sheet stability.

Patternation. A measure of the uniformity of the circumferential distribution of liquid in a conical spray. The term *radial patternation* is also used to describe the radial distribution of liquid within a conical spray.

Patternator. Two types are available: one designed to measure the radial liquid distribution in a conical spray, the other to measure the uniformity of the circumferential liquid distribution.

Plain-orifice atomizer. Atomizer in which liquid is ejected at high velocity through a small round hole; the best known example is a diesel injector.

Polydisperse spray. Spray containing drops of different sizes.

Prefilmer. Solid surface on which a thin continuous liquid film is formed.

Pressure atomizer. Single-fluid atomizer in which the conversion of pressure into kinetic energy results in a high relative velocity between the liquid and the surrounding gas.

Relative density. Ratio of the mass of a given volume of liquid to the mass of an equal volume of water; the temperature of both liquids must be stated. For example, relative density = 0.81 at 289/277 K indicates that the mass of the liquid was measured at 289 K and divided by the mass of an equal volume of water at 277 K. Formerly called specific gravity.

Relative span factor. Indicator of the range of drop sizes relative to the mass median diameter.

Reynolds number. Dimensionless ratio of inertial force to viscous force.

Rosin-Rammler distribution. Drop size distribution described in terms of two parameters, one of which provides a measure of the spread of drop sizes.

Rotary atomizer. Atomizer in which liquid is discharged from the edge of a rotating disk, cup, or slotted wheel.

Sauter mean diameter. Diameter of a droplet whose surface-to-volume ratio is equal to that of the entire spray.

Shroud air. A flow of air over the atomizer face to prevent deposition of carbon; also used to modify spray characteristics at low flow rates.

Simplex nozzle. Nozzle that employs a single swirl chamber to produce a well-atomized spray of wide cone angle.

Skewness. The axis of the nozzle spray cone is not colinear with the central axis of the nozzle; the maximum departure is stated in degrees [18].

Slinger system. Rotary atomizer employed in some small gas turbines.

Solid-cone spray. Spray in which the droplets are fairly uniformly distributed throughout a conical spray volume.

Spatial sampling. Measurement of drops contained within a volume under conditions such that contents of volume do not change during any single measurement.

Spill-return nozzle. Basically a simplex atomizer with provision for liquid to be removed from the swirl chamber and returned to supply; provides good atomization even at lowest flow rates.

Spitting. Large, irregular drops of liquid intermittently produced by otherwise uniformly fine spray; sometimes caused by internal flow leaks within the nozzle [18].

Spray angle. Angle formed by two straight lines drawn from the discharge orifice to cut the spray contours at a specific distance from the atomizer face.

Spray axis. Intersection of two planes of symmetry of the spray [5]; for symmetrical sprays the spray axis coincides with the centerline of the angle.

Stability curve. Graph showing relationship between jet velocity and breakup length.

Streak. Very narrow sector of the spray with more or less than the average concentration of droplets.

Surface tension. Property that resists expansion of liquid surface area. Surface tension forces must be overcome by aerodynamic, centrifugal, or pressure forces to achieve atomization.

Swirl chamber. Conical or cylindrical cavity having tangential inlets that impart a swirling motion to the liquid.

Temporal sampling. Measurement of drops that pass through a fixed area during a specific time interval.

Transition range. Small flow range of a dual-stage nozzle over which atomization quality is relatively poor; occurs when the pressurizing valve first opens to allow secondary liquid flow into the nozzle.

Turndown ratio. Ratio of maximum rated liquid flow to minimum rated liquid flow.

Twin-fluid atomizer. Generic term encompassing all nozzle types in which atomization is achieved using high-velocity air, gas, or steam.

Ultrasonic atomizer. Atomizer in which a vibrating surface is used to cause a liquid film to become unstable and disintegrate into drops.

Upper critical point. Point on stability curve where breakup changes from varicose to sinuous.

Varicose. Term used to describe appearance of liquid jet during breakup without interaction with air.

Velocity coefficient. Ratio of actual discharge velocity to the theoretical velocity corresponding to the total pressure differential across the nozzle.

Viscosity. Liquid property that has a marked effect on atomization quality and spray angle and also affects pumping power requirements; very dependent on liquid temperature.

Visibility technique. Drop sizing interferometry used in conjunction with laser Dopper anemometry to measure both drop size and velocity.

Weber number. Dimensionless ratio of momentum force to surface tension force.

Web bulb temperature. Droplet surface temperature during steady-state evaporation.

Whistle atomizer. Atomizer in which sound waves are used to shatter a liquid jet into droplets.

Wide-range nozzle. Nozzle designed to provide good atomization over a wide range of liquid flow rates; best known example is dual-orifice nozzle.

Windmill atomizer. Rotary atomizer used for aerial application of pesticides; unique feature is use of wind forces to provide rotary motion.

REFERENCES

1. Lefebvre, A. H., Fuel Effects on Gas Turbine Combustion—Ignition, Stability, and Combustion Efficiency, *ASME J. Eng. Gas Turbines Power,* Vol. 107, 1985, pp. 24–37.
2. Reeves, C. M., and Lefebvre, A. H., Fuel Effects on Aircraft Combustor Emissions, ASME Paper 86-GT-212, 1986.
3. Rink, K. K., and Lefebvre, A. H., Influence of Fuel Drop Size and Combustor Operating Conditions on Pollutant Emissions, SAE Technical Paper 861541, 1986.
4. Lord Rayleigh, On the Instability of Jets, *Proc. London Math. Soc.,* Vol. 10, 1878, pp. 4–13.
5. Tate, R. W., Sprays, *Kirk-Othmer Encyclopedia of Chemical Technology,* Vol. 18, 2nd ed., John Wiley and Sons, New York, 1969, pp. 634–654.
6. Fair, J., Sprays, *Kirk-Othmer Encyclopedia of Chemical Engineering,* Vol. 21, 3rd ed., John Wiley and Sons, New York, 1983, pp. 466–483.
7. Topp, M. N., Ultrasonic Atomization—A Photographic Study of the Mechanism of Disintegration, *J. Aerosol Sci.,* Vol. 4, 1973, pp. 17–25.
8. Spillmann, J., and Sanderson, R., A Disc-Windmill Atomizer for the Aerial Application of

Pesticides, *Proceedings of the 2nd International Conference on Liquid Atomization and Spray Systems,* Madison, Wis., 1982, pp. 169–172.

9. Brown, R., and York, J. L., Sprays Formed by Flashing Liquid Jets, *AIChE J.,* Vol. 8, No. 2, 1962, pp. 149–153.

10. Sher, E., and Elata, C., Spray Formation from Pressure Cans by Flashing, *Ind. Eng. Chem. Process Des. Dev.,* Vol. 16, 1977, pp. 237–242.

11. Marek, C. J., and Cooper, L. P., U.S. Patent No. 4,189,914, 1980.

12. Solomon, A. S. P., Rupprecht, S. D., Chen, L. D., and Faeth, G. M., Flow and Atomization in Flashing Injectors, *Atomization Spray Technol.,* Vol. 1, 1985, pp. 53–76.

13. Lefebvre, A. H., Wang, X. F., and Martin, C., Spray Characteristics of Aerated-Liquid Pressure Atomizers, *AIAA J. Propulsion Power,* Vol. 4, in press.

14. Christensen, L. S., and Steely, S. L., Monodisperse Atomizers for Agricultural Aviation Applications, NACA CR-159777, February 1980.

15. Chigier, N., Drop Size and Velocity Instrumentation, *Prog. Energy Combust. Sci.,* Vol. 9, 1983, pp. 155–177.

16. Putnam, A. A., Miesse, C. C., and Pilcher, J. M., Injection and Combustion of Liquid Fuels, Battelle Memorial Institute, WADC Tech. Report 56-344, 1957.

17. Tate, R. W., and Olsen, E. O., Techniques for Measuring Droplet Size in Sprays, Delavan Manufacturing Co., West Des Moines, Iowa, 1964.

18. Anon., Atomizing Nozzles for Combustion Nozzles, Delavan Manufacturing Co., West Des Moines, Iowa, 1968.

TWO

BASIC PROCESSES IN ATOMIZATION

INTRODUCTION

The atomization process is essentially one in which bulk liquid is converted into small drops. Basically, it can be considered as a disruption of the consolidating influence of surface tension by the action of internal and external forces. In the absence of such disruptive forces, surface tension tends to pull the liquid into the form of a sphere, since this has the minimum surface energy. Liquid viscosity exerts a stabilizing influence by opposing any change in system geometry. On the other hand, aerodynamic forces acting on the liquid surface may promote the disruption process by applying an external distorting force to the bulk liquid. Breakup occurs when the magnitude of the disruptive force just exceeds the consolidating surface tension force.

Many of the larger drops produced in the initial disintegration process are unstable and undergo further disruption into smaller drops. Thus, the final range of drop sizes produced in a spray depends not only on the drop sizes produced in primary atomization but also on the extent to which these drops are further disintegrated during secondary atomization.

This chapter is devoted primarily to a review of the various mechanisms that have been proposed to account for the manner in which a jet or sheet emerging from an atomizer is broken down into drops. The process of jet disintegration is of primary importance for the design of plain-orifice atomizers of the types used extensively in diesel and rocket engines. The mechanisms of sheet disintegration have direct relevance to the design of rotary atomizers for spray drying and to the performance of pressure-swirl and prefilming airblast atomizers, which are widely used in oil burners and gas turbine combustors. In view of their importance in the overall atomization process, the various mechanisms of liquid particle breakup are considered first.

STATIC DROP FORMATION

Before embarking on a discussion of practical atomization processes, it is worth-while to consider briefly the most elementary form of atomization, which is the quasi-static case of the hanging or pendant drop. This is exemplified by the slow discharge of a liquid from the end of a burette or a dripping faucet. When the gravity force on the liquid exceeds the attaching surface tension force, the liquid is pulled away from its attachment and a drop forms. The mass of the drop formed is determined by equating the gravitational and surface tension forces on the drop. For the slow emission of liquid from a thin circular tube of diameter d_o, the mass of the drop formed is

$$m_D = \frac{\pi d_o \sigma}{g} \qquad (2.1)$$

The size of the spherical drop corresponding to this mass is given by

$$D = \left(\frac{6 d_o \sigma}{\rho_L g}\right)^{1/3} \qquad (2.2)$$

For a sharp-edged opening 1 mm in diameter, Eq. (2.2) would predict drop sizes of 3.6 and 2.6 mm for water and kerosine, respectively. If the hole size were reduced to 10 μm in diameter, the predicted drop sizes for the two liquids would be 784 and 560 μm, respectively.

For the breakaway of a drop from a flat horizontal wetted surface, which involves a more complex balance of gravitational and surface tension forces, Tamada and Shiback [1] derived the following expression for drop size:

$$D = 3.3\left(\frac{\sigma}{\rho_L g}\right)^{0.5} \qquad (2.3)$$

This equation predicts that drops formed slowly by the action of gravity on a liquid film would have diameters of 9 and 5 mm for water and kerosine, respectively. Thus, the formation of drops by a dripping mechanism inevitably entails large drops and low flow rates. Although common in nature, this mechanism is clearly ineffective for practical applications, most of which call for high liquid flow rates and fine atomization. Gravitational forces are significant only in the formation of large drops and become negligibly small for the range of drop sizes of practical interest, which normally lie between 1 and 300 μm.

BREAKUP OF DROPS

When atomization occurs as a result of interaction between a liquid and the surrounding air, the overall atomization process involves several interacting mechanisms, among which is the splitting up of the larger drops during the final stages of disintegration. It is of interest, therefore, to examine the various ways in which a single drop of liquid can break up under the action of aerodynamic forces.

A rigorous mathematical solution for the breakup of a drop would demand an exact knowledge of the distribution of aerodynamic pressures on the drop. However, as soon as the drop is deformed by these pressures, the pressure distribution around it also changes and either a state of equilibrium between the external aerodynamic forces and the internal forces due to surface tension and viscosity is attained or further deformation follows, leading to possible breakup of the drop.

The influence of variations in the distribution of air pressure around a drop was examined by Klüsener [2]. Under equilibrium conditions the internal pressure at any point on the drop surface, p_I, is just sufficient to balance the external aerodynamic pressure p_A and the surface tension pressure p_σ so that

$$p_I = p_A + p_\sigma = \text{constant} \tag{2.4}$$

Note that for a spherical drop

$$p_\sigma = \frac{4\sigma}{D} \tag{2.5}$$

Clearly, a drop can remain stable as long as a change in air pressure at any point on its surface can be compensated by a corresponding change in p_σ such that p_I remains constant. However, if p_A is large compared with p_σ, then any appreciable change in p_A cannot be compensated by a corresponding change in p_σ to maintain p_I constant. In this situation, the external pressure p_A may deform the drop to an extent that leads to further reduction in p_σ and finally to disruption of the drop into smaller drops [3]. For these smaller drops, the higher value of p_σ as indicated by Eq. (2.5) may be large enough to accommodate the variations in p_A. If not, further subdivision may occur until p_σ is large enough to maintain a constant value of p_I at all points on the drop surface. When this stage is reached the drop is stable and no further breakup can occur.

These considerations lead to the concept of a critical drop size. For drops slightly larger than the critical size, the breakup time increases for decreasing size, until the stable drop has an infinite breakup time [4]. The influence of liquid viscosity, by opposing deformation of the drop, is to increase the breakup time. If the aerodynamic force on the drop declines during the breakup time, no breakup may occur, although the initial aerodynamic force was large enough to produce breakup.

Drop Breakup in Flowing Air

The investigation of drop breakup in flowing air has a long history, dating back to the beginning of this century. Large free-falling drops in still air and smaller drops in a steady stream of air were first considered by Lenard [5] and Hochschwender [6]. Since then, this breakup process has been studied extensively both theoretically and experimentally [7–30].

High-speed photographs have revealed that globules or drops of liquid can split up under the action of aerodynamic forces in a number of ways, depending on the flow pattern around them. Hinze [7] identified the following three basic types of deformation, as illustrated in Fig. 2.1:

1. The drop is flattened to form an oblate ellipsoid (lenticular deformation). Subsequent deformation depends on the magnitude of the internal forces causing the deformation. It is conjectured that the ellipsoid is converted into a torus, which becomes stretched and disintegrates into small drops.
2. The initial drop becomes elongated to form a long cylindrical thread or ligament, which breaks up into small drops (cigar-shaped deformation).
3. Local deformations on the drop surface create bulges and protuberances, which eventually detach themselves from the parent drop to form smaller drops.

According to Hinze [7], deformation of type 1 occurs if the drop is subjected to aerodynamic pressures or viscous stresses produced by parallel and rotating flows. Plane hyperbolic and Couette flows are needed to produce type 2 deformation, while type 3 (described as "bulgy" deformation) occurs in irregular flow patterns. Thus, the preference for any particular type of deformation and breakup depends partly on the physical properties of the gas and liquid phases, namely their densities, viscosities, and interfacial tension, and also on the flow pattern around the drop.

In general, the breakup of a drop in a flowing stream is controlled by the dynamic pressure, surface tension, and viscous forces. For liquids of low viscosity, the deformation of a drop is determined primarily by the ratio of the aerodynamic forces, represented by $0.5\rho_A U_R^2$, and the surface tension forces, which are related to σ/D. Forming a dimensionless group from these two opposing forces yields the Weber number, $\rho_A U_R^2 D/\sigma$. The higher the Weber number, the larger

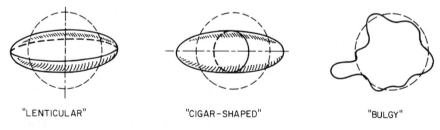

"LENTICULAR"   "CIGAR–SHAPED"   "BULGY"

Figure 2.1 Basic types of globule deformation *(Hinze [7])*.

are the deforming external pressure forces compared with the reforming surface tension forces.

For any given liquid, the initial condition for breakup is achieved when the aerodynamic drag is just equal to the surface tension force, i.e.,

$$C_D \frac{\pi D^2}{4} 0.5\rho_A U_R^2 = \pi D \sigma \tag{2.6}$$

Rearranging these terms provides the dimensionless group

$$\left(\frac{\rho_A U_R^2 D}{\sigma} \right)_{crit} = \frac{8}{C_D} \tag{2.7}$$

where the subscript crit denotes that a critical condition has been reached. As the first term in Eq. (2.7) is the Weber number, the equation may be written as

$$We_{crit} = \frac{8}{C_D} \tag{2.8}$$

For a relative velocity U_R the maximum stable drop size is obtained from Eq. (2.7) as

$$D_{max} = \frac{8\sigma}{C_D \rho_A U_R^2} \tag{2.9}$$

and the critical relative velocity at which the drop will disrupt is given by

$$U_{R_{crit}} = \left(\frac{8\sigma}{C_D \rho_A D} \right)^{0.5} \tag{2.10}$$

Many different experimental techniques have been used to investigate the breakup of large drops. They include (1) free fall from towers and stairwells, (2) suspension of drops in vertical wind tunnels with air velocity adjusted to keep the drops stationary, and (3) use of shock tubes to create supersonic velocities. The knowledge gained from these studies has direct application to the analysis of sprays.

Lane [18] has shown experimentally, and Hinze [7] has confirmed theoretically, that the mode of drop disintegration depends on whether the drop is subjected to steady acceleration or is suddenly exposed to a high-velocity gas stream. With steady acceleration the drop becomes increasingly flattened, and at a critical relative velocity it is blown out into the form of a hollow bag attached to a roughly circular rim, as illustrated in Fig. 2.2. On disintegration the bag produces a shower of very fine drops, while the rim, which contains at least 70% of the mass of the original drop, breaks up into larger drops.

A drop suddenly exposed to a fast airstream disintegrates in an entirely different manner. Instead of being blown out into a thin hollow bag anchored to a rim, the drop is deformed in the opposite direction and presents a convex surface to the flow of air. The edges of the saucer shape are drawn out into a thin sheet and then into fine filaments, which break into drops.

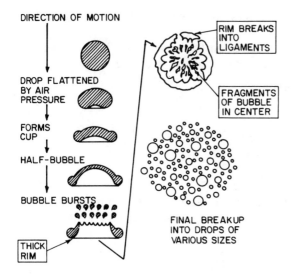

Figure 2.2 Breakup of a spherical drop by interaction with ambient air *(Lane [18], Simmons [31])*.

For the first mode of breakup, Lane [18] has established that at normal atmospheric conditions there is a limiting value of the relative velocity $U_{R_{crit}}$ below which breakup does not occur. For liquids whose surface tensions lie in the range 0.028 to 0.475 kg/s², the relationship is given by

$$U_{R_{crit}} \propto \left(\frac{\sigma}{D}\right)^{0.5} \tag{2.11}$$

This equation should be compared with Eq. (2.10). For water, Eq. (2.11) becomes

$$U_{R_{crit}} = \frac{784}{\sqrt{D}} \tag{2.12}$$

where U_R is in meters per second and D in micrometers.

From the data of Merrington and Richardson [9], Hinze [7] estimated We_{crit} to be 22 for a free-falling drop. For low-viscosity liquid drops suddenly exposed to a high-velocity airstream, the critical value of the Weber number was estimated to be 13.

Taylor [10] has provided an explanation for the observation that critical Weber numbers for drops subjected to a steady stream are almost twice as high as those for drops suddenly exposed to an airflow at constant speed. According to Taylor, the reason for this may be seen by considering the drop as a vibrating system and comparing the suddenly applied force F_T and the gradually increasing force F_S required to produce the same extension. The maximum distortion the drop can sustain without breaking occurs when $F_T = 0.5F_S$.

Hinze's value of 13 for the critical Weber number may be compared with

$8/C_D$ from Eq. (2.8), 10.6 from the data of Lane [18], 10.3 for mercury drops in air obtained by Haas [19], and 7.2 to 16.8 (with an average of about 13.0) for water, methyl alcohol, and a low-viscosity silicone oil obtained by Hanson et al. [20].

To account for the influence of liquid viscosity on drop breakup, Hinze [7] used a viscosity group defined as

$$Z = \frac{\sqrt{We}}{Re} \tag{2.13}$$

This dimensionless group represents the ratio of an internal viscosity force to an interfacial surface tension force [13]. Some idea of the significance of the group may be gained by examining these forces [21]. The interfacial tension force per unit area is represented by σ/D, while the friction force per unit area, τ, is given by the product of liquid viscosity and the velocity gradient within the drop, i.e., $\mu_L \, \delta U/\delta x$. If the drop is oscillating at its natural frequency of vibration, ω, then τ is of the order $\mu_L \omega$. The frequency is related to the properties of the drop by

$$\omega^2 = \frac{2\sigma n(n+1)(n-1)(n+2)}{\pi^2 D^3 [\rho_L(n+1) + \rho_A n]} \tag{2.14}$$

where n depends on the mode of vibration. The most important mode is the first, for which $n = 2$. Since the density of the liquid is usually much higher than that of the surrounding air, for $n = 2$ Eq. (2.14) becomes

$$\tau = \mu_L \omega \approx \frac{4\mu_L}{\pi D}\left(\frac{\sigma}{D\rho_L}\right)^{0.5} \tag{2.15}$$

The ratio of the friction force to the surface tension force is thus

$$\frac{\tau}{\sigma D} = \frac{4}{\pi}\frac{\mu_L}{(\rho_L \sigma D)^{0.5}} = \left(\frac{4}{\pi}\right)Oh \tag{2.16}$$

This equation illustrates the relevance of the Ohnesorge number Oh to the atomization process. However, as Sleicher [21] has pointed out, before the drops break up they are likely to become distorted to an extent far beyond the linear region of validity of Eq. (2.15). Therefore, the viscosity forces represented in this group may not be as important as those that resist the stretching of the drop by the flow field.

According to Hinze [7], the effect of viscosity on the critical Weber number can be expressed by a relation of the form

$$We_{crit} = \dot{We}_{crit}\,[1 + f(Oh)] \tag{2.17}$$

where $\dot{We}_{crit}$ is the critical Weber number for zero viscosity.

This form is considered acceptable because Oh goes to zero as μ_L goes to zero. Thus, when μ_L is so small that it offers no resistance to drop deformation, the initial value of the Weber number, as given by Eq. (2.17), reverts to the value for zero viscosity.

The data of Hanson et al. [20] provide qualitative support for Eq. (2.17) but do not agree in detail. Brodkey [22] has proposed the following empirical relationship:

$$We_{crit} = \dot{W}e_{crit} + 14Oh \qquad (2.18)$$

which is claimed to be accurate within 20%.

Drop Breakup in Turbulent Flow Fields

The foregoing discussion is based on the assumption of a high relative velocity between the drop and the surrounding gas. In many practical situations, a high relative velocity may not exist or may be difficult to determine. In such cases it is perhaps more logical to assume that the dynamic pressure forces of the turbulent motion determine the size of the largest drops. Kolmogorov [23] and Hinze [7] took this view and further assumed that since the breakup was to be considered local, the principles of isotropic turbulence would be valid. The kinetic energy of a turbulent fluctuation increases with increasing wavelength. Thus, velocity differences having a wavelength equal to $2D$ will produce a higher dynamic pressure than those due to fluctuations of shorter wavelength. If these fluctuations are assumed to be responsible for the breakup of drops, the critical Weber number becomes

$$We_{crit} = \frac{\rho_A \bar{u}^2 D_{max}}{\sigma} \qquad (2.19)$$

where $\bar{u}^2$ is the average value across the whole flow field of the squares of velocity differences over a distance equal to D_{max} [7]. For isotropic turbulence, the main contributions to the kinetic energy are made by fluctuations where the Kolmogorov energy distribution law is valid. For this region, where the turbulence pattern is determined solely by the energy input per unit mass per unit time, E, it can be shown that

$$\bar{u}^2 = C_1(ED)^{2/3} \qquad (2.20)$$

where, according to Batchelor [24], $C_1 = 2$.

For low-viscosity liquids (Oh $<< 1$) Eq. (2.19) may be written as

$$We_{crit} = \frac{2\rho_A}{\sigma} E^{2/3} D_{max}{}^{5/3} \qquad (2.21)$$

and

$$D_{max} = C\left(\frac{\sigma}{\rho_A}\right)^{3/5} E^{-2/5} \qquad (2.22)$$

where C is a constant to be determined experimentally. This simple result can also be derived directly from dimensional analysis on the assumption that only σ, ρ_A, and E determine the size of the largest drops.

Hinze [7] used Clay's [25] experimental data to calculate the value of the constant C in Eq. (2.22). Clay's apparatus consisted of two coaxial cylinders, of which the inner one rotated. The space between the cylinders was filled with two immiscible fluids, one of which formed discrete drops. Clay determined drop size distributions as a function of energy input and from these calculated the value of D_{95}, the value of D for which 95% of the total dispersed liquid is contained in drops of smaller diameter. By assuming $D_{max} = D_{95}$, Hinze used Clay's data to derive a value for C of 0.725 (see Fig. 2.3). Equation (2.22) thus becomes

$$D_{max} = 0.725 \left(\frac{\sigma}{\rho_A}\right)^{0.6} E^{-0.4} \qquad (2.23)$$

Substitution of this relation for D_{max} into Eq. (2.21) gives

$$We_{crit} = 1.18 \qquad (2.24)$$

Experimental data on breakup in an isotropic field are nonexistent, so direct verification of Eqs. (2.23) and (2.24) is not possible. Sleicher [21] has shown that these equations are not valid for pipe flow. In a pipe system, breakup occurs as a result of a balance between surface forces, velocity fluctuations, pressure fluctuations, and the steep velocity gradients.

The work of Kolmogorov [23] and Hinze [7] on the splitting of drops in turbulent flows has been modified by Sevik and Park [26]. They suggest that resonance can cause drop breakup in turbulent flow fields when the characteristic turbulence frequency matches the lowest or natural frequency mode of an entrained fluid particle. Since damping is very small, such drops will deform very violently if the existing frequency corresponds to one of their resonant frequencies. By setting a characteristic frequency of the turbulence equal to such a res-

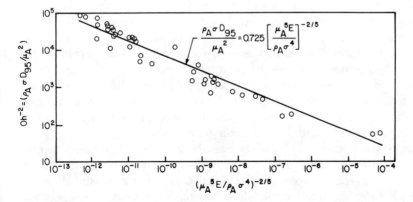

Figure 2.3 Maximum drop size as a function of energy input (Hinze [7]).

onant frequency, Sevik and Park were able to predict theoretically the critical Weber numbers corresponding to both Clay's drop-splitting experiments and their own bubble-splitting experiments. For drops it was found that

$$We_{crit} = 1.04 \qquad (2.25)$$

as compared with Hinze's value of 1.18.

Drop Breakup in Viscous Flow Fields

This mechanism of breakup applies to fluid globules surrounded by viscous fluid where there is a strong velocity gradient in the vicinity of the globules. In this situation the Reynolds numbers that are characteristic of the flow field may be so small that dynamic forces are no longer important, and breakup is controlled entirely by viscous and surface tension forces. If the viscous forces are large enough, the interfacial tension forces are unable to prevent the globule from splitting.

The first basic experiments on the splitting of drops under the action of viscous and surface tension forces were made by Taylor [10] in 1934. His test apparatus was designed to generate carefully controlled flow patterns, and several liquids with different viscosities were used. Among the observations made by Taylor, many of which were subsequently explained by Tomotika [27], are the following:

1. Under the action of viscous shear, a drop elongates into the shape of a prolate ellipsoid.
2. The deformation is governed by a dimensionless group, $\mu_C SD/\sigma$, where μ_C is the viscosity of the continuous phase and S is the maximum velocity gradient in the external flow field.
3. Breakup of the drop occurs at a critical value of the Weber number that depends on the continuous flow field.

The experimental data obtained by Meister and Scheele [28] demonstrate the effects of the viscosities of both the dispersed and continuous phases, as well as their viscosity ratio, on atomization. At this point it is appropriate to make a distinction between the *dispersed* and *continuous* phases. The dispersed phase refers to the liquid to be atomized, whereas the continuous phase represents the medium, usually gaseous, in which the atomization occurs. By injecting heptane into various concentrations of aqueous glycerine solutions, Meister and Scheele showed that variation of continuous-phase viscosity has no appreciable effect on the atomization process. The viscosity of the dispersed phase, however, delays breakup at higher velocities, thereby impeding atomization.

The analysis of Tomotika [27], as reviewed by Miesse [29], suggests that the perturbation wavelength has a minimum value (for maximum growth rate) when the viscosity ratio μ_L/μ_A is almost unity. Either decreasing or increasing the value of this ratio reduces the tendency for drop breakup. The same conclusion was reached by Hinze [7], who showed from Taylor's experimental data that a minimum value for We_{crit} of less than 1 is obtained for viscosity ratios between 1 and 5.

Taylor's theory has been modified by several workers over the years. For example, Rumscheidt and Mason [30] proposed that breakup occurs when

$$We_{crit} = \frac{1 + (\mu_L/\mu_A)}{1 + (19/16)(\mu_L/\mu_A)} \tag{2.26}$$

which varies only between 1 and 0.84 as μ_L/μ_A varies from zero to infinity.

Sevik and Park [26] pointed out that the Taylor mechanism of globule deformation applies only if both the undeformed and elongated drops are small compared with local regions of viscous flow. When the Reynolds number of the external flow field is large, as it is in most practical applications, the spatial dimensions of such local regions are very small compared with the drop sizes. Under these circumstances, the determining factor is the dynamic pressure caused by the velocity changes over distances of the order of the globule diameter. It is also clear from Taylor's experiments that it is very difficult to atomize liquids that show a high viscosity ratio. This explains why, in practice, dispersion by pure viscous flow is restricted to emulsification processes.

DISINTEGRATION OF LIQUID JETS

When a liquid jet emerges from a nozzle as a continuous body of cylindrical form, the competition set up on the surface of the jet between the cohesive and disruptive forces gives rise to oscillations and perturbations. Under favorable conditions the oscillations are amplified and the liquid body disintegrates into drops. This process is sometimes referred to as *primary* atomization. If the drops so formed exceed the critical size, they further disintegrate into drops of smaller size, a process known as *secondary* atomization.

The phenomenon of jet disintegration has been subjected to theoretical and experimental investigation for more than 100 years. A comprehensive review of jet flow was made by Krzywoblocki [32], and other reviews of relevance to liquid jet flow are also available [29, 33–38]. The following discussion is restricted to liquid jets injected into a gaseous atmosphere. It applies to any combination of liquid (the dispersed phase) and gas (the continuous phase), but for convenience the surrounding gas is usually referred to as air.

The properties of jets of most interest are the continuous length (which provides a measure of the growth rate of the disturbance) and the drop size (which is a measure of the wave number of the most unstable disturbance). Also of interest is the manner in which the jet is disrupted.

The earliest investigations into jet flow phenomena appear to have been carried out by Bidone [39] and Savart [40]. Bidone's research was concerned with the geometric forms of jets produced by nozzles of noncircular cross section, while Savart supplied the first quantitative data related to jet disintegration. His results showed that if the jet diameter is kept constant, the length of the continuous part of the jet is directly proportional to jet velocity. He also observed that, for constant jet velocity, the length of a jet is directly proportional to its diameter.

Plateau [41] is credited with the first theoretical investigation into jet instability. He showed that a cylindrical column of liquid is unstable if its length exceeds its perimeter; otherwise, two drops would be formed whose total surface area would be less than that of the original cylinder. Plateau's work helped to explain the results of Savart and provided the basis for Rayleigh's more widely known theory of jet stability.

In an early mathematical analysis, Rayleigh [11] employed the method of small disturbances to predict the conditions necessary to cause the collapse of a liquid jet issuing at low velocity, for example, a low-speed water jet in air. This comprehensive analysis has been reviewed and summarized by McCarthy and Molloy [42].

Rayleigh compared the surface energy (directly proportional to the product of surface area and surface tension) of the disturbed configuration with that of the undisturbed column. He then calculated the potential energy of the disturbed configuration (relative to the equilibrium value) as

$$E_s = \frac{\pi\sigma}{2d}(\gamma^2 + n^2 - 1)b_n^2 \tag{2.27}$$

where E_s = potential surface energy
   d = jet diameter
   b_n = constant in Fourier series expansion
   γ = dimensionless wave number = $2\pi/\lambda$
   λ = wavelength of disturbance
   n = any positive integer (including zero)

For nonsymmetrical disturbances, $n \gg 1$ and E_s is always positive, indicating that the system is always stable to this class of disturbance. When $n = 0$ and $\gamma < 1$, which is the case for symmetrical disturbances, Eq. (2.27) shows that E_s is negative and the system is unstable to this class of disturbance. Hence, a liquid jet that is affected by surface tension forces only will become unstable to any axisymmetrical disturbance whose wavelength satisfies the inequality

$$\lambda > \pi d$$

which corresponds to

$$\gamma < 1$$

The conclusion to be drawn from Rayleigh's analysis of the breakup of non-viscous liquid jets under laminar flow conditions is that all disturbances on a jet with wavelengths greater than its circumference will grow. Furthermore, his results show that one class of disturbance will grow fastest and eventually control the breakup. Although actual liquid jets are viscous, turbulent, and subjected to surrounding air influences, the conclusions of Rayleigh have found general acceptance in later theories as valid first approximations.

Rayleigh's contribution to the mathematical treatment of the breakup of jets stemmed from his recognition that jet breakup is a dynamic problem and that the rate of collapse is important. By assuming that b_n in Eq. (2.27) is proportional to $\exp(qt)$, where q is the exponential growth rate of disturbance, Rayleigh showed that the exponential growth rate of the fastest-growing disturbance is given by

$$q_{max} = 0.97 \left(\frac{\sigma}{\rho_L d^3} \right)^{0.5} \tag{2.28}$$

and λ_{opt}, corresponding to q_{max}, is

$$\lambda_{opt} = 4.51d \tag{2.29}$$

After breakup, the cylinder of length $4.51d$ becomes a spherical drop, so that

$$4.51d \times \frac{\pi}{4} d^2 = (\pi/6)D^3$$

and hence

$$D = 1.89d \tag{2.30}$$

Thus for the Rayleigh mechanism of breakup the average drop size is nearly twice the diameter of the undisturbed jet.

Figure 2.4 shows an idealization of Rayleigh breakup for a falling liquid jet and breakup as it actually occurs. It may be noted in Fig. 2.4a that this mode of drop breakup is quite regular; the liquid jet first starts to run down and eventually

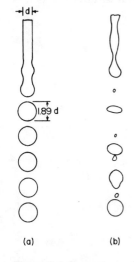

(a)   (b)

Figure 2.4 Comparison of (a) idealized jet breakup with (b) actual breakup as indicated by high-speed photographs.

collapses to form drops of uniform size and spacing. Figure 2.4b, which is based on high-speed photographs, shows the appearance of the jet in this region. The formation and growth of axisymmetric disturbances are evident, as is the characteristic dumbbell shape just prior to breakup. The jet pinches off at the ends of the connecting cylinder, which in turn becomes unstable and breaks down into drops. The drops formed from this cylinder usually coalesce, so the end result is a pattern of large drops with small single "satellite" drops between them [35].

Tyler [43] later measured the frequency of formation of drops as a jet disintegrated and related it to the wavelength of the disturbance. By assuming that the volume of the spherical drops formed by disintegration of the jet is equal to the volume of a cylinder with diameter equal to that of the undisturbed jet and with wavelength equal to the most rapidly growing disturbance, Tyler obtained the following results:

$$\frac{D}{d} = \left(1.5\frac{\lambda}{d}\right)^{1/3} \tag{2.31}$$

$$\lambda_{opt} = 4.69d \tag{2.32}$$

$$D = 1.92d \tag{2.33}$$

In view of the close agreement between his experimental results and the predictions of Rayleigh's mathematical analysis [see Eqs. (2.29) and (2.30)], Tyler concluded that cylindrical jets do break up under the conditions required for maximum instability, as predicted by Rayleigh's theory.

A more general theory for disintegration at low jet velocities was developed by Weber [44], who extended Rayleigh's analysis to include viscous liquids. He assumed that any disturbance causes rotationally symmetrical oscillations of the jet, as illustrated in Fig. 2.5a. If the wavelength of the initial disturbance is less

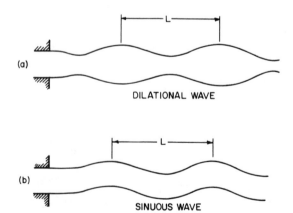

(a)

DILATIONAL WAVE

(b)

SINUOUS WAVE

Figure 2.5 (a) Jet with rotationally symmetric disturbance. (b) Jet disturbance causing wave formation.

than λ_{min}, the surface forces tend to damp out the disturbance. If λ is greater than λ_{min}, the surface tension forces tend to increase the disturbance, which eventually leads to disintegration of the jet. There is, however, one particular wavelength, λ_{opt}, that is most favorable for drop formation. For nonviscous liquids

$$\lambda_{\text{min}} = \pi d \tag{2.34}$$

$$\lambda_{\text{opt}} = \sqrt{2}\pi d = 4.44d \tag{2.35}$$

For viscous liquids

$$\lambda_{\text{min}} = \pi d \tag{2.36}$$

$$\lambda_{\text{opt}} = \sqrt{2}\pi d \left(1 + \frac{3\mu_L}{\sqrt{\rho_L \sigma d}} \right)^{0.5} \tag{2.37}$$

Thus, for nonviscous liquids the value of λ/d required to produce maximum instability is 4.44, which is close to the value of 4.51 predicted by Rayleigh for this case. It is of interest to note that the minimum wavelength is the same for both viscous and nonviscous liquids, but the optimum wavelength is greater for viscous liquids.

Weber next examined the effect of air resistance on the disintegration of jets into drops. He found that air friction shortens both the minimum wavelength and the optimum wavelength for drop formation. Recalling that for nonviscous liquids and zero air velocity

$$\lambda_{\text{min}} = 3.14d \tag{2.34}$$

$$\lambda_{\text{opt}} = 4.44d \tag{2.35}$$

at 15 m/s relative air velocity

$$\lambda_{\text{min}} = 2.2d \tag{2.38}$$

$$\lambda_{\text{opt}} = 2.8d \tag{2.39}$$

Thus, the effect of relative air velocity is to reduce the optimum wavelength for jet breakup.

Weber also considered the case where the air motion induces wave formation and showed that this can occur only at relative air velocities above a certain minimum value. For glycerin the minimum velocity for wave formation is about 20 m/s. At this velocity the theoretical breakup distance is infinite. Increasing the velocity reduces the breakup distance; for example, at 25, 30, and 35 m/s, the corresponding wavelengths are $3.9d$, $3.0d$, and $1.3d$, respectively.

Haenlein [12] presented experimental evidence in support of Weber's theoretical analysis. Using nozzles with a length-to-bore ratio of 10 and liquids of various viscosities and surface tensions, he showed that for liquids of high viscosity (0.85 kg/m s) the ratio of wavelength to jet diameter producing maximum instability could range from 30 to 40, in contrast to the value of 4.5 predicted by the Rayleigh theory for nonviscous jets.

Haenlein identified four distinct regimes of breakup in the disintegration of a liquid jet.

1. Drop formation without the influence of air. This is the mechanism studied by Rayleigh. The term "varicose" is sometimes used to describe the appearance of the jet in this regime. Radially symmetric waves, as illustrated in Fig. 2.6a, are formed by the interaction of primary disturbances in the liquid and surface tension forces. This regime is characterized by a linear relationship between the length of the jet prior to breakup and the jet velocity. Weber calculated the breakup time to be proportional to $d_0^{1.5}$ for nonviscous jets and proportional to d_0 for viscous jets.
2. Drop formation with air influence (Fig. 2.6b). As the jet velocity is increased, the aerodynamic forces of the surrounding air are no longer negligible and tend to accentuate the waves formed under regime 1.
3. Drop formation due to waviness of the jet (Fig. 2.6c). This regime is associated with increasing effectiveness of aerodynamic forces and lessened relative influence of surface tension. The term "sinuous" has been used to describe the jet in this regime.
4. Complete disintegration of the jet, i.e., atomization. The liquid is broken up at the nozzle in a chaotic and irregular manner.

Although these four separate regimes can be clearly identified, there is no sharp demarcation between them. From a practical viewpoint it is unfortunate that regime 4, which is the normal operating condition for the plain-orifice atomizer, is not easily described.

Perhaps the most commonly quoted criteria for classifying jet disintegration

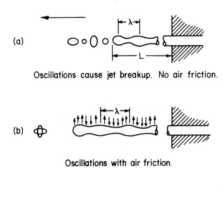

(a)

Oscillations cause jet breakup. No air friction.

(b)

Oscillations with air friction.

(c)

Wave-like breakup caused by air friction.

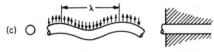

Figure 2.6 Mechanisms of drop formation *(Haenlein [12])*.

are those proposed by Ohnesorge [13]. From photographic records of jet disintegration, Ohnesorge classified the data according to the relative importance of gravitational, inertial, surface tension, and viscous forces. He used dimensionless analysis with good effect to show that the breakup mechanism of a jet could be expressed in three stages, each stage characterized by the magnitudes of the Reynolds number and a dimensionless number Z, which is obtained as

$$Z = \left(\frac{U_L^2 \rho_L d_o}{\sigma} \right)^{0.5} \left(\frac{U_L d_o \rho_L}{\mu_L} \right)^{-1} = \frac{\mu_L}{(\rho_L \sigma d_o)^{0.5}} \tag{2.40}$$

This group is sometimes referred to as the stability number, the viscosity group [7], or the Ohnesorge number (Oh).

Ohnesorge showed that the various mechanisms of jet breakup could be divided into three regions on a graph of Ohnesorge number versus Reynolds number according to the rapidity of drop formation:

1. At low Reynolds numbers, the jet disintegrates into large drops of fairly uniform size. This is the Rayleigh mechanism of breakup.
2. At intermediate Reynolds numbers, the breakup of the jet is by jet oscillations with respect to the jet axis. The magnitude of these oscillations increases with air resistance until complete disintegration of the jet occurs. A wide range of drop sizes is produced.
3. At high Reynolds numbers, atomization is complete within a short distance from the discharge orifice.

Ohnesorge's chart is shown in Fig. 2.7. Since for a given liquid and orifice size the Oh number is constant, a variation of Reynolds number on the chart follows a horizontal line. Thus, at low Reynolds numbers (region I) the jet structure is predominantly varicose and the mode of breakup follows the Rayleigh mechanism. With increase in Reynolds number the mode moves into region II, where the jet oscillates about its axis, having a twisted or sinuous appearance (see Fig. 2.5b). Passing through this narrow band, region III is reached, in which atomization occurs at the orifice from which the jet emerges.

For region III, Castleman [8] proposed a mechanism for jet disruption based on the observations of Sauter [14] and Scheubel [15]. According to Castleman, the most important factor in the process of jet disintegration is the effect of the relative motion between the outer jet layer and the air; this, combined with air friction, causes irregularities in the previously smooth liquid surface and the production of unstable ligaments. As the relative air speed increases, the size of the ligaments decreases and their life becomes shorter; upon their collapse, much smaller drops are formed, in accordance with Rayleigh's theory.

Miesse [29] found that his experimental data would fall in the appropriate area in Fig. 2.7 only if the boundary line between regions II and III was translated to the right, as indicated by the dashed line in Fig. 2.7. The equation of the modified boundary line was determined by Miesse as

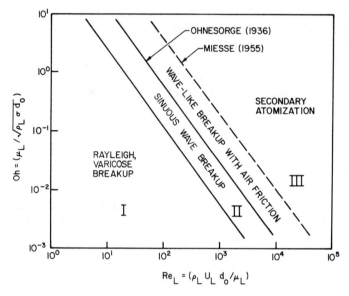

Figure 2.7 Classification of modes of disintegration *(Ohnesorge [13])*.

$$Oh = 100Re^{-0.92} \tag{2.41}$$

or

$$We^{0.5} = 150Oh^{-0.087} \tag{2.42}$$

In a more recent study, Reitz [45] attempted to resolve some of the uncertainties surrounding the Ohnesorge chart. His analysis was based on interpretation of the data on diesel sprays obtained by himself and other workers, including Giffen and Muraszew [3] and Haenlein [12]. According to Reitz [45], the following four regimes of breakup are encountered as the liquid injection velocity is progressively increased.

1. Rayleigh jet breakup. This is caused by the growth of axisymmetric oscillations of the jet surface, induced by surface tension. Drop diameters exceed the jet diameter.
2. First wind-induced breakup. The surface tension effect is now augmented by the relative velocity between the jet and the ambient gas, which produces a static pressure distribution across the jet, thereby accelerating the breakup process. As in regime 1, breakup occurs many jet diameters downstream of the nozzle. Drop diameters are about the same as the jet diameter.
3. Second wind-induced breakup. Drops are produced by the unstable growth of short-wavelength surface waves on the jet surface caused by the relative motion of the jet and the ambient gas. This wave growth is opposed by surface tension. Breakup occurs several diameters downstream of the nozzle exit. Average drop diameters are much less than the jet diameter.

4. Atomization. The jet disrupts completely at the nozzle exit. Average drop diameters are much less than the jet diameter.

These four regimes are shown in Fig. 2.8 and described in Table 2.1.

Influence of Jet Velocity Profile

It is probable that some of the peculiarities and anomalies associated with experimental work on jet stability arise from differences in the velocity profile and turbulence properties of the jet as it emerges from the nozzle. Schweitzer [33] provided an excellent qualitative description of hydraulic turbulence in jets and orifices. The jet emerges from the nozzle in either a laminar or turbulent state. When the liquid particles flow in streams parallel to the axis of the tube, the flow is laminar. When the paths of the liquid particles cross each other in a random manner, the particles having various transverse velocity components, the flow is turbulent. Laminar flow is promoted by the absence of any disturbances to the flow, a rounded entrance to the tube, and high liquid viscosity, whereas turbulent flow is promoted by high flow velocity, large tube size, surface roughness, rapid changes in tube cross-sectional area, and protuberances projecting into the flowing stream. Irrespective of the length of the tube, an initially laminar flow will remain laminar if its Reynolds number is lower than the critical Reynolds number. Conversely, an originally turbulent flow will remain turbulent if its Reynolds number remains above the critical value.

In a smooth tube with no flow disturbances, an initially laminar flow can

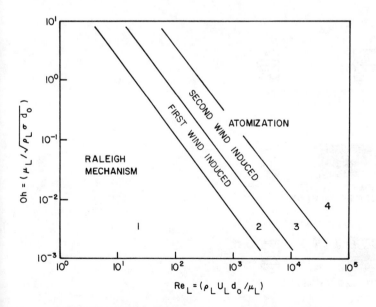

Figure 2.8 Classification of modes of disintegration *(Reitz [45])*.

Table 2.1 Classification of jet breakup regimes

Regime	Description	Predominant drop formation mechanism	Criteria for transition to next regime
1	Rayleigh breakup	Surface tension force	$We_A > 0.4$ $We_A > 1.2 + 3.4Oh^{0.9}$
2	First wind-induced breakup	Surface tension force; dynamic pressure of ambient air	—
3	Second wind-induced breakup	Dynamic pressure of ambient air opposed by surface tension force initially	$We_A > 40.3$ $We_A > 13$
4	Atomization	Unknown	—

Source: Reitz [45].

remain laminar up to Reynolds numbers much higher than the critical value, but if the Reynolds number exceeds the critical value, only a small disturbance to the flow is required to initiate a transition to turbulent flow.

As Schweitzer [33] pointed out, there is a general tendency to associate turbulent flow with a Reynolds number higher than critical, although in reality the flow is sometimes laminar at Reynolds numbers higher than critical and may be turbulent or semiturbulent at Reynolds numbers below the critical value. A semiturbulent flow comprises a turbulent core and a laminar envelope, as illustrated in Fig. 2.9.

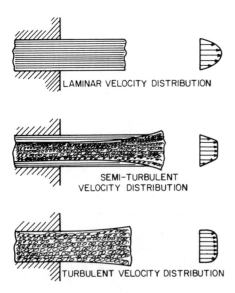

Figure 2.9 Various velocity distributions in jets as suggested by Giffen and Muraszew [3].

The critical Reynolds number may be defined as the value below which, in a long straight cylindrical tube, any disturbances in the flow will damp out. Above the critical Reynolds number, disturbances in the flow never damp out, no matter how long the tube is [33]. The critical Reynolds number so defined was found by Schiller [46] to be around 2320. In attempts to standardize the velocity profile in the emerging jet, many workers have used long tubes as nozzles to ensure that the jet initially possesses either a fully developed laminar (parabolic) profile or a fully developed turbulent velocity profile.

Spray nozzles are usually of compact size, with short orifices to minimize pressure losses. The state of turbulence in the flow at the nozzle exit is determined by the flow state upstream of the orifice and by disturbances generated in the approach passages and within the orifice itself. For a given nozzle all these factors are constant, so the nature of the flow—laminar, turbulent, or semiturbulent—is dictated by the Reynolds number.

The strong influence of the velocity profile of the emerging jet on its subsequent disintegration has been described by Schwietzer [33], McCarthy and Molloy [42], and Sterling [35, 47]. With laminar flow, the velocity distribution in the jet immediately downstream of the orifice varies in a parabolic manner, rising from zero at the outer surface to a maximum at the jet axis. If the jet is injected into quiescent or slow-moving air, there is no appreciable velocity difference between the outer skin of the jet and the adjacent air. Consequently the necessary conditions for atomization by air friction forces do not exist. However, after a certain distance, the combined effects of air friction and surface tension forces create surface irregularities that ultimately lead to disintegration of the jet.

If the flow at the orifice is fully turbulent, the radial velocity component soon leads to disruption of the surface film followed by general disintegration of the jet. It should be noted that when the issuing jet is fully turbulent, no aerodynamic forces are needed for breakup. Even when injected into a vacuum, the jet will disintegrate solely under the influence of its own turbulence.

If the jet is semiturbulent, as illustrated in Fig. 2.9, the annulus of laminar flow surrounding the turbulent core tends to prevent the liquid particles in the core from reaching and disrupting the jet surface. At the same time, the influence of air friction is minimal, due to the very low relative velocity between the jet surface and the surrounding air. Thus, jet disintegration does not occur close to the orifice exit. However, farther downstream, the faster turbulent core outpaces its protective laminar layer and then disintegrates in the normal manner of a turbulent jet. Alternatively, and to some extent simultaneously, a redistribution of energy takes place between the laminar and turbulent components of the total flow, which produces a flattening of the jet velocity profile. This process brings to the jet surface liquid particles with radial components of velocity. These particles disrupt the jet surface so that ultimately it disintegrates into drops.

The change in velocity profile (usually termed velocity profile "relaxation") that occurs downstream of the nozzle exit can have an important influence on the stability of the jet and on its subsequent breakup into drops [42]. As soon as a jet leaves a nozzle, i.e., as soon as the physical constraint of the nozzle wall is removed, the process of velocity profile relaxation occurs by a mechanism of momentum transfer between transverse layers within the jet. Thus, in addition to the normal jet-destabilizing forces discussed above, there exists an additional disruptive mechanism arising from the internal motions associated with profile relaxation.

It is well known that the kinetic energy per unit mass of flowing gas or liquid is very dependent on the fluid velocity profile. For example, in pipe flow with a fully developed parabolic velocity profile (i.e., laminar flow) the kinetic energy per unit mass of fluid is exactly twice what it would be for a flat velocity profile (plug flow) with the same average velocity. Following McCarthy and Molloy [42], a quantity ϵ can be defined such that

$$\epsilon = \int_0^A U_r^3 dA / U^3 A \qquad (2.43)$$

where U_r is the local fluid velocity and U the average fluid velocity over area A. For the three different flow situations we have

1. For plug flow, $\epsilon = 1$.
2. For fully developed turbulent flow, $\epsilon = 1.1$ to 1.2.
3. For fully developed laminar flow, $\epsilon = 2.0$.

Thus, for flow in a pipe, the quantity ϵ represents the ratio of the kinetic energy to the equivalent kinetic energy under plug flow conditions.

When a fully developed laminar jet emerges from a nozzle, its parabolic profile relaxes into a flat profile at the same average velocity. This process is accompanied by a reduction in ϵ from 2 to 1 that involves a considerable redistribution of energy within the jet, leading to the creation of forces that can be quite violent, causing the jet to burst. The phenomenon of "bursting breakup" was first observed and explained by Eisenklam and Hooper [48] and was also noted by Rupe [49]. Obviously, jets with fully developed turbulent profiles on exit ($\epsilon = 1.1$ to 1.2) are only slightly susceptible to profile relaxation effects.

For high-velocity jets it is now generally believed that the action of the surrounding air or gas is the primary cause of atomization, although jet turbulence is a contributing factor because it ruffles the surface of the jet, making it more susceptible to aerodynamic effects. The different roles played by liquid turbulence and air friction allow the distinction to be made between *primary* and *secondary* atomization. Primary atomization is related to jet breakup by the action of internal forces, such as turbulence, inertial effects, or those arising from velocity profile relaxation and surface tension. Secondary atomization always involves the action of aerodynamic forces in addition to those present in primary atomization [42]. Aerodynamic forces promote atomization by acting directly on the surface of the jet and by splitting the drops formed in primary atomization into smaller droplets.

Stability Curve

Many investigators have characterized jet behavior by determining experimentally the relationship between jet velocity and breakup length. The latter is defined as the length of the continuous portion of the jet, measured from the nozzle to the breakup point where drop formation occurs.

Laminar flow region. The general shape of the length-velocity curve is shown in Fig. 2.10. The initial dashed portion of this curve below A corresponds to drip flow. Point A denotes the lower critical velocity at which the drip flow changes to jet flow. From A to B the breakup length L increases linearly with velocity. This portion of the curve corresponds to disintegration of the jet due to surface forces as studied by Rayleigh and Weber. Weber [44] has shown that a small axisymmetric disturbance δ_0 will grow at an exponential rate q_{max} until δ acquires a value equal to the jet radius. If t_b is the time required for breakup, it is assumed that

$$r_0 = \delta_0 \exp(q_{max} t_b) \tag{2.44}$$

Hence

$$t_b = \frac{\ln(d/2\delta_0)}{q_{max}} \tag{2.45}$$

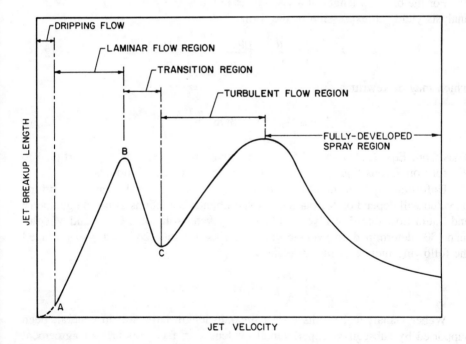

Figure 2.10 Jet stability curve indicating change of breakup length with jet velocity.

As

$$t_b = \frac{L}{U}$$

$$L = \frac{U}{q_{max}} \ln \left(\frac{d}{2\delta_0} \right) \tag{2.46}$$

From Eq. (2.28) we have

$$q_{max} = 0.97 \left(\frac{\sigma}{\rho_L d^3} \right)^{0.5} \tag{2.28}$$

Substituting for q_{max} from Eq. (2.28) into Eq. (2.46) gives

$$\frac{L}{d} = 1.03 U \left(\ln \frac{d}{2\delta_0} \right) \left(\frac{\rho_L d}{\sigma} \right)^{0.5} \tag{2.47}$$

or

$$L = 1.03 d \, We^{0.5} \ln(d/2\delta_0) \tag{2.48}$$

Equation (2.48) is thus a prediction of the breakup length of a liquid jet subjected only to inertial and surface tension forces.

For the breakup length of a viscous jet in the absence of air friction, Weber's analysis yields an expression of the form

$$L = U \left(\ln \frac{d}{2\delta_0} \right) \left[\left(\frac{\rho_L d^3}{\sigma} \right)^{0.5} + \frac{3\mu_L d}{\sigma} \right] \tag{2.49}$$

which may be rewritten as

$$L = d \, We^{0.5} (1 + 3Oh) \ln \left(\frac{d}{2\delta_0} \right) \tag{2.50}$$

Thus from Eqs. (2.48) and (2.50) we see that breakup length is proportional to $d^{1.5}$ for nonviscous liquids and proportional to d for viscous liquids.

Unfortunately, the initial disturbance $(d/2\delta_0)$ cannot be determined a priori. Its value will depend on the particular experimental conditions of nozzle geometry and liquid flow rate. For glycol and glycerol/water solutions, Grant and Middleman [50] determined an average value of 13.4 for $\ln(d/2\delta_0)$ and also provided the following more general correlation:

$$\ln \left(\frac{d}{2\delta_0} \right) = 7.68 - 2.66 Oh \tag{2.51}$$

Weber's analytical results for jet breakup length have not, in general, been supported by subsequent experimental evidence. In fact, this lack of agreement has caused some investigators to discount Weber's theory entirely. It was left to Sterling and Sleicher [47] to show that reported discrepancies between theory and experiment could be due to relaxation of the velocity profile in the jet, as discussed earlier. They also showed that, in the absence of velocity profile relaxation, Weber's theory overestimates the influence of aerodynamic forces. They then modified Weber's theory to account for the viscosity of the ambient gas. Details of their analysis are beyond the scope of this work, but the significant improvement in data prediction afforded by their approach is illustrated in Figs. 2.11 and 2.12.

Mahoney and Sterling [51] later extended the results of Sterling and Sleicher [47] to obtain a universal equation for the length of laminar Newtonian jets. Their equation is strictly valid only for jets with an initially uniform velocity profile, but it appears to be satisfactory also for jets with an initial parabolic profile when the Oh number (based on jet properties) is large. It is essentially the same as Eq. (2.50) but has an extra term that is a function of Oh and We.

$$L = d\,\mathrm{We}^{0.5}(1 + 3\mathrm{Oh})\ln\!\left(\frac{d}{2\delta_0}\right)\bigg/ f(\mathrm{Oh}, \mathrm{We}) \qquad (2.52)$$

For details of the expressions used in estimating $f(\mathrm{Oh}, \mathrm{We})$ in terms of the jet velocity and liquid properties, reference should be made to the original paper [51].

The success of Eq. (2.52) in predicting breakup length is illustrated in Fig. 2.13. This figure shows a comparison of measured breakup lengths obtained by

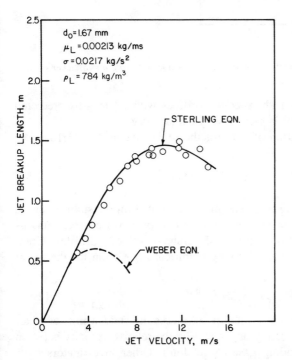

Figure 2.11 Comparison of experimental data with predictions of Weber [44] and Sterling [35].

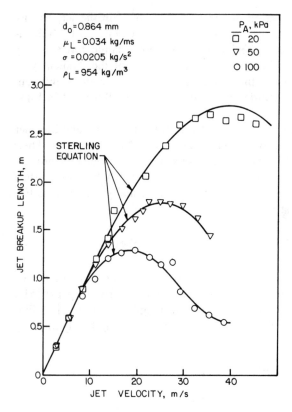

Figure 2.12 Comparison of experimental data with predicted values *(Sterling and Sleicher [47])*.

Phinney and Humphries [52] for sharp-edged orifices with jet lengths predicted from Eq. (2.52). The agreement is clearly very satisfactory.

A further empirical correlation of data for the region AB in Fig. 2.10 is the following, due to Grant and Middleman [50].

$$L = 19.5d \, \text{We}^{0.5}(1 + 3\text{Oh})^{0.85} \tag{2.53}$$

Upper critical point. According to Haenlein [12], point B on the stability curve corresponds to the change in the breakup mechanism from varicose to sinuous (see Fig. 2.5b). For jets with fully developed parabolic velocity profiles, Grant and Middleman [50] developed the following empirical correlation for the upper critical point:

$$\text{Re}_{\text{crit}} = 3.25\text{Oh}^{-0.28} \tag{2.54}$$

According to Weber, the upper critical point marks the sudden impact of air resistance on jet stability and is solely dependent on the relative velocity between the jet surface and the surrounding gaseous medium. Other investigators have attributed this critical point to the onset of turbulence in the jet. It seems probable that both mechanisms operate together in promoting more rapid jet disintegration, but which is dominant has yet to be resolved.

Van de Sande and Smith [53] have proposed the following expression for the critical Reynolds number at which the jet changes from laminar to turbulent flow:

$$Re_{crit} = 12,000\left(\frac{1}{d}\right)^{-0.3} \tag{2.55}$$

This equation is intended to apply to the transition region, shown as BC in Fig. 2.10, where the changeover from a laminar to a turbulent jet occurs.

Turbulent jets. A turbulent jet is defined as one in which the flow is turbulent at the nozzle exit. Turbulent jets tend to have a "milky" surface. This is caused by light scattering from ruffles on the surface produced by random fluctuations of the velocity components, as opposed to the clear glassy appearance of laminar jets.

The notion that atomization is due to aerodynamic interaction between the liquid and the gas leading to unstable wave growth on the liquid jet surface, as postulated by Castleman [16], has not found universal acceptance. It is agreed that aerodynamically induced wave growth needs time to develop and, therefore, an undisturbed length should be observed at the nozzle exit. However, Reitz and Bracco [54] have pointed out that the most unstable wavelength and the undis-

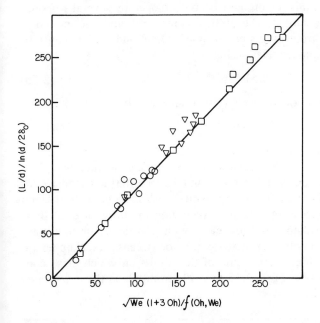

Figure 2.13 Comparison of measured jet length to jet length predicted by Eq. (2.52) *(Mahoney and Sterling [51])*.

turbed length could be much smaller than the jet diameter and therefore difficult to observe experimentally. This is borne out by the remarkably clear and detailed photograph obtained for a water jet by Taylor and Hoyt [55], shown in Fig. 2.14, which provides strong support for Castleman's hypothesis.

Alternative mechanisms for the disintegration of turbulent jets have been proposed by various workers. For example, DeJuhasz [56] has suggested that the jet breakup process starts within the nozzle itself and is strongly influenced by turbulence. As discussed earlier, Schweitzer [33] took the view that the radial velocity components generated in turbulent pipe flow could produce immediate disruption of the jet at the nozzle exit. Bergwerk [57] hypothesized that liquid cavitation phenomena inside the nozzle could create large-amplitude pressure disturbances in the flow, leading to atomization. Sadek [58] also considered that cavitation bubbles could influence the atomization process. Other workers, including Eisenklam and Hooper [48] and Rupe [49], have suggested that jet breakup is due to velocity profile relaxation, which also accounts for the greater stability of turbulent jets. From their review of experimental work on turbulent jets and their own experimental observations, Reitz and Bracco [54] concluded that no single mechanism is responsible for jet breakup in all cases and often a combination of factors is involved.

Regardless of the detailed mechanisms of the atomization process, for turbulent jet flow with strong interaction between the liquid and the ambient gaseous medium, the jet breakup length increases with increase in velocity, as indicated by the region on the stability curve beyond point C in Fig. 2.10. Empirical correlations for breakup data in this region have been developed by several workers, including Vitman [59], Lienhard and Day [60], Phinney [36], and Lafrance [61]. For turbulent jet emerging from long smooth tubes, Grant and Middleman [50] suggested the following empirical relationship:

$$L = 8.51 d_o \, \mathrm{We}^{0.32} \qquad (2.56)$$

while Baron's [62] correlation of Miesse's [29] water data gives

$$L = 538 d_o \, \mathrm{We}^{0.5} \, \mathrm{Re}^{-0.625} \qquad (2.57)$$

It is uncertain what happens to the shape of the stability curve if the jet velocity is increased indefinitely. In their review of experimental data in this region, McCarthy and Molloy [42] note that Tanasawa and Toyoda [63] assert that L/d_o increases continually with increasing velocity, while Yoshizawa et al. [64] claim that L/d_o decreases with increasing velocity. According to Hiroyasu et al. [65] and Arai et al. [66], the breakup length increases with jet velocity up to a maximum value, beyond which any further increase in velocity causes breakup length to decline, as illustrated in Fig. 2.10. Some of the results on which these conclusions are based are shown in Figs. 2.15 to 2.19, from which the data points have been removed for clarity.

Figure 2.14 (a) High-speed photograph of water jet showing surface wave instabilities and spray detachment; (b) enlarged view of water jet just downstream of nozzle exit *(Taylor and Hoyt [55]).*

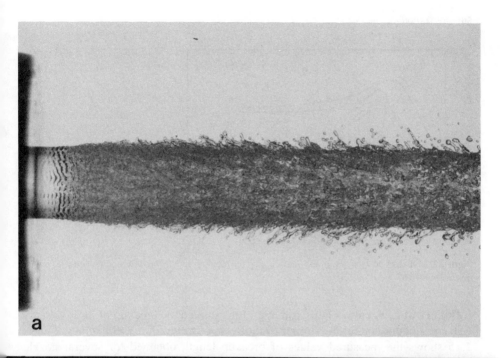

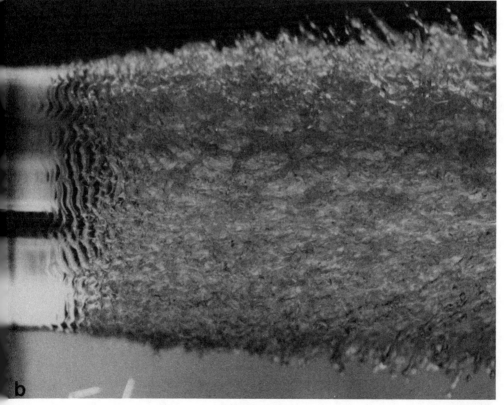

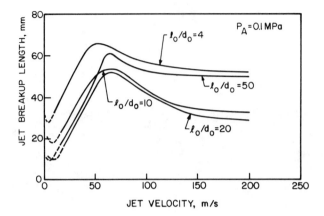

Figure 2.15 Effect of l_o/d_o ratio and jet velocity on breakup length for low ambient pressure *(Hiroyasu et al. [65])*.

Influence of l_o/d_o ratio. Hiroyasu et al. [65] studied the breakup of high-velocity water jets under conditions similar to those encountered in diesel engines. Figure 2.15 shows the measured values of breakup length obtained for several nozzle l_o/d_o ratios at jet velocities up to 200 m/s when injecting into air at normal atmospheric pressure. This figure shows breakup length increasing with injection velocity up to a maximum at around 60 m/s, beyond which any further increase in jet velocity causes breakup length to decline. The influence of l_o/d_o on breakup length shows no clear trend. In the range of jet velocities of most practical interest, i.e., >50 m/s, it is of interest to note that for l_o/d_o ratios of 10 and 20, breakup length may be increased by either reducing the l_o/d_o ratio to 4 or increasing it to 50. Similar data on the effect of l_o/d_o on L, obtained at an ambient air pressure of 3 MPa (30 atm) are shown in Fig. 2.16. At this high pressure the influence of

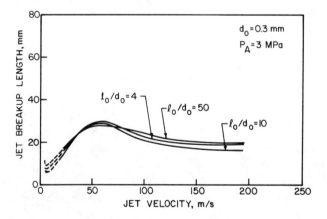

Figure 2.16 Effect of l_o/d_o ratio and jet velocity on breakup length for high ambient pressure *(Hiroyasu et al. [65])*.

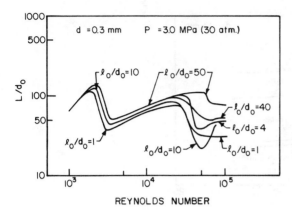

Figure 2.17 Effect of l_o/d_o ratio and Reynolds number on breakup length *(Arai et al. [66])*.

l_o/d_o is clearly much less pronounced, presumably because the aerodynamic effects on the jet surface now far outweigh the hydrodynamic instabilities generated within the liquid upstream of the nozzle exit.

Arai et al. [66] carried out a similar series of tests on the influence of l_o/d_o on breakup length for high ambient air pressures (3 MPa). Diesel-type nozzles were used with water as the test fluid. Their results are shown in Fig. 2.17. For Reynolds numbers higher than 30,000, L is seen to diminish with increase in Re, pass through a minimum, and then reach an almost constant value when Re exceeds 50,000. According to Arai et al. [66], when the nozzle l_o/d_o ratio is much higher than 10, the strong turbulence created by separated flow at the nozzle entrance is reduced and the velocity profile of the internal flow changes to that of fully developed turbulent flow. Thus the breakup length of the $l_o/d_o = 50$ nozzle is longer than that of the $l_o/d_o = 10$ nozzle in the spray flow region. However, for small l_o/d_o ratios, the turbulence in the flow through the nozzle is not fully developed. This causes the breakup length to increase with l_o/d_o in the range from 1 to 4 in the spray flow region.

Influence of ambient pressure. In diesel engines, fuel is injected from plain-orifice atomizers into air at pressures of 3 MPa and higher. Some perception of the strong influence of ambient gas pressure on breakup length may be gained by comparison of Figs. 2.15 and 2.16. It is clear from these figures that an increase in pressure causes the breakup length to diminish and at the same time reduces the influence of l_o/d_o on breakup length, as discussed above.

The effect of ambient pressure on breakup length is shown more directly in Fig. 2.18. The curves drawn in this figure indicate a strong effect of pressure in the range 0.1 to 3 MPa (1 to 30 atm). Within this range, a 30-fold increase in gas pressure produces a 3-fold reduction in breakup length. Increase in ambient gas pressure from 3 to 4 MPa appears to have little effect on breakup length. For all pressures, breakup length increases with injection velocity up to around 60

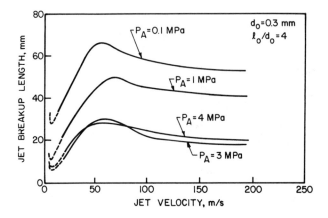

Figure 2.18 Influence of ambient pressure on breakup length *(Hiroyasu et al. [65]).*

m/s, beyond which any further increase in velocity causes breakup length to decline.

The results obtained by Arai et al. [66] on the influence of pressure on breakup length are shown in Fig. 2.19. In the laminar and transition flow regions the effect of ambient pressure is quite small, but in the turbulent flow region its influence is more pronounced. The results shown in Fig. 2.19 for the fully developed spray region are in broad agreement with those of Hiroyasu et al. [65]; both sets of data indicate a marked reduction in breakup length with increase in gas pressure. However, the results of Arai et al. generally conform more closely to the stability curve drawn in Fig. 2.10.

An interesting feature of Fig. 2.19 is that for a pressure of 0.4 MPa (4 atm) there is a range of Reynolds numbers over which the breakup length has two values. Starting from a Reynolds number below this region, increase in Reynolds number causes the breakup length to assume the smaller value. However, if the

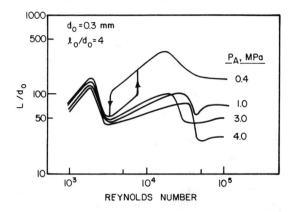

Figure 2.19 Influence of ambient pressure on breakup length *(Arai et al. [66]).*

Reynolds number is falling as it approaches and passes through this region, higher values of breakup length are obtained. Arai et al. attribute this phenomenon to the effects of flow separation and reattachment within the nozzle discharge orifice.

Influence of transverse airflow. Some atomizing devices produce jets of liquid that are disintegrated by exposure to a cross-flowing airstream. Kitamura and Takahashi [67] have studied this type of atomization. The breakup length was measured from film negatives exposed by a high-speed flash. The distance along the jet axis, the shape of which is a curved line, was used as the breakup length. Their results for water are shown in Fig. 2.20. Similar results were obtained for ethanol and aqueous glycerol solutions. They demonstrate that an increase in air velocity increases the aerodynamic destabilizing effects at the jet surface, thereby accelerating jet breakup. High-speed photographs reveal that the liquid jet is disintegrated by liquid disturbances at low air velocities, as illustrated in Fig. 2.6b, and by sinuous wave formation at high jet velocities, of the type shown in Fig. 2.6c.

DISINTEGRATION OF LIQUID SHEETS

Many atomizers do not form jets of liquid, but rather form flat or conical sheets. Impingement of two liquid streams can produce a flat sheet. Conical sheets can be obtained if a liquid flowing in a pipe is deflected through an annular orifice, the form of the sheet being governed by the angle of deflection. Conical sheets are also generated in pressure-swirl and prefilming airblast nozzles, where the

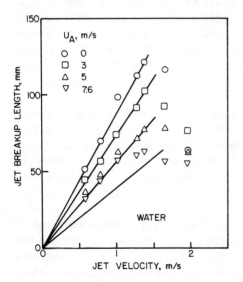

Figure 2.20 Effect of transverse airflow on breakup length *(Kitamura and Takahashi [67]).*

liquid issues from an orifice with a tangential velocity component resulting from its passage through one or more tangential or helical slots. Another widely used method of producing a flat circular sheet is by feeding the liquid to the center of a rotating disk or cup.

When a sheet of liquid emerges from a nozzle, its subsequent development is influenced mainly by its initial velocity and the physical properties of the liquid and the ambient gas. To expand the sheet against the contracting surface tension force, a minimum sheet velocity is required, which is provided by pressure, aerodynamic drag, or centrifugal forces, depending on whether the atomizer is pressure-swirl, prefilming-airblast, or rotary, respectively. Increasing the initial velocity expands and lengthens the sheet until a leading edge is formed where equilibrium exists between surface tension and inertial forces.

Fraser and Eisenklam [68] defined three modes of sheet disintegration, described as *rim, wave,* and *perforated-sheet* disintegration. In the rim mode, forces created by surface tension cause the free edge of a liquid sheet to contract into a thick rim, which then breaks up by a mechanism corresponding to the disintegration of a free jet. When this occurs, the resulting drops continue to move in the original flow direction, but they remain attached to the receding surface by thin threads that also rapidly break up into rows of drops. This mode of disintegration is most prominent where the viscosity and surface tension of the liquid are both high. It tends to produce large drops, together with numerous small satellite droplets.

In perforated-sheet disintegration, holes appear in the sheet and are delineated by rims formed from the liquid that was initially included inside. These holes grow rapidly in size until the rims of adjacent holes coalesce to produce ligaments of irregular shape that finally break up into drops of varying size.

Disintegration can also occur in the absence of perforations through the generation of a wave motion on the sheet whereby areas of the sheet, corresponding to a half or full wavelength of the oscillation, are torn away before the leading edge is reached. These areas rapidly contract under the action of surface tension, but they may suffer disintegration by air action or liquid turbulence before a regular network of threads can be formed. This type of wavy-sheet disintegration is evident in the photograph shown in Fig. 2.21.

As Fraser [69] has pointed out, the orderliness of the disintegration process and the uniformity of production of threads have a large influence on the drop size distribution. Perforations occurring in the sheet at the same distance from the orifice have a similar history, and thus thread diameters tend to be uniform and drop sizes fairly constant in perforated-sheet disintegration. However, wavy-sheet disintegration is highly irregular, and consequently drop sizes are much more varied.

Atomizers that discharge the liquid in the form of a sheet are usually capable of exhibiting all three modes of sheet disintegration. Sometimes two different modes occur simultaneously, and their relative importance can greatly influence both the mean drop size and the drop size distribution. Dombrowski, Eisenklam, Fraser, and co-workers [68–74] made numerous studies on the mechanisms of

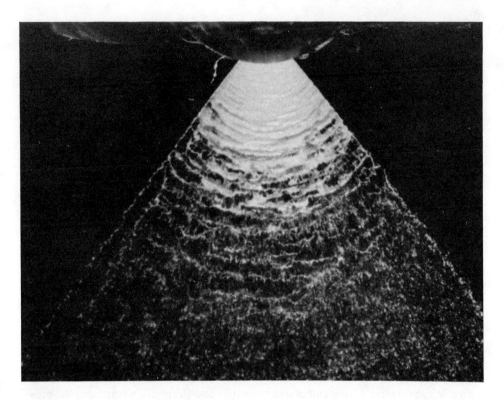

Figure 2.21 Photograph illustrating atomization by wavy-sheet disintegration *(courtesy of Ransburg Gema, Inc.).*

sheet disintegration. In the early 1950s Dombrowski and Fraser provided useful insight into the manner of liquid sheet breakup, using an improved photographic technique and a specially designed source of lighting combining high intensity with very short duration. They established that ligaments are caused principally by perforations in the liquid sheet. If the holes are caused by air friction, the ligaments break up very rapidly. However, if the holes are created by other means, such as turbulence in the nozzle, the ligaments are broken more slowly. From a large number of tests on a wide variety of liquids, Dombrowski and Fraser [72] concluded that (1) liquid sheets with high surface tension and viscosity are most resistant to disruption and (2) the effect of liquid density on sheet disintegration is negligibly small.

Flat Sheets

The mechanism for the disintegration of flat liquid sheets has been studied theoretically and experimentally by York et al. [17], who concluded that instability and wave formation at the interface between the continuous and discontinuous phases are the major factors in the breakup of a sheet of liquid into drops. They considered the system of forces acting on the slightly disturbed surface of a liquid sheet moving in air, as illustrated in Fig. 2.22. Surface tension forces try to return the protuberance back to its original position, but the air experiences a local decrease in static pressure (corresponding to the local increase in velocity) that tends to expand the protuberance farther outward. This corresponds to the normal pattern of wind-induced instability, where surface tension forces oppose any movement of the interface from its initial plane and attempt to restore equilibrium, while the aerodynamic forces increase any deviation from the interface and thereby promote instability.

At the boundary between a liquid and the ambient air the balance of forces may be expressed by the equation

$$p_L - p_A = -\sigma \frac{d^2h}{dx^2} \qquad (2.58)$$

where h is the displacement of liquid from the equilibrium position and x is distance along the liquid sheet.

To formulate the problem, York et al. [17] considered a two-dimensional infinite sheet of liquid of finite thickness with air on both sides. By neglecting viscous effects and assuming irrotational flow, the velocities can be obtained from a velocity potential. Using calculated velocities and Bernoulli's equation, the pressures may then be estimated and the displacements h determined. As in the case of a liquid jet, an exponential increase in wave amplitude occurs under certain conditions. The amplitude increase is given by

$$h_t = A \exp(\beta t) \qquad (2.59)$$

AERODYNAMIC FORCES

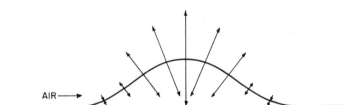

AIR →

LIQUID →

SURFACE TENSION FORCES

Figure 2.22 System of forces acting on the distributed interface of a liquid sheet moving in air.

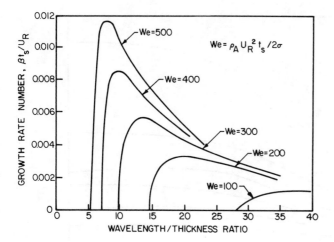

Figure 2.23 Effect of Weber number on wave growth for air and water *(York et al. [17])*.

where h_t is the amplitude at time t, A the amplitude of the initial disturbance, and β a number that determines the growth rate of the disturbance. If the amplitude of the initial disturbance, A, is known, then the time for disintegration can be calculated from the relation

$$t = \beta^{-1} \ln\left(\frac{t_s}{2A}\right) \tag{2.60}$$

where t_s is the thickness of the liquid sheet. A graph based on the analysis of York et al. is plotted in Fig. 2.23 and illustrates the effect of Weber number on growth rate for a liquid sheet in air at atmospheric pressure.

Figure 2.23 shows that the growth rate has a clearly defined maximum for a given Weber number, especially at high Weber numbers. A disturbance of that wavelength will dominate the interface and rapidly disintegrate the sheet. Figure 2.23 also indicates a fairly precise lower limit for λ/t_s below which the interface is stable. This demonstrates that short wavelengths of disturbance on thick sheets are stable unless wind velocities or Weber numbers are very high.

If the wavelength λ^* corresponding to the maximum growth rate for a given Weber number $(U_R^2 t_s \rho_A / \sigma)$ is used to calculate the corresponding Weber number $(U_R^2 \lambda^* \rho_A / \sigma)$, the two Weber numbers can be plotted on a single graph to show the effect of varying the sheet thickness alone. This is done in Fig. 2.24, which demonstrates that large variations in sheet thickness create only small changes in λ^*, especially at low density ratios. For example, for water and air at normal atmospheric pressure, a hundredfold variation in sheet thickness changes the wavelength λ^* only 10%.

Using photographs of the flat liquid sheet, Hagerty and Shea [75] obtained experimental values of the growth rate factor (β) for sinuous waves and compared them with the theoretically predicted curve calculated from the equation

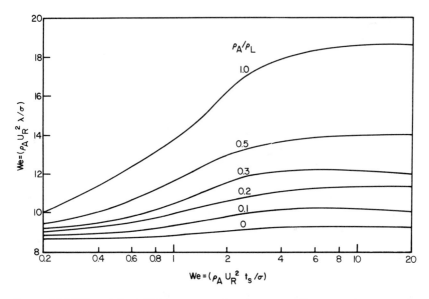

Figure 2.24 Effect of sheet thickness on wavelength corresponding to maximum growth rate *(York et al. [17])*.

$$\beta = \left[\frac{n^2 U_R^2 (\rho_A/\rho_L) - n^3 \sigma/\rho_L}{\tanh n(t_s/2)} \right]^{0.5} \tag{2.61}$$

where n is the wave number of the disturbing wave $(=\omega/U)$, U_R the relative velocity between liquid and air, and ω the wave frequency. The results of this comparison are shown in Fig. 2.25.

Their analysis also showed that the lowest stable frequency is given by

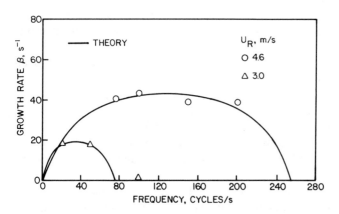

Figure 2.25 Comparison of predicted growth rate with experimental data *(Hagerty and Shea [75])*.

$$f_{min} = \frac{\omega}{2\pi} = \frac{\rho_A U_R^3}{2\pi\sigma} = \frac{U_R We}{2\pi t_s} \tag{2.62}$$

where $We = \rho_A U_R^2 t_s/\sigma$.

Frequency, velocity, and wavelength are related by

$$f = \frac{U}{\lambda} \tag{2.63}$$

Hence, from Eqs. (2.62) and (2.63), the minimum wavelength for an unstable system is obtained as

$$\lambda_{min} = \frac{2\pi\sigma}{\rho_A U_R^2} \tag{2.64}$$

From an analysis of the unstable oscillations of a liquid sheet moving in still air, Squire [76] derived the following expression for minimum wavelength:

$$\lambda_{min} = \frac{2\pi t_s \rho_L}{\rho_A(We - 1)} \tag{2.65}$$

where $We = \rho_L U_R^2 t_s/\sigma$. Since usually $We \gg 1$,

$$\lambda_{min} = \frac{2\pi\sigma}{\rho_A U_R^2}$$

which is identical to Eq. (2.64) as derived by Hagerty and Shea [75].

The optimum wavelength for sheet disintegration, λ_{opt}, is the one that has the maximum growth rate β_{max}. For $We \gg 1$, Squire obtained

$$\lambda_{opt} = \frac{4\pi\sigma}{\rho_A U_R^2} \tag{2.66}$$

and

$$\beta_{max} = \frac{\rho_A U_R^2}{\sigma(\rho_L t_s)^{0.5}} \tag{2.67}$$

Influence of surrounding gaseous medium. According to Dombrowski and Johns [71], the instability of thin liquid sheets resulting from interaction with the surrounding gaseous medium gives rise to rapidly growing surface waves. Disintegration occurs when the wave amplitude reaches a critical value; fragments of the sheet are torn off and rapidly contract into unstable ligaments under the action of surface tension, and drops are produced as the ligaments break down according to theories of varicose instability (see Fig. 2.21).

Fraser et al. [73] investigated the mechanism of disintegration of liquid sheets impinged on by high-velocity airstreams. They used a special system in which flat circular liquid sheets were produced from a spinning cup, while the atomizing airstream was admitted through an annular gap located axially symmetrically to the cup. Photographic examination showed circumferential waves initating at the position of impact of the airstream in the sheet, and the liquid sheet was observed to break down into drops through the formation of unstable ligaments.

Rizk and Lefebvre [77] studied the influence of initial liquid film thickness on spray characteristics. They used two specially designed airblast atomizers that were constructed to produce a flat liquid sheet across the centerline of a two-dimensional air duct, with the liquid sheet exposed on both sides to high-velocity air. From analysis of the processes involved and correlations of the experimental data, it was found that high values of liquid viscosity and liquid flow rate result in thicker films. It was also observed that thinner liquid films break down into smaller drops according to the relationship SMD $\propto t^{0.4}$. This is an interesting result since, if other parameters are held constant, liquid film thickness is directly proportional to nozzle size, which implies that SMD should be proportional to (atomizer linear scale)$^{0.4}$. This, in fact, is precisely the result obtained by El-Shanawany and Lefebvre [78] in their study of the effect of nozzle size on SMD.

Previous workers had noted a similar relationship. For example, the analyses of York et al. [17], Hagerty and Shea [75], and Dombrowski and Johns [71] all suggest that mean drop diameter is roughly proportional to the square root of the film thickness. In addition, the photographic studies of film disintegration carried out by Fraser et al. [73] show that, for sheets breaking down through the formation of unstable ligaments, the ligament diameter depends mainly on the sheet thickness.

Rizk and Lefebvre [77] also examined the mechanism of sheet disruption and drop formation using very high speed flash photography of 0.2 μs duration. Some typical photographs are shown in Figs. 2.26 and 2.27. Figure 2.26 was obtained with water at an air velocity of 55 m/s. It shows clearly the type of atomization process postulated by Dombrowski and Johns [71], in which the liquid/air interaction produces waves that become unstable and disintegrate into fragments. These fragments then contract into ligaments, which in turn break down into drops. With increase in air velocity, the liquid sheet disintegrates earlier and ligaments are formed nearer the lip. These ligaments tend to be thinner and shorter and disintegrate into smaller drops. With liquids of high viscosity, the wavy-surface mechanism is no longer present. Instead, the liquid is drawn out from the atomizing lip in the form of long ligaments. When atomization occurs, it does so well downstream of the atomizing lip in regions of relatively low velocity. In consequence, drop sizes tend to be higher. This type of atomization, as illustrated in Fig. 2.27 for a liquid of viscosity 0.017 kg/m s and an air velocity of 91 m/s, conforms closely to the ligament theory of Castelman [8].

The fact that thicker liquid sheets produce thicker ligaments, which disintegrate into larger drops, highlights the importance of spreading the liquid into a very thin sheet to achieve the finest atomization. Rizk and Lefebvre [77] found that the thickness of the sheet depends on both air and liquid properties. High values of liquid viscosity and/or liquid flow rate result in thicker films, while variations in surface tension appear to have no effect on the thickness of the sheet. However, for liquids of low surface tension, the sheet disintegrates more readily under the action of the airflow and the resulting ligaments are shorter.

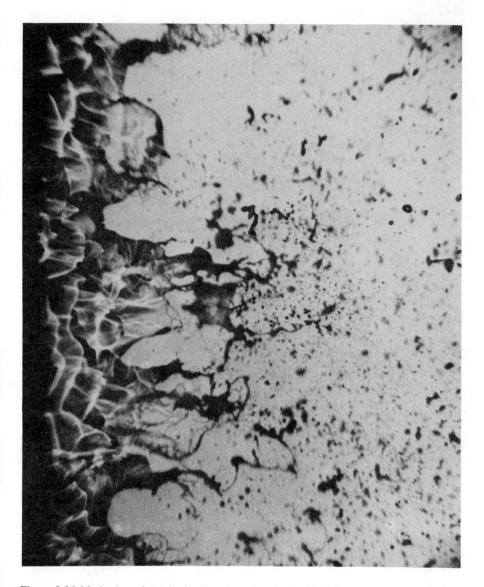

Figure 2.26 Mechanism of atomization for a low-viscosity liquid; airflow is from left to right *(Rizk and Lefebvre [77])*.

Breakup length. Arai and Hashimoto [79] studied the disintegration of liquid sheets injected into a coflowing airstream. Breakup lengths were determined by averaging values obtained from many photographs. For a constant liquid sheet thickness, Fig. 2.28 shows that breakup length decreases with increase in the relative velocity between the air and the liquid. Also indicated in the figure is that breakup length increases as the liquid sheet velocity increases or as the liquid viscosity decreases. An empirical equation for breakup length was derived as

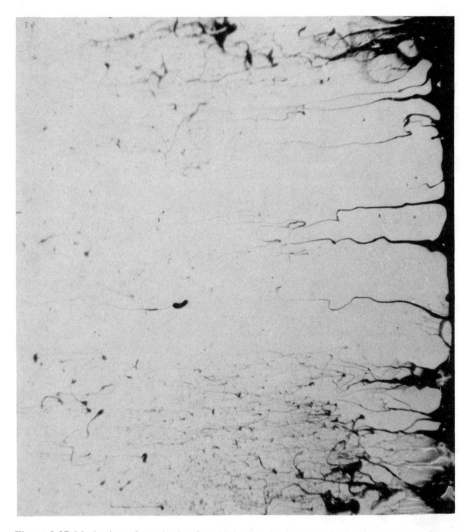

Figure 2.27 Mechanism of atomization for a high-viscosity liquid; airflow is from left to right *(Rizk and Lefebvre [77])*.

$$L = 0.123t_s^{0.5}\ We^{-0.5}\ Re^{0.6} \qquad (2.68)$$

where $We = t_s\rho_A U_R^2/2\sigma$ and $Re = t_s U_L\rho_L/\mu_L$.

Conical Sheets

The theory of attenuating conical sheets in hollow-cone, pressure-swirl nozzles has yet to be developed, but there is evidence that the radius of curvature has a destabilizing effect on the fluctuations, so that conical sheets tend to be shorter than flat sheets [80].

By making certain approximations in their analysis, York et al. [17] were able to make a rough estimate of the size of the drops produced by a pressure-swirl nozzle. Waves form near the nozzle, and those with the wavelength for maximum growth cause periodic thickening of the liquid sheet in a direction normal to the flow. Rings break off from the conical sheet, and the liquid volume contained in the rings can be estimated as the volume of a ribbon cut out of the sheet with a thickness equal to that of the sheet at the breakup distance and a width equal to one wavelength. These cylindrical ligaments then disintegrate into drops according to the Rayleigh mechanism. The resulting mean drop diameter is estimated as

$$D = 2.13(t_s\lambda^*)^{0.5} \tag{2.69}$$

where t_s is the sheet thickness and λ^* is the wavelength for maximum growth rate, as estimated from Fig. 2.24.

York et al. [17] found that the infinite flat sheet model used in their theoretical analysis was not approximated closely by the conical spray in their experiments, so the agreement between theory and experiment was only qualitative. Hagerty and Shea [75] overcame this problem by selecting for study a system that produced a flat sheet of liquid which could be subjected to waves of any desired frequency. The plane sheet was 15 cm wide and 1.6 mm thick, and its velocity could be varied continuously up to 7.6 m/s. Their analysis included both sinuous and dilation waves, as illustrated in Fig. 2.29. Sinuous waves are produced when the two surfaces of the sheet oscillate in phase, whereas dilation waves are the result of out-of-phase motion of the two surfaces. According to Fraser et al. [70], dilation waves in a sheet may be neglected since their degree of instability is always less than that of sinuous ones.

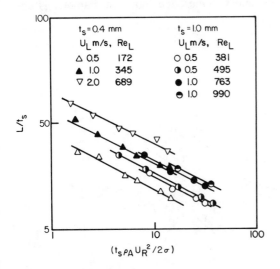

Figure 2.28 Breakup length of a flat liquid sheet in a coflowing airstream *(Arai and Hashimoto [79])*.

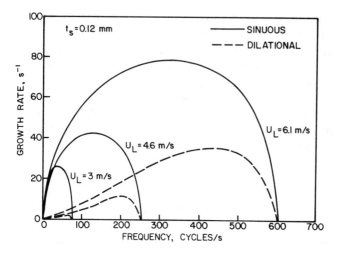

Figure 2.29 Growth rate versus frequency for sinuous and dilational waves *(Hagerty and Shea [75])*.

Fan Sheets

Fraser et al. [70] extended the theories of Hagerty and Shea [75] and Squire [76] to derive an expression for the drop sizes produced by the breakup of a low-viscosity, fan spray sheet. Their model assumes that the most rapidly growing (β_{max}) wave is detached at the leading edge in the form of a ribbon a half-wavelength wide $(\lambda_{opt}/2)$. This ribbon immediately contracts into a filament of diameter D_L, which subsequently disintegrates into drops of equal diameter. The mechanism is represented schematically in Fig. 2.30.

By equating the ribbon and ligament volumes, the ligament diameter is obtained as

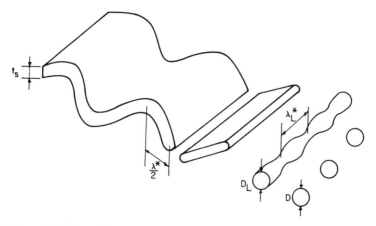

Figure 2.30 Successive stages in the idealized breakup of a wavy sheet *(Fraser et al. [70])*.

$$D_L = \left(\frac{2}{\pi} \lambda_{opt} t_s\right)^{0.5} \tag{2.70}$$

In accordance with Rayleigh's analysis [11], the collapse of a ligament produces drops of diameter

$$D = 1.89 D_L \tag{2.30}$$

and hence

$$D = \text{const}(\lambda_{opt} t_s)^{0.5} \tag{2.71}$$

The sheet thickness at breakup is derived as

$$t_s = \left(\frac{1}{2H^2}\right)^{1/3} \left(\frac{k^2 \rho_A^2 U_R^2}{\rho_L \sigma}\right)^{1/3} \tag{2.72}$$

where k = spray nozzle parameter, $H = \ln(h^0/h_0)$, h^0 = amplitude at breakup, and h_0 = initial amplitude at orifice.

From tests carried out with varying density and liquid velocity, Fraser et al. [70] determined that H remains sensibly constant. Their final expression for mean drop diameter is obtained by combining Eqs. (2.66), (2.71), and (2.72) to yield

$$D \propto \left(\frac{\rho_L}{\rho_A}\right)^{1/6} \left(\frac{k\sigma}{\rho_L U_R^2}\right)^{1/3} \tag{2.73}$$

For a given pressure-swirl or fan spray nozzle, k is constant and U_R is equal to the sheet velocity U_L, which is related to the nozzle pressure differential by

$$\Delta P_L = 0.5 \rho_L U_L^2 \tag{2.74}$$

Thus, for a given nozzle

$$D \propto \left(\frac{k\sigma \rho_L^{0.5}}{\Delta P_L \rho_A^{0.5}}\right)^{1/3} \tag{2.75}$$

The dimensions of k in Eq. (2.73) are meters squared, and from the description of k provided by Fraser et al. it appears to be proportional to the nozzle flow number. Substituting $k = FN$ in Eq. (2.73) gives

$$D \propto \left(\frac{\rho_L}{\rho_A}\right)^{1/6} \left(\frac{FN\sigma}{\rho_L U_L^2}\right)^{1/3} \tag{2.76}$$

By substituting in the above equation

$$FN = \frac{\dot{m}_L}{\sqrt{\rho_L \Delta P_L}} \tag{2.77}$$

it becomes

$$D \propto \left(\frac{\dot{m}_L \sigma}{\rho_A^{0.5} \Delta P_L^{1.5}}\right)^{1/3} \tag{2.78}$$

This equation shows how mean drop size is influenced by the nozzle operating conditions of injection pressure differential and liquid flow rate. Equation (2.78) also indicates that drop sizes decrease with increase in ambient air density, a result that was confirmed by Fraser et al.'s measurements of the drop sizes produced by the disintegration of water sheets at subatmospheric pressure. Subsequent research by other workers has generally supported their conclusion that an increase in ambient air pressure leads to a diminution of mean drop size.

Equations (2.75)–(2.78) apply solely to flow situations where the liquid viscosity is negligibly small. Dombrowski and Johns [71] and Hasson and Mizrahi [81] examined the more realistic case where the liquid has finite viscosity and the thickness of the sheet diminishes as it moves away from the orifice. The latter workers were able to correlate their measurements of Sauter mean diameter for a fan-spray nozzle by using Eq. (2.75), which was developed theoretically for inviscid flow, empirically correlated for viscosity,

$$\text{SMD} = 0.071 \left(\frac{t_s x \sigma \mu_L^{0.5}}{\rho_L^{0.5} U_L^2} \right)^{1/3} \text{cm} \tag{2.79}$$

where x is the distance downstream from the nozzle and cgs units are employed.

This equation provided a satisfactory correlation of the experimental data over the range of viscosities from 0.003 to 0.025 kg/m s.

SUMMARY

The theoretical work described in this chapter has made a valuable contribution to our understanding of the fundamental mechanisms involved in atomization, but it has not culminated in quantitative descriptions that can be used to design and predict the performance of practical atomizers. That is why most of the experimental data have been accumulated in the form of empirical and semiempirical equations of the type presented in Chapter 6. Nevertheless, from the theoretical, experimental, and photographic evidence now available, certain general conclusions can be drawn regarding the key factors that govern jet and sheet disintegration and the manner and extent of their influence on the drop sizes produced in the spray. It is now clear that important spray characteristics, such as mean drop size and drop size distribution, are dependent on a large number of variables such as nozzle geometry, the physical properties of the liquid being atomized, and the physical properties, turbulence characteristics, and flow conditions in the surrounding gas.

Theoretical and experimental studies on jet and sheet disintegration have both demonstrated the need to group these variables into nondimensional parameters to clarify their effects on the atomization process. As in most flow systems, the Reynolds grouping, which represents the ratio of momentum forces to viscous drag forces, usefully describes the flow state of the emerging jet in terms of both velocity profile and the magnitude of the radial velocity components that promote jet disintegration. The flow characteristics of most relevance to atomization are velocity, velocity profile, and turbulence properties, all of which can contribute effectively to jet or sheet disintegration, especially under conditions where aerodynamic influences are relatively small.

In reality, aerodynamic forces are seldom small, and the Weber number, which represents the ratio of the disruptive aerodynamic forces to the restoring surface tension forces, becomes very significant. In fact, for low-viscosity liquids subjected to ambient gas at high relative velocities, the Weber grouping becomes dominant.

Another important dimensionless grouping is the Ohnesorge number. This group contains only the properties of the globules formed in primary atomization before they split up into smaller drops during secondary atomization. It is sometimes called a stability group because it provides an indication of the resistance of a globule to further disintegration, but it is also called a viscosity group because it accounts for the effect of liquid viscosity on the globule. Other dimensionless groups are obtained as the density ratios of the gas and liquid phases and various geometric ratios of the spray generating system.

The fundamental principle of the disintegration of a liquid consists of increasing its surface area, usually in the form of a cylindrical rod or sheet, until it becomes unstable and disintegrates into drops. If the issuing jet is in laminar flow, a vibration in the jet or an external disturbance causes disintegration as predicted by the Rayleigh breakup mechanism. If the jet is fully turbulent, it can break up without the application of external aerodynamic forces. Jet or sheet disintegration is promoted by increases in relative velocity and is impaired by increases in liquid viscosity. Regardless of all other effects, the breakup process is always accelerated by air resistance, which increases with air density. Where the air resistance is high, as in most practical applications, the mutually opposing aerodynamic and surface tension forces acting on the liquid surface give rise to oscillations and perturbations. Under favorable conditions these oscillations are amplified and the liquid body breaks up. Jets break up into drops, while sheets disintegrate into ligaments and then into drops. If the drops so formed exceed the stable maximum limit, they further disintegrate into smaller drops until all the drops are below the critical size.

The random nature of the atomization process, as summarized above, means that most sprays are characterized by a wide range of drop sizes. The methods used to describe and quantify drop size distributions form the subject of the next chapter.

NOMENCLATURE

A	area, m^2
C_D	drag coefficient
D	drop diameter, m

D_L	ligament diameter, m
d	jet diameter, m
d_o	orifice diameter, m
E	rate of energy input per unit mass
E_s	potential surface energy
f	frequency
g	acceleration due to gravity
h	displacement (amplitude) of liquid from equilibrium position, m
h_0	amplitude at orifice, m
h^0	amplitude at breakup, m
h_t	amplitude at time t, m
L	breakup length, m
l	length of jet, m
l_o	orifice length, m
m_D	mass of drop, kg
$\dot{m}$	mass flow rate, kg/s
n	order of mode of vibration, or wave number of disturbing wave
Oh	Ohnesorge number
ΔP_L	nozzle liquid pressure differential, Pa
p_I	internal pressure, Pa
p_σ	surface tension pressure, Pa
Re	Reynolds number
r	jet radius, m
S	maximum velocity gradient in external flow field
t_b	breakup time, s
t_s	sheet thickness, m
U	axial velocity, m/s
$\bar{u}$	rms value of fluctuating velocity component, m/s
x	distance downstream of nozzle, m
We	Weber number
α	dimensionless wave number ($=2\pi/\lambda$)
β	number that determines growth rate of disturbance
δ	amplitude of disturbance, m
λ	wavelength of disturbance, m
λ^*	wavelength for maximum growth rate, m
μ	dynamic viscosity, kg/m s
ρ	density, kg/m^3
σ	surface tension, kg/s^2
ω	natural frequency of vibration

Subscripts

A	air
L	liquid
R	relative value
0	initial value

crit   critical value
max   maximum value
min   minimum value
opt   optimum value

REFERENCES

1. Tamada and Shiback, cited in Atomization—a Survey and Critique of the Literature, by C. E. Lapple, J. P. Henry, and D. E. Blake, Stanford Research Institute Report No. 6, 1966.
2. Klüsener, O., The Injection Process in Compressorless Diesel Engines, *VDI Z.*, Vol. 77, No. 7, February 1933.
3. Giffen, E., and Muraszew, A., *The Atomization of Liquid Fuels*, John Wiley and Sons, New York, 1953.
4. Gordon, D. G., Mechanism and Speed of Breakup of Drops, *J. Appl. Phys.*, Vol. 30, No. 11, 1959, pp. 1759–1761.
5. Lenard, P., *Meteorol. Z.*, Vol. 21, 1904, p. 249.
6. Hochschwender, E., Ph.D. thesis, University of Heidelberg, 1949.
7. Hinze, J. O., Fundamentals of the Hydrodynamic Mechanism of Splitting in Dispersion Processes, *AIChE J.*, Vol. 1, No. 3, 1955, pp. 289–295.
8. Castelman, R. A., The Mechanism of the Atomization of Liquids, *J. Res. Natl. Bur. Stand.*, Vol. 6, No. 281, 1931, pp. 369–376.
9. Merrington, A. C., and Richardson, E. G., The Breakup of Liquid Jets, *Proc. Phys. Soc. London*, Vol. 59, No. 331, 1947, pp. 1–13.
10. Taylor, G. I., The Function of Emulsion in Definable Field Flow, *Proc. R. Soc. London Ser. A*, Vol. 146, 1934, pp. 501–523.
11. Rayleigh, Lord, On the Instability of Jets, *Proc. London Math. Soc.*, Vol. 10, 1878, pp. 4–13.
12. Haenlein, A., Disintegration of a Liquid Jet, NACA TN 659, 1932.
13. Ohnesorge, W., Formation of Drops by Nozzles and the Breakup of Liquid Jets, *Z. Angew. Math. Mech.*, Vol. 16, 1936, pp. 355–358.
14. Sauter, J., Determining Size of Drops in Fuel Mixture of Internal Combustion Engines, NACA TM 390, 1926.
15. Scheubel, F. N., On Atomization in Carburettors, NACA TM 644, 1931.
16. Castleman, R. A., Jr., The Mechanism of the Atomization Accompanying Solid Injection, NACA Report 440, 1932.
17. York, J. L., Stubbs, H. F., and Tek, M. R., The Mechanism of Disintegration of Liquid Sheets, *Trans. ASME*, Vol. 75, 1953, pp. 1279–1286.
18. Lane, W. R., Shatter of Drops in Streams of Air, *Ind. Eng. Chem.*, Vol. 43, No. 6, 1951, pp. 1312–1317.
19. Haas, F. C., Stability of Droplets Suddenly Exposed to a High Velocity Gas Stream, *AIChE J.*, Vol. 10, 1964, pp. 920–924.
20. Hanson, A. R., Domich, E. G., and Adams, H. S., Shock Tube Investigation of the Breakup of Drops by Air Blasts, *Phys. Fluids*, Vol. 6, 1963, pp. 1070–1080.
21. Sleicher, C. A., Maximum Drop Size in Turbulent Flow, *AIChE J.*, Vol. 8, 1962, pp. 471–477.
22. Brodkey, R. A., *The Phenomena of Fluid Motions*, Addison-Wesley, Reading, Mass., 1967.
23. Kolmogorov, A. N., On the Disintegration of Drops in a Turbulent Flow, *Dokl. Akad. Nauk SSSR*, Vol. 66, 1949, p. 825–828.
24. Batchelor, G. K., *The Theory of Homogeneous Turbulence*, University Press, Cambridge, 1956.
25. Clay, P. H., *Proc. R. Acad. Sci. (Amsterdam)*, Vol. 43, 1940, p. 852.
26. Sevik, M., and Park, S. H., The Splitting of Drops and Bubbles by Turbulent Fluid Flow, *J. Fluids Eng.*, Vol. 95, 1973, pp. 53–60.

27. Tomotika, S., Breaking Up of a Drop of Viscous Liquid Immersed in Another Viscous Fluid Which Is Extending at a Uniform Rate, *Proc. R. Soc. London Ser. A*, Vol. 153, 1936, pp. 302–320.
28. Meister, B. J., and Scheele, G. F., Drop Formation from Cylindrical Jets in Immiscible Liquid System, *AIChE J.*, Vol. 15, No. 5, 1969, pp. 700–706.
29. Miesse, C. C., Correlation of Experimental Data on the Disintegration of Liquid Jets, *Ind. Eng. Chem.*, Vol. 47, No. 9, 1955, pp. 1690–1701.
30. Rumscheidt, F. D., and Mason, S. G., Particle Motion in Sheared Suspensions. Deformation and Burst of Fluid Drops in Shear and Hyperbolic Flows, *J. Colloid Sci.*, Vol. 16, 1967, pp. 238–261.
31. Simmons, H. C., The Atomization of Liquids; Principles and Methods, Parker Hannifin Report No. 7901/2-0, 1979.
32. Krzywoblocki, M. A., Jets—Review of Literature, *Jet Propul.*, Vol. 26, 1957, pp. 760–779.
33. Schweitzer, P. H., Mechanism of Disintegration of Liquid Jets, *J. Appl. Phys.*, Vol. 8, 1937, pp. 513–521.
34. Marshall, W. R., *Atomization and Spray Drying*, Chem. Eng. Prog. Monogr. Ser., No. 2, Vol. 50, 1954.
35. Sterling, A. M., The Instability of Capillary Jets, Ph.D. thesis, University of Washington, 1969.
36. Phinney, R. E., The Breakup of a Turbulent Liquid Jet in a Gaseous Atmosphere, *J. Fluid Mech.*, Vol. 60, 1973, pp. 689–701.
37. Bixson, L. L., and Deboi, H. H., Investigation of Rational Scaling Procedure for Liquid Fuel Rocket Engines, Technical Documentary Report SSD-TDR-62-78, Rocket Research Laboratories, Edwards Air Force Base, Calif., 1962.
38. Kocamustafaogullari, G., Chen, I. Y., and Ishii, M., Unified Theory for Predicting Maximum Fluid Particle Size for Drops and Bubbles, Argonne National Laboratory Report NUREG/CR-4028, 1984.
39. Bidone, G., Expériences sur la Forme et sur la Direction des Veines et des Courants d'Eau Lances par Diverses Ouvertures, Imprimerie Royale, Turin, 1829, pp. 1–136.
40. Savart, F., *Ann. Chim. Phys.*, Vol. 53, 1833, pp. 337–386.
41. Plateau, J., Statique Expérimentale et Théorique des Liquides Soumis aux Seules Forces Moléculaires, cited by Lord Rayleigh, *Theory of Sound*, Vol. II, Dover Publications, New York, 1945.
42. McCarthy, M. J., and Molloy, N. A., Review of Stability of Liquid Jets and the Influence of Nozzle Design, *Chem. Eng. J.*, Vol. 7, 1974, pp. 1–20.
43. Tyler, F., Instability of Liquid Jets, *Philos. Mag. (London)*, Vol. 16, 1933, pp. 504–518.
44. Weber, C., Disintegration of Liquid Jets, *Z. Angew. Math. Mech.*, Vol. 11, No. 2, 1931, pp. 136–159.
45. Reitz, R. D., Atomization and Other Breakup Regimes of a Liquid Jet, Ph.D. thesis, Princeton University, 1978.
46. Schiller, L., Untersuchungen ueber Laminare und Turbulente Stromung, *VDI Forschungsarbeit.*, Vol. 248, 1922.
47. Sterling, A. M., and Sleicher, C. A., The Instability of Capillary Jets, *J. Fluid Mech.*, Vol. 68, 1975, pp. 477–495.
48. Eisenklam, P., and Hooper, P. C., The Flow Characteristics of Laminar and Turbulent Jets of Liquid, Ministry of Supply D.G.G.W. Report/EMR/58/10, September 1958.
49. Rupe, J. H., Jet Propulsion Laboratory Report No. 32-207, January 1962.
50. Grant, R. P., and Middleman, S., Newtonian Jet Stability, *AIChE J.*, Vol. 12, No. 4, 1966, pp. 669–678.
51. Mahoney, T. J., and Sterling, M. A., The Breakup Length of Laminar Newtonian Liquid Jets in Air, *Proceedings of the 1st International Conference on Liquid Atomization and Spray Systems*, Tokyo, 1978, pp. 9–12.
52. Phinney, R. E., and Humphries, W., Stability of a Viscous Jet—Newtonian Liquids, NOLTR 70-5, January 1970, U.S. Naval Ordnance Laboratory, Silver Spring, Md.

53. Van de Sande, E., and Smith, J. M., Jet Breakup and Air Entrainment by Low-Velocity Turbulent Jets, *Chem. Eng. Sci.,* Vol. 31, No. 3, 1976, pp. 219–224.
54. Reitz, R. D., and Bracco, F. V., Mechanism of Atomization of a Liquid Jet, *Phys. Fluids,* Vol. 25, No. 2, 1982, pp. 1730–1741.
55. Taylor, J. J., and Hoyt, J. W., Water Jet Photography—Techniques and Methods, *Exp. Fluids,* Vol. 1, 1983, pp. 113–120.
56. DeJuhasz, K. J., *Trans. ASME,* Vol. 53, 1931, p. 65.
57. Bergwerk, W., Flow Pattern in Diesel Nozzle Spray Holes, *Proc. Inst. Mech. Eng.,* Vol. 173, 1959, pp. 655–660.
58. Sadek, R., Communication on Flow Pattern in Nozzle Spray Holes and Discharge Coefficient of Orifices, *Proc. Inst. Mech. Eng.,* Vol. 173, No. 25, 1959, pp. 671–672.
59. Vitman, L. A., Problems of Heat Transfer and Hydraulics in Two-Phase Media, collection of articles edited by S. S. Kutateladze, Moscow, 1961, p. 374.
60. Lienhard, J. H., and Day, J. B., The Breakup of Superheated Liquid Jets, *Trans. ASME J. Basic Eng., Ser. D,* Vol. 92, No. 3, 1970, pp. 515–522.
61. Lafrance, P., The Breakup Length of Turbulent Liquid Jets, *Trans. ASME J. Fluids Eng.,* June 1977, pp. 414–415.
62. Baron, T., Technical report No. 4, University of Illinois, 1949; cited by Miesse [29].
63. Tanasawa, Y., and Toyoda, S., *Trans. Jpn. Soc. Mech.,* Vol. 20, 1954, p. 300.
64. Yoshizawa, Y., Kawashima, T., and Yanaida, K., *Tohoku Kozan (J. Tohoku Mining Soc.),* Vol. 11, 1964, p. 37; cited by McCarthy and Molloy [42].
65. Hiroyasu, H., Shimizu, M., and Arai, M., The Breakup of High Speed Jet in a High Pressure Gaseous Atmosphere, *Proceedings of the 2nd International Conference on Liquid Atomization and Spray Systems,* Madison, Wis., 1982, pp. 69–74.
66. Arai, M., Shimizu, M., and Hiroyasu, H., Breakup Length and Spray Angle of High Speed Jet, *Proceedings of the 3rd International Conference on Liquid Atomization and Spray Systems,* London, 1985, pp. 1B/4/1–10.
67. Kitamura, Y., and Takahashi, T., Stability of a Liquid Jet in Air Flow Normal to the Jet Axis, *J. Chem. Eng. Jpn.,* Vol. 9, No. 4, 1976, pp. 282–286.
68. Fraser, R. P., and Eisenklam, P., Research into the Performance of Atomizers for Liquids, *Imp. Coll. Chem. Eng. Soc. J.,* Vol. 7, 1953, pp. 52–68.
69. Fraser, R. P., Liquid Fuel Atomization, *Sixth Symposium (International) on Combustion,* Reinhold, New York, 1957, pp. 687–701.
70. Fraser, R. P., Eisenklam, P., Dombrowski, N., and Hasson, D., Drop Formation from Rapidly Moving Sheets, *AIChE J.,* Vol. 8, No. 5, 1962, pp. 672–680.
71. Dombrowski, N., and Johns, W. R., The Aerodynamic Instability and Disintegration of Viscous Liquid Sheets, *Chem. Eng. Sci.,* Vol. 18, 1963, pp. 203–214.
72. Dombrowski, N., and Fraser, R. P., A Photographic Investigation into the Disintegration of Liquid Sheets, *Philos. Trans. R. Soc. London Ser. A, Math. Phys. Sci.,* Vol. 247, No. 924, 1954, pp. 101–130.
73. Fraser, R. P., Dombrowski, N., and Routley, J. H., The Atomization of a Liquid Sheet by an Impinging Air Stream, *Chem. Eng. Sci.,* Vol. 18, 1963, pp. 339–353.
74. Crapper, G. D., and Dombrowski, N., A Note on the Effect of Forced Disturbances on the Stability of Thin Liquid Sheets and on the Resulting Drop Size, *Int. J. Multiphase Flow,* Vol. 10, No. 6, 1984, pp. 731–736.
75. Hagerty, W. W., and Shea, J. F., A Study of the Stability of Plane Fluid Sheets, *J. Appl. Mech.,* Vol. 22, No. 4, 1955, pp. 509–514.
76. Squire, H. B., Investigation of the Instability of a Moving Liquid Film, *Br. J. Appl. Phys.,* Vol. 4, 1953, pp. 167–169.
77. Rizk, N. K., and Lefebvre, A. H., Influence of Liquid Film Thickness on Airblast Atomization, *Trans. ASME J. Eng. Power,* Vol. 102, 1980, pp. 706–710.
78. El-Shanawany, M. S. M. R., and Lefebvre, A. H., Airblast Atomization: The Effect of Linear Scale on Mean Drop Size, *J. Energy,* Vol. 4, No. 4, 1980, pp. 184–189.

79. Arai, T., and Hashimoto, H., Disintegration of a Thin Liquid Sheet in a Cocurrent Gas Stream, *Proceedings of the 3rd International Conference on Liquid Atomization and Spray Systems*, London, 1985, pp. V1B/1/1–7.

80. Eisenklam, P., Recent Research and Development Work on Liquid Atomization in Europe and the U.S.A., Paper presented at the 5th Conference on Liquid Atomization, Tokyo, 1976.

81. Hasson, D., and Mizrahi, J., The Drop Size of Fan Spray Nozzle, Measurements by the Solidifying Wax Method Compared with Those Obtained by Other Sizing Techniques, *Trans. Inst. Chem. Eng.*, Vol. 39, No. 6, 1961, pp. 415–422.

THREE

DROP SIZE DISTRIBUTIONS OF SPRAYS

INTRODUCTION

A spray is generally considered as a system of drops immersed in a gaseous continuous phase. Examples of natural sprays include rain, drizzle, fog, and waterfall mists. Figure 3.1 indicates the range of drop sizes as they occur in certain natural phenomena and also as commonly produced by atomizers.

Most practical atomizers generate drops in the size range from a few micrometers up to around 500 μm. Owing to the heterogeneous nature of the atomization process, the threads and ligaments formed by the various mechanisms of jet and sheet disintegration vary widely in diameter, and the resulting main drops and satellite drops vary in size correspondingly. Practical nozzles do not, therefore, produce sprays of uniform drop size at any given operating condition; instead, the spray can be regarded as a spectrum of drop sizes distributed about some arbitrarily defined mean value. Only under certain special conditions (such as are obtained, for example, with a rotary cup atomizer operating within a limited range of liquid flow rates and rotational speeds) can a fairly homogeneous spray be produced. Thus, in addition to mean drop size, another parameter of importance in the definition of a spray is the distribution of drop sizes it contains.

GRAPHICAL REPRESENTATION
OF DROP SIZE DISTRIBUTIONS

An instructive picture of drop size distribution may be obtained by plotting a histogram of drop size, each ordinate representing the number of drops whose

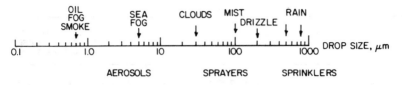

Figure 3.1 Spectrum of drop sizes [1].

dimensions fall between the limits $D - \Delta D/2$ and $D + \Delta D/2$. A typical histogram of this type is shown in Fig. 3.2, in which $\Delta D = 17$ μm. If, instead of plotting the number of drops, the volume of the spray corresponding to a range of drop sizes between $D - \Delta D/2$ and $D + \Delta D/2$ is plotted as a histogram of drop size, the resulting distribution is skewed to the right, as shown in Fig. 3.3, due to the weighting effect of the larger drops.

As ΔD is made smaller, the histogram assumes the form of a frequency curve that may be regarded as a characteristic of the spray, provided it is based on sufficiently large samples. Such a curve, shown in Fig. 3.4, is usually referred to as a *frequency distribution* curve. The ordinate values may be expressed in several alternative ways: as the number of drops with a given diameter, the relative number or fraction of the total, or the fraction of the total number per size class. If the ordinate is expressed in the last manner, the area under the frequency distribution curve must be equal to 1.0.

It is evident that incremental frequency plots may be constructed directly from drop size distribution data by plotting $\Delta N_i/N \Delta D_i$ versus D or $\Delta Q_i/Q \Delta D_i$ versus D, where ΔN_i is the number increment within ΔD_i and ΔQ_i is the volume increment within ΔD_i. The ΔQ_i can be obtained as

$$\Delta Q_i = \Delta N_i \left(\frac{\pi}{6}\right)[0.5(D_{i_1} + D_{i_2})]^3 \tag{3.1}$$

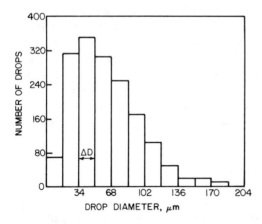

Figure 3.2 Typical drop size histogram.

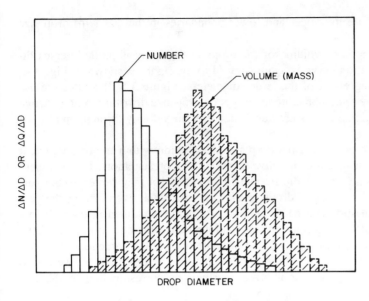

Figure 3.3 Drop size histograms based on number and volume.

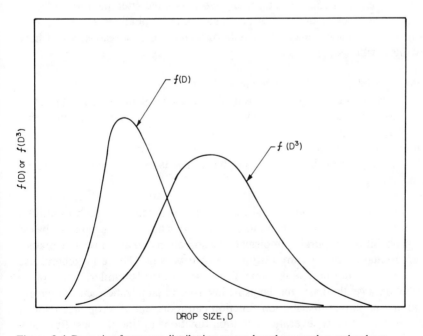

Figure 3.4 Drop size frequency distribution curves based on number and volume.

where D_{i_1} and D_{i_2} are the upper and lower boundaries in the ΔD_i (ith) drop size class.

If the surface area or volume of the drops in the spray is plotted versus diameter, the distribution curve is again skewed to the right, as shown in Fig. 3.4, due to the weighting effect of the larger diameters. Figure 3.5 illustrates the use of this type of curve to demonstrate the effect on drop size distribution of a change in operating conditions, in this case an increase in air velocity through an airblast atomizer.

In addition to representing the drop size distribution with a frequency plot, it is also informative to use a *cumulative distribution* representation. This is essentially a plot of the integral of the frequency curve, and it may represent the percentage of the total number of drops in the spray below a given size or the percentage of the total surface area or volume of a spray contained in drops below a given size. Cumulative distribution curves plotted on arithmetic coordinates have the general shape shown in Fig. 3.6; the ordinate may be the percentage of drops by number, surface area, or volume whose diameter is less than a given drop diameter. Figure 3.7 shows cumulative distributions corresponding to the frequency distribution curves of Fig. 3.5.

MATHEMATICAL DISTRIBUTION FUNCTIONS

Because the graphical representation of drop size distribution is laborious and not easily related to experimental results, many workers have attempted to replace it with mathematical expressions whose parameters can be obtained from a limited number of drop size measurements. Suitable mathematical expressions would have the following attributes [3]:

1. Provide a satisfactory fit to the drop size data.
2. Allow extrapolation to drop sizes outside the range of measured values.
3. Permit easy calculation of mean and representative drop diameters and other parameters of interest.
4. Provide a means of consolidating large amounts of data.
5. Ideally, furnish some insight into the basic mechanisms involved in atomization.

In the absence of any fundamental mechanism or model on which to build a theory of drop size distributions, a number of functions have been proposed, based on either probability or purely empirical considerations, that allow the mathematical representation of measured drop size distributions. Those in general use include normal, log-normal, Nukiyama-Tanasawa, Rosin-Rammler, and upper-limit distributions. As the basic mechanisms involved in atomization are not clearly understood and no single distribution function can represent all drop size data, it is usually necessary to test several distribution functions to find the best fit to any given set of experimental data.

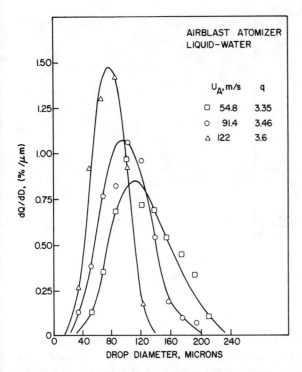

Figure 3.5 Graphs illustrating the effect of atomizing air velocity on drop size distribution [2].

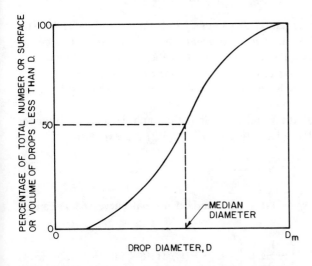

Figure 3.6 Typical shape of cumulative drop size distribution curve.

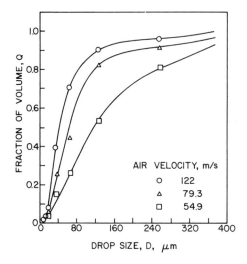

Figure 3.7 Effect of atomizing air velocity on cumulative volume distribution.

Normal Distribution

This distribution function is based on the random occurrence of a given drop size. It is comparatively simple to use, but its application is limited to processes that are random in nature and where no specific bias is present. It is usually expressed in terms of a number distribution function $f(D)$ that gives the number of particles of a given diameter D. We have

$$\frac{dN}{dD} = f(D) = \frac{1}{\sqrt{2\pi}\,s_n} \exp\left[-\frac{1}{2s_n^2}(D - \bar{D})^2\right] \tag{3.2}$$

where s_n is a measure of the deviation of values of D from a mean value $\bar{D}$. The term s_n is usually referred to as the *standard deviation,* and s_n^2 is the variance, as defined in standard textbooks on statistics. A plot of the distribution function is shown in Fig. 3.8. It is usually described as the *standard normal* curve. The area under the curve from $-\infty$ to $+\infty$ is equal to 1, and the areas on either side of the y axis are equal.

The integral of the standard normal curve is the cumulative standard number distribution function $F(D)$. By substituting into Eq. (3.2) the expression

$$t = \frac{D - \bar{D}}{s_n} \tag{3.3}$$

and noting that $\bar{D} = 0$ and $s_n = 1$ for the standard normal curve, $F(D)$ can be derived as

$$F(D) = \left(\frac{1}{\sqrt{2\pi}}\right) \int_{-\infty}^{D} \exp -(t^2/2)\, dt \tag{3.4}$$

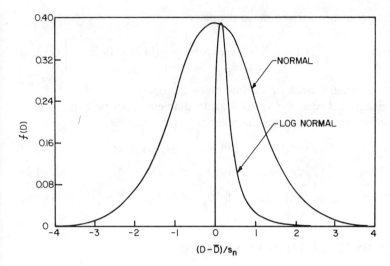

Figure 3.8 Normal and log-normal distributions.

Tabulated values of this integral may be found in most mathematical handbooks. Equation (3.4) indicates that if the data conform to a normal distribution, they will lie on a straight line when plotted on arithmetic-probability graph paper. This type of graph can be used in conjunction with Eq. (3.4) to define the mean diameter and the standard deviation.

Log-normal Distribution

Many particle size distributions that occur in nature have been found to follow the *Gaussian* or *normal* distribution law if the logarithm of the particle diameter is used as the variable. With this modification Eq. (3.2) becomes

$$\frac{dN}{dD} = f(D) = \frac{1}{\sqrt{2\pi}Ds_g} \exp -\left[\frac{1}{2s_g^2}(\ln D - \ln \bar{D}_{ng})^2\right] \tag{3.5}$$

where $\bar{D}_{ng}$ is the number geometric mean drop size and s_g is the geometric standard deviation. The terms s_g and $\bar{D}_{ng}$ have the same significance on a log-probability graph as s_n and $\bar{D}$ do on an arithmetic-probability graph.

From inspection of Eq. (3.5) it is evident that when drop size data fit this type of function, the logarithm of the diameter is distributed normally. Thus letting $y = \ln(\bar{D}/\bar{D}_{ng})$ reduces Eq. (3.5) to the normal distribution form, Eq. (3.2). Equation (3.5) is also plotted in Fig. 3.8 to compare it with the normal distribution function. Log-normal functions can also be written for surface and volume distributions as

Surface distribution: $\quad f(D^2) = \dfrac{1}{\sqrt{2\pi}Ds_g} \exp -\left[\dfrac{1}{2s_g^2}(\ln D - \ln \bar{D}_{sg})^2\right] \tag{3.6}$

where $\bar{D}_{sg}$ is the geometric surface mean diameter

Volume distribution: $\qquad f(D^3) = \dfrac{1}{\sqrt{2\pi}Ds_g} \exp -\left[\dfrac{1}{2s_g^2}(\ln D - \ln \bar{D}_{vg})^2\right]$ $\qquad$ (3.7)

where $\bar{D}_{vg}$ is the geometric mass or volume mean diameter.

The relationships between the various mean diameters can be expressed in terms of the number geometric mean diameter $\bar{D}_{ng}$. For example,

$$\text{Surface:} \qquad \ln \bar{D}_{sg} = \ln \bar{D}_{ng} + 2s_g^2 \qquad (3.8)$$

$$\text{Volume:} \qquad \ln \bar{D}_{vg} = \ln \bar{D}_{ng} + 3s_g^2 \qquad (3.9)$$

$$\text{Volume/surface (SMD):} \qquad \ln \bar{D}_{vsg} = \ln \bar{D}_{ng} + 2.5s_g^2 \qquad (3.10)$$

EMPIRICAL DISTRIBUTION FUNCTIONS

Several empirical relationships have been proposed to characterize the distribution of drop sizes in a spray. None of these is universally better than any other, and the extent to which any particular function matches any given set of data depends largely on the mechanism of disintegration involved. Some of the functions most frequently used in the analysis and correlation of drop size data are given below.

Nukiyama and Tanasawa

One relatively simple mathematical function that adequately describes the actual distribution is that due to Nukiyama and Tanasawa [4]:

$$\frac{dN}{dD} = aD^p \exp -(bD)^q \qquad (3.11)$$

This expression contains four independent constants, namely a, b, p, and q. Most of the commonly used size distribution functions represent either simplifications or modifications of this function. One example is the Nukiyama-Tanasawa [4] equation in which $p = 2$:

$$\frac{dN}{dD} = aD^2 \exp -(bD)^q \qquad (3.12)$$

Dividing this equation through by D^2 and taking logarithms of both sides gives

$$\ln\left(\frac{1}{D^2}\frac{dN}{dD}\right) = \ln a - bD^q \qquad (3.13)$$

For any given set of data a value of q may be assumed and a graph plotted of $\ln(D^{-2}\,dN/dD)$ against D^q. If the assumed value of q is correct, this plot will yield a straight line from which values of a and b may be determined.

Rosin-Rammler

At present the most widely used expression for drop size distribution is one that was originally developed for powders by Rosin and Rammler [5]. It may be expressed in the form

$$1 - Q = \exp -(D/X)^q \qquad (3.14)$$

where Q is the fraction of the total volume contained in drops of diameter less than D, and X and q are constants. Thus, by applying the Rosin-Rammler relationship to sprays, it is possible to describe the drop size distribution in terms of the two parameters X and q. The exponent q provides a measure of the spread of drop sizes. The higher the value of q, the more uniform is the spray. If q is infinite, the drops in the spray are all the same size. For most sprays the value of q lies between 1.5 and 4. However, for rotary atomizers q can be as high as 7.

Although it assumes an infinite range of drop sizes, the Rosin-Rammler expression has the virtue of simplicity. Moreover, it permits data to be extrapolated into the range of very fine drops, where measurements are most difficult and least accurate.

A typical Rosin-Rammler plot is shown in Fig. 3.9. The value of q is obtained as the slope of the line, while X, which is a representative diameter of some kind, is given by the value of D for which $1 - Q = \exp -1$. Solution of this equation yields the result that $Q = 0.632$; that is, X is the drop diameter such that 63.2% of the total liquid volume is in drops of smaller diameter.

Modified Rosin-Rammler

From analysis of a considerable body of drop size data obtained with pressure-swirl nozzles, Rizk and Lefebvre [6] found that although the Rosin-Rammler

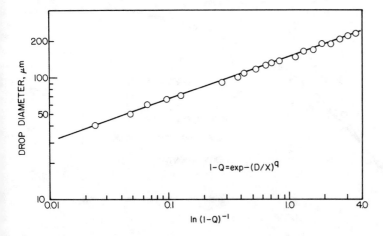

Figure 3.9 Typical Rosin-Rammler plot.

expression provides an adequate data fit over most of the drop size range, there is occasionally a significant deviation from the experimental data for the larger drop sizes. By rewriting the Rosin-Rammler equation in the form

$$1 - Q = \exp -\left(\frac{\ln D}{\ln X}\right)^q \tag{3.15}$$

the volume distribution equation is given by

$$\frac{dQ}{dD} = q\,\frac{(\ln D)^{q-1}}{D(\ln X)^q}\exp -\left(\frac{\ln D}{\ln X}\right)^q \tag{3.16}$$

which gives a much better fit to the drop size data, as illustrated in Figs. 3.10 and 3.11 from reference [7]. However, many more comparative assessments must be performed before the modified version can be claimed superior to the original Rosin-Rammler formula.

Upper-Limit Function

Mugele and Evans [8] analyzed the various functions used to represent drop size distribution data by computing mean diameters for the experimental data and comparing them with means calculated from the distribution functions given above. The empirical constants used in these functions were determined from the experimental distributions. As a result of their analysis, Mugele and Evans proposed a so-called *upper-limit function* as being the best way of representing the drop size distributions of sprays. This is a modified form of the log-probability equation that is based on the normal distribution function. The volume distribution equation is given by

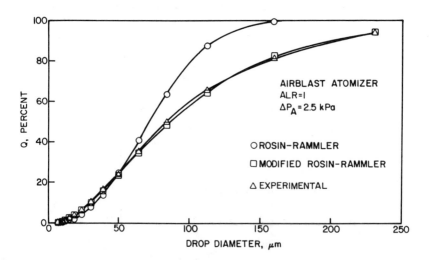

Figure 3.10 Comparison of Rosin-Rammler and modified Rosin-Rammler distributions *(Rizk [7])*.

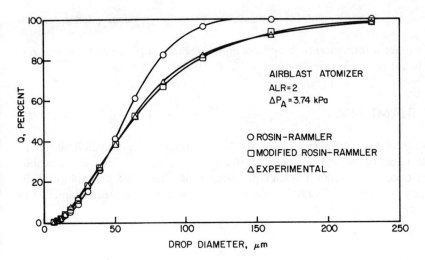

Figure 3.11 Comparison of Rosin-Rammler and modified Rosin-Rammler distributions *(Rizk [7])*.

$$\frac{dQ}{dy} = \delta \exp \frac{-\delta^2 y^2}{\sqrt{\pi}} \qquad (3.17)$$

where

$$y = \ln \frac{aD}{D_m - D} \qquad (3.18)$$

As y goes from $-\infty$ to $+\infty$, D goes from D_0 (minimum drop size) to D_m (maximum drop size), while δ is related to the standard deviation of y and, hence, of D; a is a dimensionless constant. The Sauter mean diameter is given by

$$\text{SMD} = \frac{D_m}{1 + a \exp(1/4\delta^2)} \qquad (3.19)$$

It follows that a reduction in δ implies a more uniform distribution.

The upper-limit distribution function assumes a realistic spray of finite minimum and maximum drop sizes, but it involves difficult integration that requires the use of log-probability paper. The value of D_m must be assumed, and usually many trials are needed to find the most suitable value.

Mugele and Evans [8] have summarized the various statistical formulas for computing mean diameters and variances. The principal conclusion appears to be that the drop size distribution should be given in each case by the best empirical representation obtainable. Until atomization mechanisms can be suitably related to one or more distribution functions, there seems to be no theoretical justification for a belief (or expectation) that one function is generally superior to another for representing drop size distribution. Probably the best reasons for selecting a given distribution function would be (1) mathematical simplicity, (2) ease of manipulation in computations, and (3) consistency with the physical phenomena involved.

For further information on empirical equations for drop size distribution, reference should be made to references [1, 3, 8, 9, 10].

MEAN DIAMETERS

In many calculations of mass transfer and flow processes it is convenient to work only with mean or average diameters instead of the complete drop size distribution. The concept of mean diameter has been generalized and its notation standardized by Mugele and Evans [8]. One of the most common mean diameters is D_{10}, where

$$D_{10} = \frac{\int_{D_0}^{D_m} D(dN/dD)\, dD}{\int_{D_0}^{D_m} (dN/dD)\, dD} \tag{3.20}$$

Other mean diameters of interest include the

$$\text{Surface mean:} \quad D_{20} = \left[\frac{\int_{D_0}^{D_m} D^2(dN/dD)\, dD}{\int_{D_0}^{D_m} (dN/dD)\, dD}\right]^{1/2} \tag{3.21}$$

$$\text{Volume mean:} \quad D_{30} = \left[\frac{\int_{D_0}^{D_m} D^3(dN/dD)\, dD}{\int_{D_0}^{D_m} (dN/dD)\, dD}\right]^{1/3} \tag{3.22}$$

In general, we have [8]

$$(D_{ab})^{a-b} = \frac{\int_{D_0}^{D_m} D^a(dN/dD)\, dD}{\int_{D_0}^{D_m} D^b(dN/dD)\, dD} \tag{3.23}$$

where a and b may take on any values corresponding to the effect investigated, and the sum $a + b$ is called the order of the mean diameter.

Equation (3.23) may also be written as

$$D_{ab} = \left[\frac{\Sigma N_i D_i^a}{\Sigma N_i D_i^b}\right]^{1/(a-b)}$$

(3.24)

where i denotes the size range considered, N_i is the number of drops in size range i, and D_i is the middle diameter of size range i. Thus, for example, D_{10} is the linear average value of all the drops in the spray; D_{30} is the diameter of a drop whose volume, if multiplied by the number of drops, equals the total volume of the sample; and D_{32} (SMD) is the diameter of the drop whose ratio of volume to surface area is the same as that of the entire spray. These and other important mean diameters are listed in Table 3.1, along with their fields of application as suggested by Mugele and Evans [8].

REPRESENTATIVE DIAMETERS

For most engineering purposes the distribution of drop sizes in a spray may be represented concisely as a function of two parameters (as in the Rosin-Rammler expression, for example), one of which is a representative diameter and the other a measure of the range of drop sizes. In some instances it may be advantageous to introduce another term, such as a parameter to express minimum drop size, but basically there must be at least two parameters to describe the drop size distribution.

Table 3.1 Mean diameters and their applications

a	b	$a + b$ (order)	Symbol	Name of mean diameter	Expression	Application
1	0	1	D_{10}	Length	$\dfrac{\Sigma N_i D_i}{\Sigma N_i}$	Comparisons
2	0	2	D_{20}	Surface area	$\left(\dfrac{\Sigma N_i D_i^2}{\Sigma N_i}\right)^{1/2}$	Surface area controlling
3	0	3	D_{30}	Volume	$\left(\dfrac{\Sigma N_i D_i^3}{\Sigma N_i}\right)^{1/3}$	Volume controlling, e.g., hydrology
2	1	3	D_{21}	Surface area-length	$\dfrac{\Sigma N_i D_i^2}{\Sigma N_i D_i}$	Absorption
3	1	4	D_{31}	Volume-length	$\left(\dfrac{\Sigma N_i D_i^3}{\Sigma N_i D_i}\right)^{1/2}$	Evaporation, molecular diffusion
3	2	5	D_{32}	Sauter (SMD)	$\dfrac{\Sigma N_i D_i^3}{\Sigma N_i D_i^2}$	Mass transfer, reaction
4	3	7	D_{43}	De Brouckere or Herdan	$\dfrac{\Sigma N_i D_i^4}{\Sigma N_i D_i^3}$	Combustion equilibrium

There are many possible choices of representative diameter, each of which could play a role in defining the distribution function. The various possibilities include the following:

$D_{0.1}$ = drop diameter such that 10% of total liquid volume is in drops of smaller diameter.

$D_{0.5}$ = drop diameter such that 50% of total liquid volume is in drops of smaller diameter. This is the mass median diameter (MMD).

$D_{0.632}$ = drop diameter such that 63.2% of total liquid volume is in drops of smaller diameter. This is X in Eq. (3.14).

$D_{0.9}$ = drop diameter such that 90% of total liquid volume is in drops of smaller diameter.

$D_{0.999}$ = drop diameter such that 99.9% of total liquid volume is in drops of smaller diameter.

D_{peak} = value of D corresponding to peak of drop size frequency distribution curve.

The locations of various representative diameters on a drop size frequency curve (assuming a Rosin-Rammler distribution) are shown in Fig. 3.12.

Chin and co-workers [11, 12] have argued in support of the Rosin-Rammler distribution function because it is simple to use and can readily be obtained using a standard drop size analyzer. A further useful advantage is that all the represen-

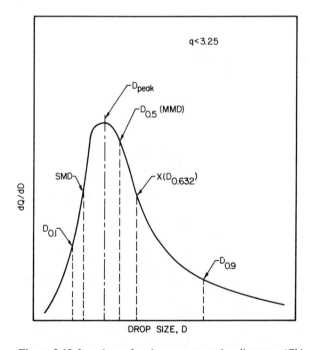

Figure 3.12 Locations of various representative diameters *(Chin and Lefebvre [11])*.

tative diameters in the spray are uniquely related to each other via the distribution parameter q. For example, we have [12]

$$\frac{\text{MMD}}{\text{SMD}} = (0.693)^{1/q} \, \Gamma\left(1 - \frac{1}{q}\right) \tag{3.25}$$

where Γ denotes the gamma function.

This equation demonstrates that the ratio MMD/SMD is not constant, as sometimes asserted, but is a unique function of q.

The definition of D_{peak} is evident from inspection of Fig. 3.12. Clearly, at D_{peak}, $d^2Q/dD^2 = 0$, so differentiation of Eq. (3.14) gives

$$\frac{d^2Q}{dD^2} = q(q-1)\frac{D^{q-2}}{X^q}\exp-\left(\frac{D}{X}\right)^q - \left(q\frac{D^{q-1}}{X^q}\right)^2\exp-\left(\frac{D}{X}\right)^q = 0 \tag{3.26}$$

Hence,

$$\frac{D_{\text{peak}}}{X} = \left(1 - \frac{1}{q}\right)^{1/q} \tag{3.27}$$

From Eq. (3.14) we have

$$\frac{D_{0.1}}{X} = (0.1054)^{1/q} \tag{3.28}$$

$$\frac{D_{0.9}}{X} = (2.3025)^{1/q} \tag{3.29}$$

Also,

$$\frac{\text{MMD}}{X} = (0.693)^{1/q} \tag{3.30}$$

and

$$\frac{\text{SMD}}{X} = \left[\Gamma\left(1 - \frac{1}{q}\right)\right]^{-1} \tag{3.31}$$

Other useful relationships are

$$\frac{D_{0.1}}{\text{MMD}} = (0.152)^{1/q} \tag{3.32}$$

$$\frac{D_{0.9}}{\text{MMD}} = (3.32)^{1/q} \tag{3.33}$$

$$\frac{D_{0.999}}{\text{MMD}} = (9.968)^{1/q} \tag{3.34}$$

It is of interest to examine where MMD and SMD lie on the distribution curve in relation to the peak diameter. From Eqs. (3.27) and (3.30) we have

$$\frac{D_{peak}}{MMD} = \frac{(1 - 1/q)^{1/q}}{(0.6931)^{1/q}} = \left(1.4428 - \frac{1.4428}{q}\right)^{1/q} \quad (3.35)$$

which shows that MMD = D_{peak} when $q = 3.2584$. Referring to Fig. 3.12, this means that MMD will appear on the left-hand side or the right-hand side of D_{peak} depending on whether q is greater or less than 3.2584, respectively.

From Eqs. (3.27) and (3.31) we have

$$\frac{D_{peak}}{SMD} = \left(1 - \frac{1}{q}\right)^{1/q} \Gamma\left(1 - \frac{1}{q}\right) \quad (3.36)$$

From this equation it is apparent that D_{peak} will always be larger than SMD, so SMD must always lie on the left-hand side of D_{peak}.

Several representative diameters are defined in Table 3.2, and the ratios of different diameters are shown in Table 3.3 and Figs. 3.13 to 3.15. Figures 3.16 and 3.17 show, respectively, the effects on the drop size frequency distribution curves and the cumulative drop size distribution curves of changes in SMD and q. These figures were calculated using Eq. (3.38).

From inspection of Table 3.3 it is evident that

1. The ratio of MMD/SMD is always greater than one. For $q \geq 3$ it changes only little, as illustrated in Fig. 3.15.
2. For a spray having $q = 3$, $D_{0.9}$ is only 50% higher than $D_{0.5}$, but for $q \leq 1.7$, $D_{0.9}$ is more than double $D_{0.5}$.
3. For many sprays q lies between 2 and 2.8, and SMD is between 80 and 84% of the peak diameter.

From Eq. (3.25) and Eqs. (3.27) to (3.36) the general conclusion is that when the Rosin-Rammler expression is used, the ratio of any two representative diameters is always a unique function of q. In Eq. (3.14), instead of using X, we can use any representative diameter and it will give the same distribution. For example, Eq. (3.14) can be rewritten as

Table 3.2 Some representative diameters

Symbol	Name	Value for Rosin-Rammler distribution	Position on Q versus D plot
$D_{0.1}$		$X(0.1054)^{1/q}$	$Q = 10\%$
D_{peak}	Peak diameter	$X\left(1 - \frac{1}{q}\right)^{1/q}$	Peak point on $\dfrac{dQ}{dD}$ versus D curve
$D_{0.5}$	Mass median diameter (MMD)	$X(0.693)^{1/q}$	$Q = 50\%$
$D_{0.632}$	Characteristic diameter	X	$Q = 63.2\%$
$D_{0.9}$		$X(2.3025)^{1/q}$	$Q = 90\%$
$D_{0.999}$	Maximum diameter	$X(6.9077)^{1/q}$	$Q = 99.9\%$

Table 3.3 Relationship between Rosin-Rammler distribution parameter q and other spray parameters

q	$\dfrac{D_{0.9}}{D_{0.5}}$	$\dfrac{D_{peak}}{D_{0.5}}$	$\dfrac{D_{peak}}{SMD}$	$\dfrac{D_{0.5}}{SMD}$	$Q_{at\,SMD}$ (%)	$\Gamma\left(1 - \dfrac{1}{q}\right)$	$\dfrac{D_{peak}}{X}$	$\dfrac{D_{0.9}}{X}$	$\dfrac{D_{0.1}}{X}$	$\dfrac{D_{0.5}}{X}$	$\dfrac{SMD}{X}$	$\Delta = \dfrac{D_{0.9} - D_{0.1}}{D_{0.5}}$	$\Delta_B = \dfrac{D_{0.999} - D_{0.5}}{D_{0.5}}$
1.2	2.71952	0.30494	1.2506	4.1013	11.965	5.5673	0.22467	2.00376	0.15331	0.73681	0.17965	2.51143	5.7939
1.4	2.35134	0.53100	1.2870	2.4238	18.18	3.1496	0.40868	1.81437	0.20041	0.76967	0.31755	2.09695	4.1671
1.6	2.11772	0.68120	1.2841	1.8851	22.23	2.3707	0.54171	1.68417	0.24500	0.79527	0.42187	1.80966	3.2081
1.8	1.94832	0.78125	1.2701	1.6257	25.10	1.9930	0.63730	1.58939	0.28645	0.81577	0.50180	1.59719	2.5871
2.0	1.82262	0.84935	1.2534	1.4757	27.26	1.7727	0.70711	1.51743	0.32459	0.83255	0.56418	1.43275	2.1570
2.2	1.72582	0.89683	1.2367	1.3790	28.95	1.6291	0.75918	1.46098	0.35955	0.84654	0.61388	1.30110	1.8437
2.4	1.64910	0.93068	1.2212	1.3122	30.31	1.5288	0.79885	1.41554	0.39155	0.85837	0.65415	1.19295	1.6065
2.6	1.58685	0.95529	1.2071	1.2636	31.48	1.4550	0.82967	1.37820	0.42083	0.86852	0.68788	1.1023	1.4213
2.8	1.53536	0.97348	1.1943	1.2269	32.36	1.3956	0.85402	1.34698	0.44767	0.87731	0.71506	1.02507	1.2731
3.0	1.49210	0.98712	1.1830	1.1984	33.15	1.3542	0.87358	1.32050	0.47231	0.88500	0.73848	0.95841	1.1520
3.2	1.45524	0.99747	1.1726	1.1756	33.84	1.3183	0.88950	1.29775	0.49498	0.89178	0.75857	0.90019	1.0514
3.4	1.42348	1.00539	1.1633	1.1571	34.43	1.2889	0.90263	1.27801	0.51588	0.89781	0.77591	0.84888	0.9665
3.6	1.39582	1.0115	1.1549	1.1418	34.95	1.2642	0.91357	1.26071	0.5352	0.90320	0.79103	0.80327	0.8940
3.8	1.37154	1.01624	1.1473	1.1290	35.40	1.2434	0.92278	1.24543	0.55311	0.90805	0.80430	0.76242	0.8314
4.0	1.35004	1.01992	1.1403	1.118	35.83	1.2253	0.93060	1.23184	0.56973	0.91244	0.81613	0.72564	0.7768

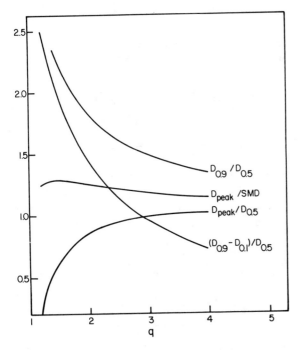

Figure 3.13 Relationship between Rosin-Rammler distribution parameter q and various spray characteristics *(Chin and Lefebvre [11])*.

$$Q = 1 - \exp -\left[0.693\left(\frac{D}{\text{MMD}}\right)^{q}\right] \tag{3.37}$$

or

$$Q = 1 - \exp -\left[\Gamma\left(1 - \frac{1}{q}\right)^{-q}\left(\frac{D}{\text{SMD}}\right)^{q}\right] \tag{3.38}$$

Although Eq. (3.14) is simpler than Eq. (3.38), the latter is strongly recommended because it shows clearly both the fineness and the spread of drop sizes in the spray.

From Table 3.3 it can be seen that, for a constant SMD of 50 μm, changing q from 2 to 3 will produce the following changes in MMD and X:

q	2	2.2	2.4	2.6	2.8	3.0
$D_{0.5}$ (MMD), μm	73.78	68.95	65.61	63.18	61.35	59.90
X, μm	88.62	81.45	76.45	72.69	69.93	67.70

Some drop size analyzers give a direct estimate of $D_{0.9}$, $D_{0.1}$, $D_{0.5}$, SMD, etc., but others can only estimate X and q. The above equations, along with Table 3.3 and Figs. 3.13–3.15, can then be used to obtain SMD, $D_{0.5}$, $D_{0.9}$, and $D_{0.1}$.

It is of interest to note in Table 3.3 that as q increases (toward a more uniform spray), the volume fraction Q at which D = SMD also increases. This means that, relatively speaking, SMD is increasing. This somewhat curious trend occurs because increase in q causes all the drops in the spray to shift closer to $D_{0.5}$, which eliminates many of the smallest drops. For many combustion systems, where very small drops are needed to provide high initial fuel evaporation rates for rapid ignition, a more uniform distribution of drop sizes in the spray may not always be desirable.

It is clearly important to distinguish between the concept of a representative diameter and a diameter that provides an indication of atomization quality. Any of the diameters listed in Table 3.2 can be used as a representative diameter to describe drop size distribution, but only the SMD can properly indicate the fineness of the spray from a combustion viewpoint. Even if the mass median diameter ($D_{0.5}$) is decreasing, it is still not certain whether the spray is finer, because if q also changes it is possible that SMD will either increase or remain sensibly constant. For the same reason, other representative diameters such as D_{peak}, X, and $D_{0.9}$ cannot indicate the fineness of the spray. It is strongly recommended, therefore, that for combustion applications the Sauter mean diameter be used to describe atomization quality, since the use of any representative diameter could lead to erroneous conclusions about the fineness of the spray.

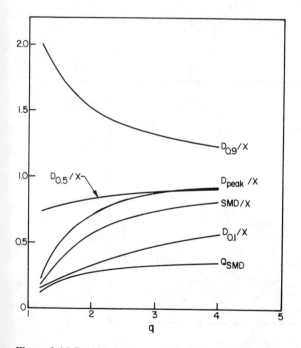

Figure 3.14 Relationship between Rosin-Rammler distribution parameter q and various spray characteristics *(Chin and Lefebvre [11])*.

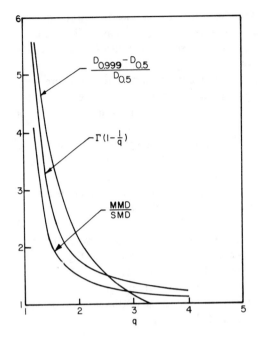

Figure 3.15 Relationship between Rosin-Rammler distribution parameter q and various spray characteristics *(Chin and Lefebvre [11])*.

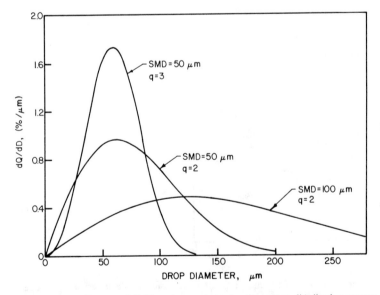

Figure 3.16 Influence of SMD and q on drop size frequency distribution curves.

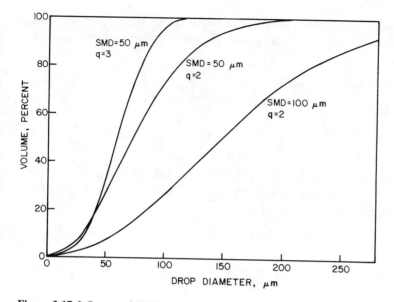

Figure 3.17 Influence of SMD and q on cumulative drop size distribution curves.

DROP SIZE DISPERSION

The term *dispersion* is sometimes used as an alternative to *distribution* to express the range of drop sizes in a spray. The concept of *width* is also employed to indicate drop size range, but it can be misleading. For example, the width of a spray having a certain SMD may be less than that of another spray having a higher SMD, but it does not necessarily follow that the former has a smaller drop size dispersion than the latter, because their average values are different.

It is generally recognized that the ratio MMD/SMD provides a good indication of drop size dispersion (see, for example, reference [13]) but, as discussed above, a slight change in this ratio could correspond to a large change in dispersion (or q).

Droplet Uniformity Index

Tate [14] has proposed a *droplet uniformity index* to describe the spread of drop sizes in a spray. It is defined as

$$\text{Droplet uniformity index (volumetric basis)} = \frac{\sum_i V_i(D_{0.5} - D_i)}{D_{0.5}} \tag{3.39}$$

where D_i is the midpoint of size class i and V_i is the volume fraction in the size class.

This expression indicates the spread relative to the mass median diameter and takes into account all the discrete size classes. It can be shown that the droplet uniformity index depends mainly on q and to a lesser extent on mean diameter.

Relative Span Factor

This parameter, defined by

$$\Delta = \frac{D_{0.9} - D_{0.1}}{D_{0.5}} \tag{3.40}$$

has much to commend it. First, it provides a direct indication of the range of drop sizes relative to the mass median diameter. Second, from Eqs. (3.28)–(3.30) we have, for a Rosin-Rammler distribution,

$$\Delta = (3.322)^{1/q} - (0.152)^{1/q} \tag{3.41}$$

which is also a unique function of q.

The manner and extent of change in the relative span factor with q is shown in Table 3.3.

Dispersion Index

Another parameter that is used to express the range of drop sizes is the *dispersion index*, which is defined for a Rosin-Rammler distribution as

$$\delta = \int_0^{D_m} D \frac{dQ}{dD} dD = \int_0^{D_m} q \left(\frac{D}{X}\right)^q \exp\left[-\left(\frac{D}{X}\right)^q\right] dD \tag{3.42}$$

which is essentially the same as the *uniformity index* proposed by Tate [14].

It should be noted that the dispersion index depends not only on q but also to a lesser degree on X. Due to its more complicated form, δ has no advantage over the relative span factor Δ. Thus for most engineering purposes the relative span factor provides an adequate description of dispersion.

Dispersion Boundary Factor

To give some indication of the maximum possible drop size, it is useful to define a dispersion boundary factor given by

$$\Delta_B = \frac{D_{0.999} - D_{0.5}}{D_{0.5}} \tag{3.43}$$

By assuming a Rosin-Rammler distribution, from Eq. (3.14) we have

$$\frac{D_{0.999}}{X} = (6.90775)^{1/q} \tag{3.44}$$

Hence, from Eqs. (3.34) and (3.44) we have

$$\Delta_B = (9.9665)^{1/q} - 1 \qquad (3.45)$$

which is also a unique function of q. Values of Δ_B are listed in Table 3.3.

From Table 3.3, which shows the variation of dispersion boundary factor with q, it is possible to estimate the maximum drop size in a spray for any given value of q. For example, for

$$q = 2, \qquad D_{0.999} = 3.16 D_{0.5}$$

$$q = 3, \qquad D_{0.999} = 2.15 D_{0.5}$$

$$q = 4, \qquad D_{0.999} = 1.77 D_{0.5}$$

The rationale employed by Chin and his colleagues [11, 12] in developing methods for defining drop size distributions solely in terms of a representative diameter and the Rosin-Rammler dispersion parameter q could easily be extended to more complicated distribution functions such as the upper-limit function. However, this would add complexity without necessarily improving accuracy and would be difficult to justify unless some situation arose where another distribution function demonstrated a clear superiority over the Rosin-Rammler distribution parameter.

CONCLUDING REMARKS

Much of this chapter has been devoted to a review of available methods for characterizing drop size distributions in sprays. It is difficult to judge whether any one method is superior to all others; such a conclusion must await theoretical determinations based on a more complete understanding of the basic mechanisms involved in drop formation. In the meantime, the following observations may be of guidance in avoiding some of the more commonly encountered pitfalls:

1. No single parameter can completely define a drop size distribution. For example, two sprays are not necessarily similar just because they have the same Sauter mean diameter (SMD) or the same mass median diameter (MMD). In many practical applications it is the smallest drop sizes in a spray or the largest drop sizes that are of paramount importance, and neither SMD nor MMD can provide this information.
2. There is no universal correlation between the mean diameter (or representative diameter) of a spray and its drop size distribution. They are completely independent of each other.
3. Mean diameters and representative diameters are different in nature. The MMD is not a mean diameter; it is a representative diameter. In particular, MMD ($D_{0.5}$) should not be confused with the mass mean diameter D_{30}.
4. If the Rosin-Rammler distribution is used, the distribution of drop sizes in a spray is defined by two parameters, a representative diameter and a measure of drop size dispersion.

5. Relative span factor can be used to indicate the spread of drop sizes in a spray.
6. Dispersion boundary factor can be used to define a meaningful maximum drop size.

NOMENCLATURE

D	drop diameter
$\bar{D}$	mean drop diameter
$D_{0.1}$	drop diameter such that 10% of total liquid volume is in drops of smaller diameter
$D_{0.5}$	drop diameter such that 50% of total liquid volume is in drops of smaller diameter
$D_{0.632}$	drop diameter such that 63.2% of total liquid volume is in drops of smaller diameter
$D_{0.9}$	drop diameter such that 90% of total liquid volume is in drops of smaller diameter
$D_{0.999}$	drop diameter such that 99.9% of total liquid volume is in drops of smaller diameter
D_{peak}	value of D corresponding to peak of drop size frequency distribution curve
D_m	maximum drop diameter
D_0	minimum drop diameter
D_{ng}	geometric number mean diameter
D_{sg}	geometric surface mean diameter
D_{vg}	geometric volume mean diameter
MMD	mass median diameter ($D_{0.5}$)
N	number of drops
Q	liquid volume fraction (or percentage) containing drops of smaller diameter than D
q	Rosin-Rammler drop size distribution parameter
SMD	Sauter mean diameter (D_{32})
s_g	geometric standard deviation
s_n	standard deviation
V	volume
X	characteristic diameter in Rosin-Rammler equation
δ	dispersion index
Δ	relative span factor
Δ_B	dispersion boundary factor

REFERENCES

1. Fraser, R. P., and Eisenklam, P., Liquid Atomization and the Drop Size of Sprays, *Trans. Inst. Chem. Eng.*, Vol. 34, 1956, pp. 294–319.

2. Rizk, N. K., and Lefebvre, A. H., Airblast Atomization: Studies on Drop-Size Distribution, *J. Energy*, Vol. 6, No. 5, 1982, pp. 323–327.

3. Miesse, C. C., and Putnam, A. A., Mathematical Expressions for Drop-Size Distributions, *Injection and Combustion of Liquid Fuels*, Section II, WADC Technical Report 56-344, Battelle Memorial Institute, March 1957.

4. Nukiyama, S., and Tanasawa, Y., Experiments on the Atomization of Liquids in an Air Stream, Report 3, On the Droplet-Size Distribution in an Atomized Jet, Defense Research Board, Department National Defense, Ottawa, Canada; translated from *Trans. Soc. Mech. Eng. Jpn.*, Vol. 5, No. 18, 1939, pp. 62–67.

5. Rosin, P., and Rammler, E., The Laws Governing the Fineness of Powdered Coal, *J. Inst. Fuel*, Vol. 7, No. 31, 1933, pp. 29–36.

6. Rizk, N. K., and Lefebvre, A. H., Drop-Size Distribution Characteristics of Spill-Return Atomizers, *AIAA J. Propul. Power*, Vol. 1, No. 3, 1985, pp. 16–22.

7. Rizk, N. K., Spray Characteristics of the LHX Nozzle, Allison Gas Turbine Engines Report Nos. AR 0300-90 and AR 0300-91, 1984.

8. Mugele, R., and Evans, H. D., Droplet Size Distributions in Sprays, *Ind. Eng. Chem.*, Vol. 43, No. 6, 1951, pp. 1317–1324.

9. Brodkey, R. A., *The Phenomena of Fluid Motions*, Addison-Wesley, Reading, Mass., 1967.

10. Marshall, W. R., Jr., Mathematical Representations of Drop-Size Distributions of Sprays, *Atomization and Spray Drying*, Chapter VI, Chem. Eng. Prog. Monogr. Ser., Vol. 50, No. 2, American Institute of Chemical Engineers, 1954.

11. Chin, J. S., and Lefebvre, A. H., Some Comments on the Characterization of Drop-Size Distributions in Sprays, *Proceedings of the 3rd International Conference on Liquid Atomization and Spray Systems*, London, Paper No. IVA/1, 1985.

12. Zhao, Y. H., Hou, M. H., and Chin, J. S., Dropsize Distributions from Swirl and Airblast Atomizers, *Atomization Spray Technol.*, Vol. 2, 1986, pp. 3–15.

13. Martin, C. A., and Markham, D. L., Empirical Correlation of Drop Size/Volume Fraction Distribution in Gas Turbine Fuel Nozzle Sprays, ASME Paper ASME 79-WA/GT-12, 1979.

14. Tate, R. W., Some Problems Associated with the Accurate Representation of Drop-Size Distributions, *Proceedings of the 2nd International Conference on Liquid Atomization and Sprays (ICLASS)*, Madison, Wis., 1982, pp. 341–351.

CHAPTER

FOUR

ATOMIZERS

INTRODUCTION

Atomization is often accomplished by discharging the liquid at high velocity into a relatively slow-moving stream of air or gas. Notable examples include the various forms of pressure atomizers and also rotary atomizers that eject the liquid at high velocity from the periphery of a rotating cup or disk. An alternative approach is to expose a relatively slow-moving liquid to a high-velocity airstream. The latter method is generally known as twin-fluid, air-assist, or airblast atomization.

In this chapter the general features of the main types of atomizers in industrial and laboratory use are described. Primary emphasis is placed on the atomizers employed in combustion equipment, since this is by far their most widespread application. Most of the fuels employed in heat engines and industrial furnaces are liquids that must be atomized before being injected into the combustion zone. The process of atomization produces a very high ratio of surface to mass in the liquid phase, thereby promoting rapid evaporation and high rates of combustion.

The fuel injection process plays a major role in many aspects of combustion performance. It seems likely to assume even greater importance in the future, as all types of combustion equipment will be called on to burn fuels of diminishing quality, while being subjected to increasingly severe regulative standards for exhaust pollutant emissions. Fuel nozzles having multifuel capability will also be in increasing demand.

ATOMIZER REQUIREMENTS

An ideal atomizer would possess all the following characteristics:

1. Ability to provide good atomization over a wide range of liquid flow rates.
2. Rapid response to changes in liquid flow rate.
3. Freedom from flow instabilities.
4. Low power requirements.
5. Capability for scaling, to provide design flexibility.
6. Low cost, light weight, ease of maintenance, and ease of removal for servicing.
7. Low susceptibility to damage during manufacture and installation.

Fuel nozzles should have all the above features, plus the following:

1. Low susceptibility to blockage by contaminants and to carbon buildup on the nozzle face.
2. Low susceptibility to gum formation by heat soakage.
3. Uniform radial and circumferential fuel distribution.

PRESSURE ATOMIZERS

As their name suggests, pressure atomizers rely on the conversion of pressure into kinetic energy to achieve a high relative velocity between the liquid and the surrounding gas. Most of the atomizers in general use are of this type. They include plain-orifice and simplex nozzles, as well as various wide-range designs such as variable-geometry, duplex, and dual-orifice injectors. These various types of pressure atomizers are discussed below in turn.

Plain Orifice

The atomization of a low-viscosity liquid is most easily accomplished by passing it through a small circular hole. If the velocity is low, the liquid emerges as a thin distorted pencil, but if the liquid pressure exceeds the ambient gas pressure by about 150 kPa, a high-velocity liquid jet is formed that rapidly disintegrates into a well-atomized spray. Disintegration of the jet is promoted by an increase in flow velocity, which increases both the level of turbulence in the liquid jet and the aerodynamic drag forces exerted by the surrounding medium, and is opposed by increases in liquid viscosity and surface tension, which resist breakup of the ligaments.

The sprays produced by plain-orifice atomizers have a cone angle that usually lies between 5° and 15°. This cone angle is only slightly affected by the diameter and length/diameter ratio of the orifice and is mainly dependent on the viscosity and surface tension of the liquid and the turbulence of the issuing jet. An increase in turbulence increases the ratio of the radial to the axial component of velocity in the jet and thereby increases the cone angle.

Diesel injectors. Perhaps the best known example of a plain-orifice atomizer is the diesel injector, which is designed to provide a pulsed or intermittent supply of fuel to the combustion space for each power stroke of the piston. As the air into which the fuel is injected is already at a high pressure due to compression by the piston, very high injection pressures must be used to achieve the desired atomization and penetration of the spray.

The formation of a suitable mixture of fuel and air in diesel engines is accomplished by a variety of designs of fuel injectors and combustion chambers. In all cases the main objectives are to achieve the desired spray characteristics in terms of penetration, angle, and atomization quality and to provide for relative motion between the air and fuel drops or vapor.

The first stage of mixing fuel and air is accomplished as the continuous jet of fuel, moving at high velocity into air at high pressure, becomes finely atomized at its outer edges. At this point the spray structure comprises a "solid" fuel jet surrounded by a mantle containing drops of fuel mixed with air. The velocity of the fuel jet is highest at the center of the jet and decreases to almost zero at the interface between the zone of disintegration and the ambient air. In the vicinity of this interface where the velocity of drops relative to air is low, further mixing of fuel and air is accomplished by the air movement provided in the design of the combustion chamber or by random turbulence in the combustion chamber. In the disintegrating jet, concentrations of fuel range from zero at the periphery to 100% at the center [1].

As soon as drops are formed, vaporization of fuel begins. Spontaneous ignition of the vaporized fuel occurs in a region somewhere between the injector and the tip of the spray. A flame then spreads rapidly along the progressing spray, and burning occurs in a mixture containing air, fuel vapor, and partially evaporated fuel drops. During the later stages of injection, fuel is injected into the spreading flame, and combustion takes place under turbulent, fuel-rich conditions. During this phase the fuel spray is entraining air by its own motion, and it is now well recognized that the exhaust gas concentrations of unburned hydrocarbon, nitric oxide, and particulates are very dependent on the aerodynamic motion of the spray and the rate of fuel-air mixing.

A schematic layout of a Cummins diesel injector and its associated drive train is shown in Fig. 4.1. At the beginning of the injection cycle the plunger is at its maximum lift and the fuel feed port is open. The cam then drives the injector plunger down, closing off the feed orifice. Injection of fuel into the combustion chamber does not start until the plunger reaches the "solid fuel height," that is, the plunger displacement at which the volume remaining in the injector body is completely filled with the fuel quantity metered into the injector during the filling period [2]. Further displacement of the injector then drives the fuel out of the spray holes into the combustion chamber, the mechanical drive system (cam, push tube, rocker, and link), providing the force to generate the very high injection

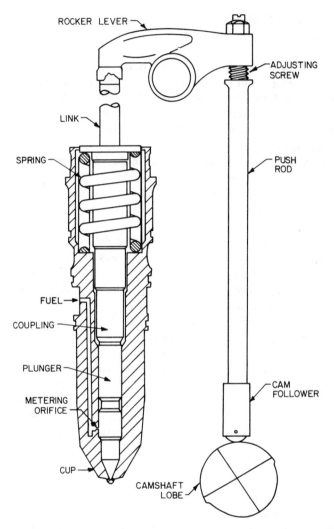

Figure 4.1 Diesel injector and associated drive system. *(Courtesy of Cummins Engine Company.)*

pressures required, typically around 83–103 MPa (12,000–15,000 psi). The end of injection is controlled by driving the tip of the plunger into the mating conical section of the injector cup, as illustrated in Fig. 4.2. The number and angular location of the injection holes are designed to provide a spray pattern that matches the combustion chamber geometry, as illustrated in Fig. 4.3.

Mechanically actuated diesel injectors have been in use for many years, and a great deal of practical knowledge of the characteristics of the system has been accumulated. In recent years attention has turned toward electromagnetic fuel injection. An example of this type of injector, manufactured by the Diesel Equipment Division of General Motors, is shown schematically in Fig. 4.4. This type

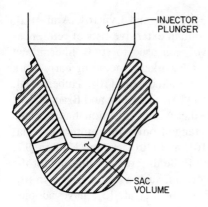

Figure 4.2 Enlarged view of sac volume and injector holes.

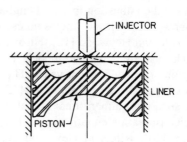

Figure 4.3 Typical combustion chamber geometry for direct-injection diesel engine.

of nozzle permits injection pressure and duration to be varied independently of each other and of engine speed.

The mode of operation of the electromagnetic injector, which requires a constant high-pressure fuel source, has been described by Sinnamon et al. [3]. An electromagnet actuates a pilot valve that modulates fuel pressure in the modulation chamber above the injector needle, as shown in Fig. 4.4. Energizing the electromagnet causes the pilot valve to open, the pressure in the modulation chamber to drop, and the injector needle to lift. Fuel from the constant-pressure source then exits the nozzle. The nozzle is closed by deenergizing the electromagnet. The pilot valve then closes and the pressure buildup in the modulation chamber returns the injector needle to the closed position. The orifice between the fuel supply passage and the modulation chamber is sized to ensure rapid closing of the injector needle, yet prevent rapid pressure fluctuations in the modulation chamber from affecting the nozzle pressure.

Many studies have been carried out on fuel sprays in diesel engines. The

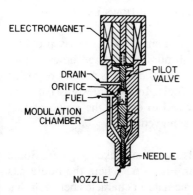

Figure 4.4 Schematic diagram of electromagnetic diesel injector. *(Courtesy of General Motors Corp.)*

studies of Elkotb [4], Borman and Johnson [5], Adler and Lyn [6], Arai et al. [7], and Hiroyasu [8] are frequently cited and contain extensive lists of references on diesel sprays. Visualization techniques have been used to study intermittent sprays and combustion in diesel engines by several workers, including early work by Schweitzer [9] and later studies by Lyn and Valdamanis [10], Huber et al. [11], Rife and Heywood [12], Hiroyasu and Arai [13], and Reitz and Bracco [14].

The use of computer models to simulate diesel fuel injection has become common practice in recent years. These models range from highly empirical equations such as those of Chiu et al. [15], Dent [16], and Hiroyasu and Kadota [17] to detailed multidimensional models that require finite difference solutions for the basic conservation equations [18, 19]. Other workers, such as Adler and Lyn [6], Melton [20], Rife and Heywood [12], and Sinnamon et al. [3], have adopted intermediate approaches that generally use integral-type continuity and momentum equations and rely heavily on experimental data.

The objective of these models is mainly to characterize and predict fuel spray penetration and trajectory within a diesel engine combustion chamber. These predictions can then be applied to engine design or to the analysis of experimental data acquired from existing engines. For example, the model developed by Sinnamon et al. [3] permits injection from any position within the chamber and in any direction, both with and without air swirl. Comparisons with experiments have shown that the model can predict spray penetration and trajectory with reasonable accuracy and responds properly to changes in nozzle diameter, injection pressure, direction of injection, cylinder air density, and air swirl rate.

Afterburner injectors. Another important application of plain-orifice atomizers in the combustion field is to jet engine afterburners, where the fuel injection system normally consists of one or more circular manifolds supported by struts inside the jet pipe. Fuel is supplied to the manifolds by feed pipes in the support struts and is sprayed into the flame zone from holes drilled in the manifolds. Sometimes "stub pipes" of the type shown in Fig. 4.5 are used instead of manifolds, and many fuel injector arrays consist of stub pipes mounted radially on circular manifolds. In all cases the objective is to provide a uniform distribution of well-atomized fuel throughout the portion of the gas stream that flows into the combustion zone.

For any given liquid flow rate, a large number of small holes clearly provides a more uniform distribution of liquid than a small number of large holes. However, owing to the risk of blockage, a hole diameter of 0.5 mm is usually regarded as the minimum practical size for kerosine-type fuels. In some designs (see, for example, Fig. 4.26) a variable hole size is provided to extend the range of fuel flow rates over which good atomization can be achieved.

Rocket injectors. Plain-orifice atomizers are also used in rocket engines. Some typical configurations are shown schematically in Fig. 4.6. Where the liquid jets are designed to impinge, the atomization is intermediate in character between that of a plain jet and that of a liquid sheet [21]. At low injection velocities and with

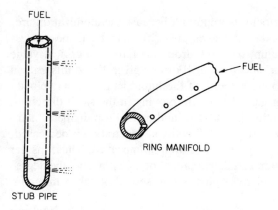

Figure 4.5 Plain-orifice atomizers for afterburners.

large impingement angles, a well-defined sheet is formed at right angles to the plane of the two jets; however, the sheet becomes progressively less pronounced as the liquid velocity is increased or the impingement angle is reduced. High-speed photography reveals that under typical operating conditions the spray is produced both by a process resembling the disintegration of a plain jet and by rather ill-defined sheet formation at the intersection of the two streams [21].

Other applications. Plain-orifice nozzles are widely used as a means of introducing liquid into a flowing stream of air or gas. Two cases of practical importance are (1) injection into a coflowing or contraflowing stream of air and (2) transverse injection across a flowing stream of air.

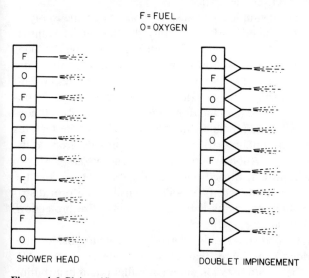

Figure 4.6 Plain-orifice atomizers for rocket engines.

The influence of air or gas velocity is important because the atomization process is not completed as soon as the jet leaves the orifice. Instead, the process continues in the surrounding medium until the drop size falls to a critical value below which no further disintegration can occur. For any given liquid this critical drop size depends not on the absolute velocity of the liquid jet but on its velocity relative to that of the surrounding gas. If both are moving in the same direction, penetration is augmented, atomization is retarded, and mean drop diameter is increased. When the movements are in opposite directions, penetration is decreased, the cone angle widens, and the quality of atomization is improved. Thus, insofar as gaseous flow affects the formation and development of the spray and the degree of atomization achieved, it is the relative velocity that should be taken into consideration.

These effects of air motion on the spray characteristics of plain-orifice atomizers are relevant only to situations in which the air velocity is not sufficiently high to change the basic nature of the atomization process. If, however, the issuing liquid jet is subjected to a high-velocity airstream, the mechanism of jet disintegration changes and corresponds to airblast atomization. Plain-jet airblast atomizers of this type are sometimes used in gas turbines and are discussed later in this chapter.

Simplex

The narrow spray cone angles exhibited by plain-orifice atomizers are disadvantageous for most practical applications. Much wider cone angles are achieved in the simplex or pressure-swirl atomizer, in which a swirling motion is imparted to the liquid so that, under the action of centrifugal force, it spreads out in the form of a conical sheet as soon as it leaves the orifice.

There are two basic types of simplex nozzles. In one design the spray is comprised of drops that are distributed fairly uniformly throughout its volume. This is generally described as a *solid-cone* spray. The other nozzle type produces a *hollow-cone* spray, in which most of the drops are concentrated at the outer edge of a conical spray pattern. These two spray structures are illustrated in Fig. 4.7. Figure 4.8 shows a Delavan solid-cone nozzle of the type used in gas scrubbing, coke quenching, chemical processing, and drenching operations [22]. It consists of a one-piece cast body with a removable vane-type core. This core features a cylindrical hole that functions as a plain-orifice atomizer to provide drops at the center of the conical spray pattern. Another type of solid-cone nozzle is shown in Fig. 4.9. This nozzle has a special orifice outlet configuration to accent the corners of the spray pattern. It is designed to provide a sensibly uniform distribution of drops in a square pattern and is used for gas washing, fire protection, foam breaking, gravel washing, and vegetable cleaning [22].

The main drawback of solid-cone nozzles is relatively coarse atomization, the drops at the center of the spray being larger than those near the periphery [23]. Hollow-cone nozzles provide better atomization and their radial liquid distribution is also preferred for many industrial purposes, especially for combustion applications.

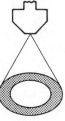

HOLLOW CONE

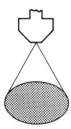

SOLID CONE

Figure 4.7 Spray produced by pressure-swirl nozzles.

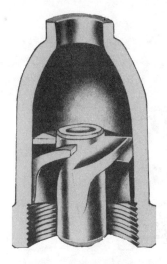

Figure 4.8 Solid cone simplex nozzle. *(Courtesy of Delavan, Inc.)*

The simplest form of hollow-cone atomizer is the so-called simplex atomizer, as illustrated in Fig. 4.10. Liquid is fed into a swirl chamber through tangential ports that give it a high angular velocity, thereby creating an air-cored vortex. The outlet from the swirl chamber is the final orifice, and the rotating liquid flows through this orifice under both axial and radial forces to emerge from the atomizer in the form of a hollow conical sheet, the actual cone angle being determined by the relative magnitude of the tangential and axial components of velocity at exit.

The development of the spray passes through several stages as the liquid injection pressure is increased from zero.

1. Liquid dribbles from the orifice.
2. Liquid leaves as a thin distorted pencil.
3. A cone forms at the orifice but is contracted by surface tension forces into a closed bubble.
4. The bubble opens into a hollow tulip shape terminating in a ragged edge, where the liquid disintegrates into fairly large drops.
5. The curved surface straightens to form a conical sheet. As the sheet expands its thickness diminishes, and it soon becomes unstable and disintegrates into ligaments and then drops in the form of a well-defined hollow-cone spray.

These five stages of spray development are illustrated in Figs. 4.11 and 4.12. Various types of hollow-cone simplex atomizers have been developed for com-

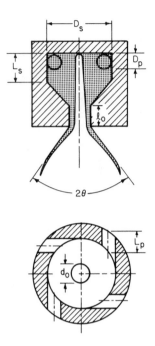

Figure 4.9 Square-spray simplex nozzle. *(Courtesy of Delavan, Inc.)*

Figure 4.10 Schematic view of a simplex swirl atomizer.

bustion applications. They differ mainly in the method used to impart swirl to the issuing jet [24]. They include swirl chambers with tangential slots or drilled holes and swirling by helical slots or vanes, as illustrated in Fig. 4.13. Where spray uniformity is not a prime consideration, economies can be achieved by using thin, removable swirl plates out of which the swirl chamber entry ports are cut or stamped.

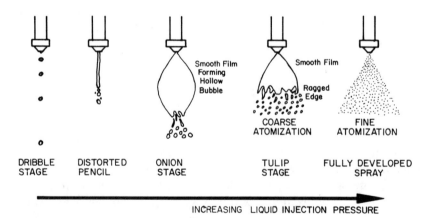

Figure 4.11 Stages in spray development with increase in liquid injection pressure.

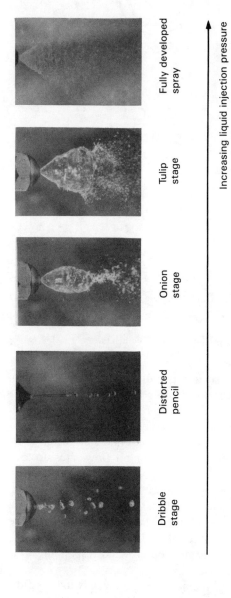

Dribble stage   Distorted pencil   Onion stage   Tulip stage   Fully developed spray

Increasing liquid injection pressure

Figure 4.12 Photographs illustrating spray development in a simplex swirl atomizer.

115

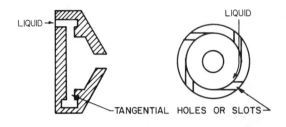

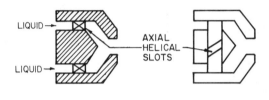

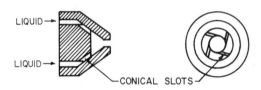

Figure 4.13 Various designs of simplex swirl atomizers.

A more rugged construction is indicated in the Delavan nozzle shown in Fig. 4.14. This nozzle features a one-piece body with a removable spiral-type core. It is used primarily for gas washing, spray cooling, rinsing, and dust suppression.

Many different types of large simplex atomizers are employed in utility boilers and industrial furnaces. A typical commercial design is shown in Fig. 4.15. Atomizers of this type are designed for oil flow rates up to around 4000 kg/h. Figure 4.16 shows a more advanced version that embodies several features for improving the quality of atomization [25]. These include (1) minimum area of wetted surface to reduce frictional losses, (2) improved mixing between the discrete jets emanating from each inlet slot, and (3) reduced losses due to flow separations at sharp corners. According to Jones [25], these design improvements yield a 12% reduction in mean drop size, corresponding to a 30% reduction in the emission of particulates.

A major drawback of the simplex atomizer is that its flow rate varies as the square root of the injection pressure differential. Thus, doubling the flow rate demands a fourfold increase in injection pressure. For low-viscosity liquids the lowest injection pressure at which atomization can be achieved is about 100 kPa. This means that an increase in flow rate to some 20 times the minimum value

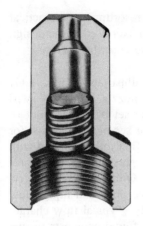

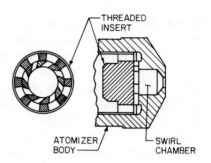

Figure 4.14 Hollow-cone simplex nozzle. *(Courtesy of Delavan, Inc.)*

Figure 4.15 Commercial design of large pressure atomizer *(Jones [25]).*

would require an injection pressure of 40 MPa. On the other hand, if the nozzle flow number were made large enough to pass the maximum flow rate at a more acceptable value of injection pressure, say around 7 MPa, then at the lowest flow rate the injection pressure would be only 17.5 kPa (i.e., around 2.5 psi) and atomization quality would be extremely poor. This basic drawback of the simplex atomizer has led to the development of various wide-range atomizers, such as duplex, dual-orifice, and spill atomizers, in which ratios of maximum to minimum flow rate in excess of 20 can readily be achieved with injection pressures not exceeding 7 MPa. These various designs have been described by Joyce [26], Mock and Ganger [27], Carey [28], Radcliffe [29, 30], Tipler and Wilson [31], and Dombrowski and Munday [32].

Wide-Range Nozzles

Although many different types of wide-range atomizers have been produced, the design objective in all cases is the same, namely to provide good atomization over

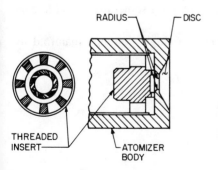

Figure 4.16 Improved design of large pressure atomizer *(Jones [25]).*

the entire operating range of liquid flow rates without resorting to impractical levels of pump pressure. Considerable ingenuity has been exercised in the design of wide-range atomizers, the most important of which are described below.

Duplex. The essential features of a duplex atomizer are illustrated in Fig. 4.17. The main factor that distinguishes it from the simplex nozzle is that its swirl chamber employs two sets of tangential swirl ports, one set being the pilot or primary ports for low flows and the other set the main or secondary passage for large flows. In operation, the primary swirl ports are the first to be supplied with liquid from the primary manifold, while a spring-loaded pressurizing valve prevents liquid from entering the secondary fuel manifold. Only when a predetermined injection pressure has been reached does the valve open and liquid flow through primary and secondary swirl ports simultaneously. Typical flow characteristics for a duplex nozzle are shown in Fig. 4.18. This figure demonstrates the superior performance of the duplex nozzle, especially at low flows. Consider, for example, a condition where the flow rate is 10% of the maximum value. Inspection of Fig. 4.18 shows that the injection pressure of the duplex nozzle is about eight times higher than that of the simplex nozzle. It will be shown later that this reduces mean drop size almost threefold. This same advantage of better atomization at low flow rates applies equally well to the dual-orifice atomizer, which is described in the next section.

A drawback of the duplex atomizer is a tendency for the spray cone angle to be smaller in the combined flow range than in the primary flow range by about 20°. This is because in going from the primary flow to the combined flow the ratio $A_p/D_s d_o$ is increased and the spray cone angle thereby reduced (see Fig. 7.4). In some designs this problem is overcome by setting the primary swirl ports on a smaller tangent circle than the secondary ports. The effect of this is to reduce the swirl component and hence the spray cone angle at low flow rates.

In the layout designed for gas turbines, as shown in Fig. 4.19, only one pressurizing valve is fitted irrespective of the number of atomizers used. This valve controls the distribution of fuel between the primary and secondary manifolds. This particular system has the disadvantage, when operating in the primary flow range, of allowing interflow between atomizers to occur via the secondary manifold. To alleviate this problem, pressurizing valves are sometimes fitted to each atomizer feed pipe, as shown in Fig. 4.20.

Another gas turbine duplex system employs a single common manifold for

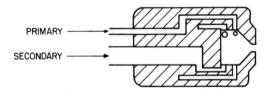

PRIMARY

SECONDARY

Figure 4.17 Duplex atomizer.

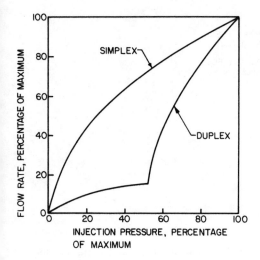

Figure 4.18 Flow characteristics of simplex and duplex atomizers.

primary and secondary fuel supply, with a pressurizing valve incorporated in each atomizer assembly. As the pump delivery pressure increases to meet rising engine fuel requirements, it eventually becomes high enough to overcome the spring loading of the pressurizing valves, whereupon the increased flow is admitted to the swirl chambers via the secondary swirl ports. The pressurizing valve is set to open at about 700 kPa, so that when the engine is operating at low power conditions all the fuel passes through the small primary slots, thereby ensuring good atomization.

The design procedure for duplex atomizers is the same as that for simplex atomizers except that special treatment is required if the primary ports are to be set on a smaller tangent circle than the main ports to give a constant spray cone angle [31].

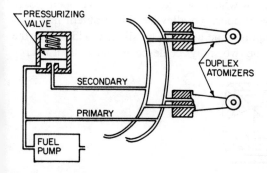

Figure 4.19 Two-manifold duplex system.

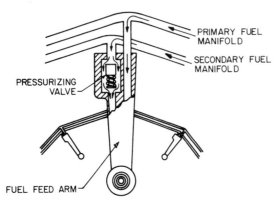

Figure 4.20 Atomizer fitted with pressurizing valve.

Dual orifice. The "dual-orifice atomizer," known in the United Kingdom as the "duple" atomizer, is shown diagrammatically in Fig. 4.21. For more than 30 years this type of nozzle has been widely used on many types of aircraft and industrial engines. Practical designs tend to be rather more complex than is suggested by the schematic diagram of Fig. 4.21, as illustrated in Fig. 4.22 for the Pratt and Whitney JT8 engine nozzle.

Essentially, a dual-orifice atomizer comprises two simplex nozzles that are fitted concentrically, one inside the other. The primary nozzle is mounted on the inside, and the juxtaposition of primary and secondary is such that the primary spray does not interfere with either the secondary orifice or the secondary spray within the orifice. When the fuel delivery is low it all flows through the primary nozzle, and atomization quality tends to be high because a fairly high fuel pressure is needed to force the fuel through the small ports in the primary swirl chamber. As the fuel supply is increased, a fuel pressure is eventually reached at which the pressurizing valve opens and admits fuel to the secondary nozzle. When this happens atomization quality is greatly impaired because the secondary fuel pressure is low. With further increase in fuel flow the secondary fuel pressure increases, and atomization quality starts to improve. However, there is inevitably a range of fuel flows, starting from the point at which the pressurizing valve opens, over which atomization quality is relatively poor. To alleviate this problem it is customary to arrange for the primary spray cone angle to be slightly wider than the secondary spray cone angle, so that the two sprays coalesce and share their energy

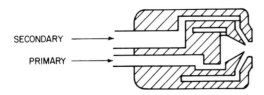

Figure 4.21 Dual-orifice atomizer.

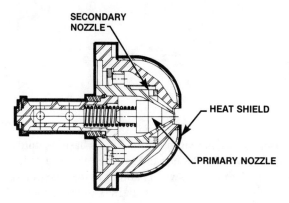

Figure 4.22 Dual-orifice atomizer used on Pratt and Whitney JT8 engine. *(Courtesy of Fuel Systems TEXTRON.)*

within a short distance from the atomizer. This helps the situation to some extent, but the atomization quality may still be unsatisfactory. Thus the designer must ensure that opening of the pressurizing valve does not coincide with an engine operating point at which high combustion efficiency and low pollutant emissions are prime requirements. D. R. Carlisle (personal communication) has analyzed the atomizing conditions in this critical flow range, starting from the assumption that the primary and secondary sprays coalesce and share momenta. This analysis shows that

$$\Delta P_e = \frac{4\Delta P_p}{R^2} [(1 + R)^{0.5} - 1]^2 \tag{4.1}$$

where ΔP_e is the equivalent injection pressure of the combined spray, ΔP_p is the primary injection pressure, and R is the ratio of secondary to primary flow numbers.

Thus for any combination of primary and secondary flow numbers and any given pressurizing-valve opening pressure, a minimum value of ΔP_e, the equivalent injection pressure of the combined spray, can be calculated and used in Eq. (6.18) for estimating mean drop size. The liquid flow rate corresponding to worst atomization occurs when the ratio of total fuel flow to primary fuel flow is equal to $(1 + R)^{0.5}$.

Carlisle has also examined the influence of gravity head on the fuel distributions obtained from multiple atomizer systems of the type employed on most aircraft engines. Maldistribution of fuel occurs when the secondary fuel pressure is low, and the effect of gravity head is to create a relatively high injection pressure in the bottom atomizer. The results of Carlisle's analysis confirm that increasing the pressurizing-valve opening pressure improves the distribution and that increasing R makes matters worse.

Spill return. This is basically a simplex atomizer, except that the rear wall of the swirl chamber, instead of being solid, contains a passage through which liquid

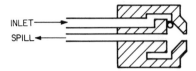

Figure 4.23 Spill-return atomizer.

can be "spilled" away from the atomizer, as shown in Fig. 4.23. Its basic features have been described by Joyce [26], Carey [28], Pilcher and Miesse [33], and Tyler and Turner [34]. Liquid is always supplied to the swirl chamber at the maximum pressure and flow rate. When the nozzle is operating at its maximum capacity, the valve located in the spill line is fully closed and all the liquid is ejected from the nozzle in the form of a well-atomized spray. Opening the valve allows liquid to be diverted away from the swirl chamber, leaving less to pass through the atomizing orifice. The spill atomizer's constant use of a relatively high pressure means that, even at extremely low flow rates, there is adequate swirl to provide efficient atomization. According to Carey [28], satisfactory atomization can be achieved even when the flow rate is as low as 1% of its maximum value, and, in general, the tendency is for atomization quality to improve as the flow rate is reduced.

The wide range of flows over which atomization quality is high is a most useful characteristic of the spill-return atomizer. Joyce [26] has stated that a flow range of 20 to 1 can be attained by spill control alone, using a constant supply pressure. If pressure is also varied over a range of 25 to 1, a flow range of 100 to 1 can readily be achieved. Other attractive features include absence of moving parts and, because the flow passages are designed to handle large flows all the time, freedom from blockage by contaminants in the liquid.

A disadvantage of the spill-return atomizer is the large variation in spray angle with change in flow rate. The effect of a reduction in flow rate is to lower the axial component of velocity without affecting the tangential component; consequently, the spray angle widens. The spray angle at minimum flow can be up to 50° wider than at maximum flow.

Another disadvantage of the spill system is that problems of metering the flow rate are more complicated than with other types of atomizer, and a larger-capacity pump is needed to handle the large recirculating flows. For these reasons interest in the spill-return atomizer for gas turbines has declined in recent years, and its main application has been in large industrial furnaces. However, if the aromatic content of gas turbine fuels continues to rise, it could pose serious problems of blockage, by gum formation, of the fine passages of conventional pressure atomizers. The spill-return atomizer, having no small passages, is virtually free from this defect. This factor, combined with its excellent atomizing capability, makes it an attractive proposition for dealing with the various alternative fuels now being considered for gas turbine applications, most of which combine high aromaticity with high viscosity.

Miscellaneous types. An interesting type of nozzle, designed by Lubbock for use on the early Whittle engine, is shown in Fig. 4.24. It is basically a simplex swirl atomizer, except that the back face of the swirl chamber is formed by the front face of a piston. The piston slides in a cylindrical sleeve in which narrow but quite deep tangential slots are cut. A helical coiled spring presses the piston in a forward direction toward the discharge orifice.

The device is quite simple in operation. At low flow rates, liquid enters the swirl chamber through the very limited depth of the narrow tangential slots not obscured by the piston in its most forward position. The small area of the inlet ports at this condition necessitates fairly high injection pressures to achieve the desired flow rate, so good atomization is automatically ensured. On progressively increasing the injection pressure to achieve a higher flow rate, the piston moves rearward, compressing the spring and thereby exposing an increasing depth of tangential slot to the flow. Finally, at maximum supply pressure, the tangential slots are open to their maximum area and the output of the nozzle attains its peak value.

According to Joyce [26], a flow range of about 25 to 1 can be achieved with this type of nozzle in the pressure range from 0.14 to 3.5 MPa (20 to 500 psi). The main advantage of the device is that good atomization can be achieved over a wide flow range, using a single feed pipe. Drawbacks of the system include variation in spray cone angle with flow rate, due to the change in A_p with flow rate (see Fig. 7.4), and the risk of the piston becoming jammed by gum or other impurities in the liquid.

Two other nozzle types that utilize liquid pressure to vary the effective flow area of the nozzle are shown schematically in Figs. 4.25 and 4.26. Figure 4.25 illustrates what is essentially a simplex swirl atomizer, except that its front face consists of a thin metallic disk. This disk, or diaphragm, distorts under the action of liquid pressure to increase the flow area, as indicated in the figure. In principle, it is exactly the same as the Lubbock nozzle described above. Both devices avoid the need for excessive hydraulic pressures by arranging for the flow number to increase as the supply pressure goes up. Special care must be taken to select the right material for the diaphragm. Control of diaphragm thickness and heat treatment are also important considerations. The main drawback of this device, and one that has militated against its more widespread adoption, is the difficulty of "matching" sets of atomizers to give closely similar progressive flow characteristics over the entire operational range of liquid pressure.

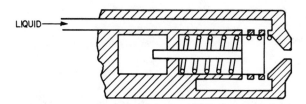

Figure 4.24 Variable-geometry pressure-swirl atomizer.

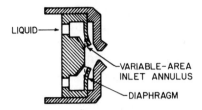

Figure 4.25 Diaphragm-type simplex atomizer.

The nozzle shown in Fig. 4.26 has been used with some success in jet engine afterburner systems. As illustrated in the figure, the fuel manifold itself forms part of the nozzle assembly. When additional fuel flow is required, the attendant increase in fuel pressure causes the oval manifold to expand away from the pintle, thereby increasing the effective discharge area of the nozzle. The pintle itself is shaped to direct the fuel away from the nozzle in the form of a conical spray.

Use of Shroud Air

When pressure-swirl nozzles are used in gas turbine combustors, a common practice is to house the nozzle within an annular passage, as shown in Fig. 4.27. A small proportion of the total combustor airflow, usually less than 1%, flows through this passage and is discharged at its downstream end, where it flows radially inward across the nozzle face. This air is usually called shroud air or anticarbon air. Its original purpose was to protect the nozzle from overheating by the flame and to prevent deposition of coke, which could interfere with the fuel spray. However, experience has shown that shroud air not only fulfills this purpose satisfactorily but also, when used judiciously, can have a marked and beneficial effect on atomization quality, particularly where the spray would otherwise be quite coarse.

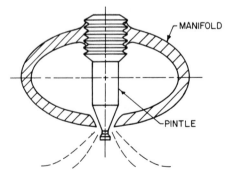

Figure 4.26 Schematic diagram of Pratt and Whitney augmentor nozzle. *(Courtesy of Fuel Systems TEXTRON.)*

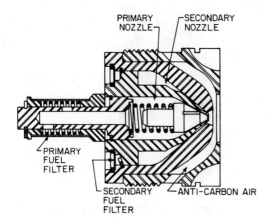

Figure 4.27 Dual-orifice assembly illustrating use of anticarbon (shroud) air. *(Courtesy of Fuel Systems TEXTRON.)*

The effectiveness of shroud air in reducing the mean drop size of a spray was examined by Clare et al. [35], using various selected swirled and unswirled air shrouds. They found that swirling shroud air improved atomization quality (in one case the SMD was reduced from 171 to 126 μm) and assisted in holding the spray tulip open at low fuel pressures; it had no effect on the cone angle of the fully developed spray. Unswirled shroud air was less effective in reducing the mean drop size (from 171 to 149 μm), and it reduced the cone angle of the spray by up to 30° at the lowest fuel flow rate. These results show that shroud air can also be used to vary and control the cone angle of the spray. Unswirled shroud air tends to narrow the spray angle, whereas swirling shroud air, by inducing gas movement in a radially outward direction, tends to widen the spray cone angle.

To summarize, in addition to its primary function of cooling the atomizer face and keeping it free of deposits, shroud air can be used to improve atomization quality at low fuel flows, thereby extending weak-extinction limits and improving lightup capability. However, excessive amounts of unswirled shroud air result in a spray cone angle that is too narrow and could give rise to combustion instability.

Fan Spray Nozzles

These atomizers are used in the coating industry, in some small annular gas turbine combustors, and in other special applications where a narrow elliptical pattern is more appropriate than the normal circular pattern. In the fan or flat spray atomizer shown in Fig. 4.28 the spray is formed by arranging for a round liquid jet to impinge on a curved surface. This arrangement produces a wide, flat, relatively coarse spray pattern containing a fairly uniform distribution of drops. The nominal spray pattern is 120° or more, depending on size [22].

In the nozzle shown in Fig. 4.29 the discharge orifice is formed by the intersection of a V groove with a hemispheric cavity communicating with a cylin-

Figure 4.28 Flat spray "flood" nozzle. *(Courtesy of Delavan, Inc.)*

Figure 4.29 Flat spray nozzle. *(Courtesy of Delavan, Inc.)*

drical liquid inlet [22, 23]. It produces a liquid sheet parallel to the major diameter of the orifice. It is the most widely used type of fan spray nozzle, producing a narrow elliptical spray pattern with tapered edges that provide uniform distribution when overlapped. Excellent atomization and patterns can be obtained with viscous and non-Newtonian materials [23, 36].

Fan sprays can also be obtained by cutting slots in plane or cylindrical surfaces and arranging for the liquid to flow into the slots from two opposite directions. In the single-hole fan spray injector there is a single shaped orifice that forces the liquid into two opposing streams within itself, so that a flat spray issues from the orifice and spreads out in the shape of a sector of a circle of about 75° angle. In engine applications an air shroud is usually fitted around the nozzle to provide both air assistance in atomization and air scavenging of the nozzle tip, using the pressure differential across the combustor liner wall.

The behavior of fan or flat sprays has been investigated by Fraser and Eisenklam [37], Dombrowski and Fraser [38], Dombrowski et al. [39], Carr [40, 41], and Lewis [21]. The results obtained by Lewis suggest that the mean drop diameter for fan spray injectors increases with increase in ambient pressure. He attributes this to sheet disintegration occurring closer to the injector face, so that drops are produced from a thicker sheet initially. Lewis also comments that ambient pressure has little effect on the spray angle obtained with fan spray injectors.

In the absence of surface tension, the edges of the sheet would travel in straight lines from the orifice so that a sector of a circle would be formed [39]. However, as a result of surface tension, the edges contract and a curved boundary is formed as the sheet develops beyond the orifice. A typical flash photograph of a fan spray is illustrated in Fig. 2.21. Breakdown of the edges of the sheet is restrained by viscous forces, which explains why these edges disintegrate into relatively large drops, as observed by Snyder et al. [36].

The work of Dombrowski et al. [39] showed that the trajectory of the liquid sheet is a function of injection pressure, sheet thickness, and surface tension and is independent of liquid density. The thickness of the sheet at any point is inversely proportional to its distance from the orifice.

Generally, drop sizes from fan spray atomizers are larger than those from swirl spray atomizers of the same flow rate. However, Carr [40, 41] has described a fan spray atomizer that produces very fine sprays (SMD < 25 μm) even with oils having viscosities as high as 8.5×10^{-6} m²/s (8.5 cS). The main design features of this nozzle are shown in Fig. 4.30. The drop size data obtained using a 6.5×10^{-6} m²/s oil are shown in Fig. 4.31. The beneficial effect of shroud air on atomization quality at low values of ΔP_L is evident in this figure.

Fan spray atomizers lend themselves ideally to small annular combustors, as they provide a good lateral spread of fuel and allow the number of injection points to be minimized. The AVCO Lycoming Company has manufactured flat-spray atomizers in several versions. One type is illustrated in a much simplified form in Fig. 4.32. More detailed information on this and other types of Lycoming flat-spray atomizers is provided in reference [42].

ROTARY ATOMIZERS

In the rotary atomizer, liquid is fed onto a rotating surface, where it spreads out fairly uniformly under the action of centrifugal force. The rotating surface may take the form of a flat disk, vaned disk, cup, or slotted wheel. A rotating cup

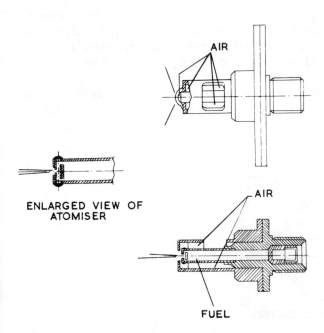

Figure 4.30 Lucas fan spray atomizer *(Carr [40])*.

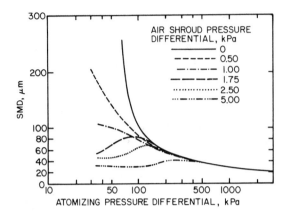

Figure 4.31 Atomizing characteristics of Lucas fan spray atomizer *(Carr [40])*.

design is shown schematically in Fig. 4.33. Diameters vary from 25 to 450 mm, the small disks rotate up to 1000 rps, the larger disks rotate up to 200 rps, and atomizing capacities are up to 1.4 kg/s [43]. Where a coaxial air jet is used to assist atomization, lower speeds of the order of 50 rps may be used. The system has extreme versatility and has been shown to atomize successfully liquids varying widely in viscosity. An important asset is that the thickness and uniformity of the liquid sheet can readily be controlled by regulating the liquid flow rate and the rotational speed.

The process of centrifugal atomization has been studied by several workers, including Hinze and Milborn [44], Dombrowski and Fraser [38], Kayano and Kamiya [45], and Tanasawa et al. [46] and has been described in detail in reviews of atomization methods by Dombrowski and Munday [32] and Matsumoto et al. [47].

Several mechanisms of atomization are observed with a rotating flat disk, depending on the flow rate of the liquid and the rotational speed of the disk. At low flow rates the liquid spreads out across the surface and is centrifuged off as

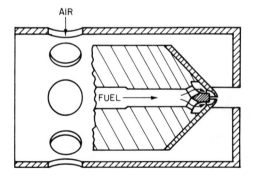

Figure 4.32 AVCO Lycoming fan spray injector *(Watkins [42])*.

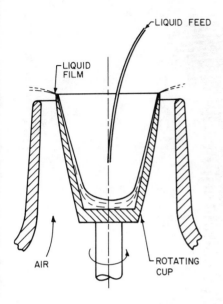

Figure 4.33 Rotating cup atomizer.

discrete drops of uniform size, each drop drawing behind it a fine ligament. The drops finally separate from the ligaments, which are themselves converted into a series of fine drops of fairly uniform size. This process is essentially a discontinuous one and occurs from place to place at the periphery of the rotating cup or disk. If the flow rate is increased the atomization process remains basically the same, except that ligaments are formed along the entire periphery and are larger in diameter. This process is illustrated in Fig. 4.34. Ligament formation can be made to produce much finer drops if an electrostatic field is applied between the rotating element and the earth.

With further increase in flow rate the condition is eventually reached where the ligaments can no longer accommodate the flow of liquid, and a thin continuous sheet is formed that extends from the lip until an equilibrium condition is achieved, at which the contraction force at the free edge due to surface tension is just equal to the kinetic energy of the advancing sheet. A thick rim is produced, which again disintegrates into ligaments and drops. However, because the rim has no controlling solid surface, the ligaments are formed in an irregular manner that results in an appreciable variation in drop size. Serrating the edge of the cup or disk delays the transition from ligament formation to sheet formation. This is well illustrated in the flash photograph of Fig. 4.35, which shows a rotating cup with part of its edge serrated and the remainder unserrated. Finally, when the peripheral speed of the rotating disk becomes very high, more than 50 m/s, say, the liquid appearing at the edge of the disk is immediately atomized by the surrounding air.

Generally, it is found that atomization quality is improved by (1) increase in rotational speed, (2) decrease in liquid flow rate, (3) decrease in liquid viscosity,

Figure 4.34 Photograph illustrating atomization by ligament formation. *(Courtesy of Ransburg Gema, Inc.)*

and (4) serration of the outer edge. Moreover, to obtain a uniform film thickness, and hence a more uniform drop size, the following conditions should be met [48]: (1) the centrifugal force should be large in comparison to the gravitational force, (2) cup rotation should be vibrationless, (3) liquid flow rate should be constant, and (4) cup surfaces should be smooth.

The main drawback of the flat disk atomizer is that slippage occurs between the liquid and the disk, especially at high rotational speeds. In consequence, the liquid is ejected from the edge of the disk at a velocity much lower than the disk peripheral speed. This problem is overcome in commercial atomizers by the use of radial vanes. The liquid supplied to the vaned disk or wheel flows over its surface until it is contained by a rotating vane. It then flows radially outward under the influence of centrifugal force, covering the vane surface in a thin film. As the vanes, whether radial or curved, prevent transverse flow of liquid over the surface, no slippage occurs once the liquid has contacted the vane, and discharge velocities approximate the wheel peripheral velocity.

Some of the wheeled designs produced by the NIRO Atomizer company are shown in Fig. 4.36. These atomizers have found a wide range of industrial applications. For any given liquid, mean drop size is controlled by a change in

Figure 4.35 Photograph illustrating the effect of edge serrations in delaying the transition from ligament formation to sheet formation. *(Courtesy of Ransburg Gema, Inc.)*

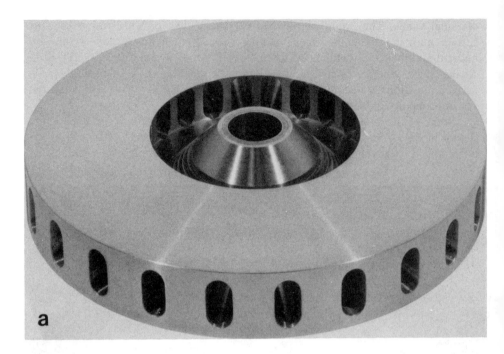

Figure 4.36 Designs of vaned atomizer wheels. (a) Straight vanes; (b) curved vanes (wheel top cover removed. *(Courtesy of NIRO Atomizer.)*

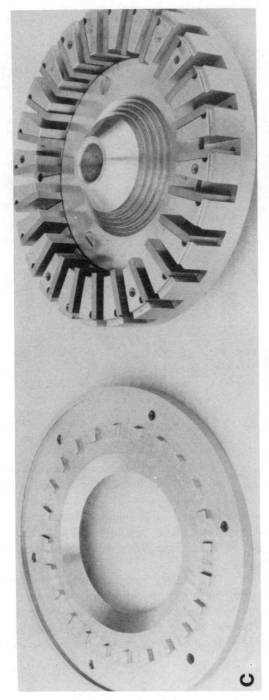

Figure 4.36 Designs of vaned atomizer wheels (*Continued*). (c) Wear-resistant vanes (inserts). (*Courtesy of NIRO Atomizer.*)

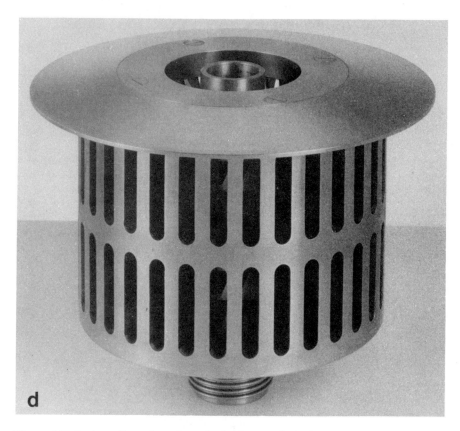

Figure 4.36 Designs of vaned atomizer wheels (*Continued*). (d) high-capacity vanes. (*Courtesy of NIRO Atomizer.*)

rotational speed. Masters [43] has provided detailed descriptions of these and other types of rotary atomizers used in spray drying. The designs shown in Fig. 4.36 can be fabricated in metals resistant to corrosion, such as hastelloy, tantalum, and titanium, to allow corrosive liquids to be handled. Problems of excessive wear on wheel surfaces when abrasive feeds are atomized are overcome by incorporating resistant inserts within the inner wheel body, as shown in Fig. 4.36c for the straight-vane design.

According to Masters [49], rotary atomizers will find increased use in industrial spray-drying operations. This is due to developments in new designs of rotary atomizers, with more reliable drive systems and more reliable wheel operation. Significant advances have been made in handling high feed rates (up to 40 kg/s) at high wheel peripheral speeds that yield very fine atomization (SMD < 20 μm).

Slinger System

Although rotary atomizers have been widely used in the chemical processing industry for almost a century and have had a broad range of applications including, recently, the aerial distribution of pesticides, they have not generally aroused much interest as fuel injectors for heat engines. A notable exception is the drum atomizer developed by the Turbomeca company in France for gas turbine engines. It is used in conjunction with a radial-annular combustion chamber, as illustrated in Fig. 4.37. Fuel is supplied at low pressure along the hollow main shaft and is discharged radially outward through holes drilled in the shaft. These injection holes vary in number from 9 to 18 and in diameter from 2.0 to 3.2 mm. The holes may be drilled in the same plane as a single row, but some installations feature a double row of holes. The holes never run full; they have a capacity that is many times greater than the required flow rate. They are made large to obviate blockage. However, it is important that the holes be accurately machined and finished, since experience has shown that uniformity of flow between one injection hole and another is dependent on their dimensional accuracy and surface finish. Clearly, if one injection hole supplies more fuel than the others, it will produce a rotating "hot spot" in the exhaust gases, with disastrous consequences for the particular turbine blade on which the hot spot happens to impinge.

Flow uniformity is also critically dependent on the flow path provided for the fuel inside the shaft, especially in the region near the holes. Where there are two rows of holes it is important to achieve the correct flow division between the two rows. Again, the internal geometry of the shaft in the vicinity of the holes is of prime importance.

The main advantages of the slinger system are its cheapness and simplicity. Only a low-pressure fuel pump is needed, and the quality of atomization depends only on engine speed. The equivalent injection pressures are very high, of the order of 34 MPa at full speed, and satisfactory atomization is claimed at speeds as low as 10% of the rated maximum. The influence of fuel viscosity is small, so the system has a potential multifuel capability.

The system seems ideally suited for small engines of low compression ratio, and this has been its main application to date. As the success of the system depends on high rotational speeds, usually greater than 350 rps, it is clearly less suitable for large engines where shaft speeds are much lower. In the United States the slinger system has been used successfully on several engines produced by

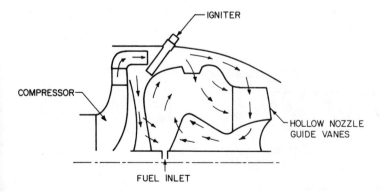

Figure 4.37 Turbomeca slinger system.

Williams Research Corporation. The amount of published information on the system is not large, but useful descriptions have been provided by Wehner [50], Maskey and Marsh [51], Nichol [52], and Burgher [53].

AIR–ASSIST ATOMIZERS

This chapter contains several references to the beneficial effect of flowing air in assisting the disintegration of a liquid jet or sheet. Examples include the use of shroud air in fan spray and pressure-swirl nozzles. As discussed earlier, a basic drawback of the simplex nozzle is that if the swirl ports are sized to pass the maximum flow rate at the maximum injection pressure, the pressure differential will be too low to give good atomization at the lowest flow rate. This problem can be overcome by using dual-orifice or duplex nozzles; however, an alternative approach is to use air or steam to augment the atomization process at low injection pressures. A wide variety of designs of this type have been produced for use in industrial gas turbines and oil-fired furnaces. Useful descriptions of these have been provided by Romp [54], Gretzinger and Marshall [55], Mullinger and Chigier [56], Bryce et al. [57], Sargeant [58], and Hurley and Doyle [59]. In all designs a high-velocity gas stream impinges on a relatively low-velocity liquid stream, either internally as shown in Fig. 4.38 or externally as shown in Fig. 4.39.

In the internal-mixing type, the spray cone angle is a minimum for maximum airflow, and the spray widens as the airflow is reduced. This type of atomizer is very suitable for highly viscous liquids, and good atomization can be obtained down to very low liquid flow rates. External-mixing types can be designed to give a constant spray angle at all liquid flow rates, and they have the advantage that there is no danger of liquid finding its way into the air line. However, their utilization of air is less efficient, and consequently their power requirements are higher.

The main drawback of air-assist atomizers from a gas turbine standpoint is the need for an external supply of high-pressure air. This virtually rules them out for aircraft applications. They are much more attractive for large industrial engines, especially as the high-pressure air is needed only during engine lightup and acceleration.

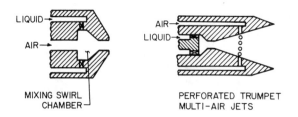

Figure 4.38 Internal-mixing air-assist atomizer.

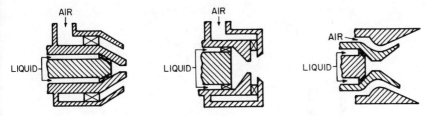

Figure 4.39 External-mixing air-assist atomizer.

For the air-assist nozzles described above, the pressure atomizing component is capable of providing good atomization over most of the operating range. Air or steam assistance is called on to supplement the atomization process only at conditions of low liquid flow rates, where the nozzle pressure differential is too low for satisfactory pressure atomization. However, in some other nozzle designs, the level of atomization quality achieved with the liquid injector alone is always so low that air assistance is required over the entire operating range. One example of this type is a Parker Hannifin nozzle designed especially for handling coal-water slurries and other liquids that would be difficult or impossible to atomize by more conventional means. The key features of this nozzle are shown schematically in Fig. 4.40. It uses both inner and outer airstreams to achieve a shearing action on an annular liquid sheet at the nozzle tip. Both airflows are swirled clockwise, with the liquid sheet swirled anticlockwise.

The Lezzon nozzle shown in Fig. 4.41 also uses inner and outer airflows to shear a liquid sheet. However, in this concept the initial contact of these flows occurs inside the nozzle, thereby allowing the air and liquid to interact before issuing from the nozzle. In one arrangement, external adjustments on the nozzle body provide continuous variations of the liquid and air exit gap heights [60].

Other examples of interest are the NGTE design [61] and the Mullinger and Chigier nozzle [56], illustrated in Figs. 4.42 and 4.43, respectively. The latter nozzle was designed to simulate, for research purposes, one injection port of the Babcock Y-jet atomizer, shown schematically in Fig. 4.44.

The Y-jet is very widely used in large oil-fired boiler plants. It generally consists of a number of jets, from a minimum of 2 to a maximum of 20, arranged

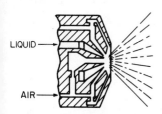

Figure 4.40 Parker Hannifin slurry nozzle.

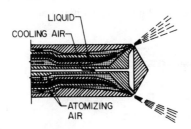

Figure 4.41 Lezzon nozzle.

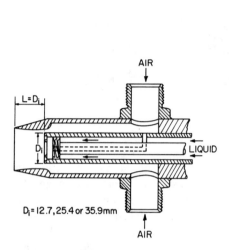

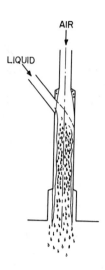

Figure 4.42 National Gas Turbine Establishment atomizer *(Wigg [61])*.

Figure 4.43 Mullinger and Chigier atomizer [56].

in an annular fashion to provide a hollow conical spray. In each individual Y, oil is injected at an angle into the exit port, where it mixes with the atomizing fluid (air or steam) admitted through the air port. The exit ports are uniformly spaced around the atomizer body at an angle to the nozzle axis, so that the individual jets of two-phase mixture issuing from the exit ports rapidly merge to form a hollow conical spray.

According to Bryce et al. [57] the most common method of operating Y-jet atomizers is to maintain the pressure of the atomizing fluid constant over the full range of oil flows, with oil flow controlled by variation of the oil pressure. This control mode results in the atomizing fluid/oil mass flow ratio increasing as the fuel flow decreases.

An alternative method of operation is to maintain the ratio of fluid to oil pressure constant over the fuel flow range. This results in higher atomizing fluid flows being available at the upper end of the oil flow range and lower atomizing fluid flows at the lower oil flows, which gives a saving in atomizing fluid at low loads without significantly reducing the quality of atomization.

A typical throughput of heavy residual fuel oil for a large utility boiler burner is around 2 kg/s. Steam is generally employed as the atomizing fluid. An obvious

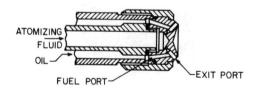

Figure 4.44 Typical Y-jet tip arrangement *(Sargeant [58])*.

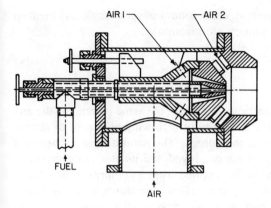

Figure 4.45 Combined pressure and twin-fluid atomizer *(Stambuleanu [62])*.

advantage of using steam is that any heat transferred from the steam to the fuel in the mixing ports will enhance atomization by reducing the fuel's viscosity and surface tension. However, comparative tests carried out by Bryce et al. [57] showed that compressed air produced a much finer spray than steam.

Another type of twin-fluid atomizer is illustrated in Fig. 4.45. The fuel is first pressure-atomized and then subjected to two stages of air (or steam) atomization. In the second stage, swirl vanes or slots are used to impart a helical motion to the airflow.

A widely used type of air-assist nozzle manufactured by the Delavan company is shown in Fig. 4.46. Air is introduced tangentially into the nozzle chamber to create primary atomization. As liquid leaves the orifice it impinges against the

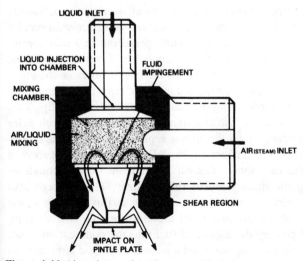

Figure 4.46 Air-assist nozzle. *(Courtesy of Delavan, Inc.)*

deflector ring, which serves a dual purpose: close control of spray angle and breakup of the spray into smaller drops (secondary atomization).

AIRBLAST ATOMIZERS

In principle, the airblast atomizer functions in exactly the same manner as the air-assist atomizer; both employ the kinetic energy of a flowing airstream to shatter the fuel jet or sheet into ligaments and then drops. The main difference between the two systems lies in the quantity of air employed and its atomizing velocity. With the air-assist nozzle, where the air is supplied from a compressor or a high-pressure cylinder, it is important to keep the airflow rate down to a minimum. However, as there is no special restriction on air pressure, the atomizing air velocity can be made very high. Thus air-assist atomizers are characterized by their use of a relatively small quantity of very high velocity air. However, because the air velocity through an airblast atomizer is limited to a maximum value (usually around 120 m/s), corresponding to the pressure differential across the combustor liner, a larger amount of air is required to achieve good atomization. However, this air is not wasted, because after atomizing the fuel it conveys the drops into the combustion zone, where it meets and mixes with the additional air needed for complete combustion.

Airblast atomizers have many advantages over pressure atomizers, especially in their application to combustion systems operating at high pressures. They require lower fuel pump pressures and produce a finer spray. Moreover, because the airblast atomization process ensures thorough mixing of air and fuel, the ensuing combustion process is characterized by very low soot formation and a blue flame of low luminosity, resulting in relatively low flame radiation and a minimum of exhaust smoke.

The merits of the airblast atomizer have led to its installation in a wide variety of aircraft, marine, and industrial gas turbines. Most of the systems now in service are of the *prefilming* type, in which the liquid is first spread out in a thin continuous sheet and then subjected to the atomizing action of high-velocity air. One example of a prefilming airblast atomizer designed for gas turbines is shown in Fig. 4.47. In this design the fuel flows through a number of equispaced tangential ports onto a prefilming surface before being discharged at the atomizing lip. Two separate airflows are provided to allow the atomizing air to impact on both sides of the liquid sheet. One airstream flows through a central circular passage containing a swirler, which causes the airflow to be deflected radially outward to strike the inner surface of the fuel sheet. The other airstream flows through an annular passage surrounding the main body of the atomizer. This passage also contains an air swirler to impart a swirling motion to the airflow impinging on the outer surface of the fuel sheet. At exit from the atomizer the two swirling airflows merge together and convey the atomized fuel into the combustion zone.

Some high-performance aircraft engines used a *piloted* or *hybrid* atomizer, as illustrated in Fig. 4.48. This is essentially a prefilming airblast atomizer with the

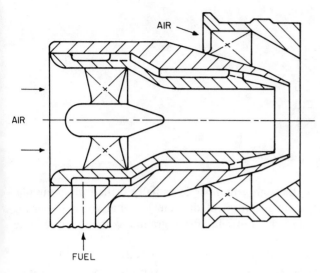

Figure 4.47 Prefilming airblast atomizer. *(Courtesy of Parker Hannifin Corp.)*

addition of a simplex nozzle. It is designed to overcome a basic weakness of the pure airblast atomizer, namely poor atomization at the low air velocities associated with low cranking speeds. The simplex nozzle supplies pressure-atomized fuel to achieve rapid lightup during engine cranking and in the event of a flameout at high altitudes.

An alternative form of airblast atomizer is one in which the fuel is injected into the high-velocity airstream in the form of one or more discrete jets. Figure 4.49 illustrates a plain-jet airblast atomizer designed for gas turbine applications.

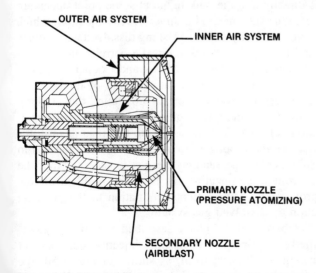

Figure 4.48 Piloted airblast atomizer. *(Courtesy of Fuel Systems TEXTRON.)*

LIQUID

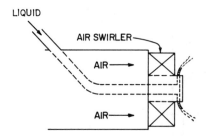

Figure 4.49 Plain-jet airblast atomizer *(Jasuja [63])*.

In this nozzle the fuel flows through a number of radially drilled plain circular holes, from which it emerges in the form of discrete jets that enter a swirling airstream. These jets then undergo in-flight disintegration without any further preparation such as prefilming.

Due to its inherent simplicity, the airblast concept lends itself to a wide variety of design configurations. However, in all cases the basic objective is the same, namely to deploy the available air in the most effective manner to achieve the best possible level of atomization.

EFFERVESCENT ATOMIZERS

All the twin-fluid atomizers described above, in which air is used either to augment atomization or as the primary driving force for atomization, have one important feature in common: the bulk liquid to be atomized is first transformed into a jet or sheet before being exposed to high-velocity air. An alternative approach is to introduce the air or gas directly into the bulk liquid at some point upstream of the nozzle discharge orifice. One such method is *supercritical* injection, which relies on the *flashing* of dissolved gas in the liquid. Flashing dissolved gas systems have been studied by Sher and Elata [64], who developed a correlation of atomization properties in terms of bubble growth rates. Marek and Cooper [65] have proposed the use of dissolved gases to improve the atomization of liquid fuels. In their study on supercritical injection, Solomon et al. [66] found that flashing even small quantities of dissolved gas (mole fractions less than 15%) had a significant effect on the atomization of Jet-A fuel. However, to be offset against this good atomization performance are the problems involved in coercing the gas to dissolve in the liquid and subsequently persuading the gas to emerge from the liquid when it is needed to promote atomization. In fact, the problem of low bubble growth rate could prove to be a fundamental limitation to the practical application of flashing injection by dissolved gas systems.

In a recent publication, Lefebvre et al. [67] have described a method of atomization that employs the same basic principles as flashing injection but does not share its practical limitations. In its simplest form, as shown in Fig. 4.50, the new concept comprises a plain-orifice atomizer with means for injecting air (or

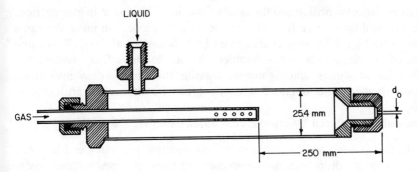

LIQUID

GAS →

25.4 mm

250 mm

d_o

Figure 4.50 Schematic drawing of effervescent flow atomizer [67].

gas) into the bulk liquid at some point upstream of the injector orifice. This gas is not intended to impart kinetic energy to the liquid stream, and it is injected at low velocity. In consequence, the pressure differential between the gas and the liquid is very small, only what is needed to induce the gas to enter the flowing liquid. Measurements of ΔP_{G-L} for various values of liquid injection pressure and gas flow rate are shown in Fig. 4.51. The injected gas forms bubbles, which produce a two-phase flow at the injector orifice. Although the basic atomization mechanism has not yet been studied in detail, it is thought that the liquid flowing through the injector orifice is "squeezed" by the gas bubbles into thin shreds and ligaments. This is an important aid to atomization because it is well established that the drop sizes produced in a spray, whether by pressure or airblast atomization, are roughly proportional to the square root of the initial thickness or diameter of the ligaments from which they are formed [68]. Thus, the larger the amount of gas used in atomization, the larger will be the number of bubbles flow-

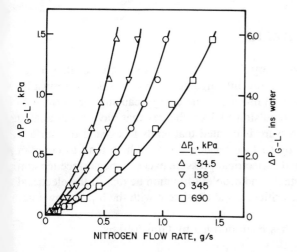

$$\frac{\Delta P_L, kPa}{}$$

△	34.5
▽	138
○	345
□	690

ΔP_{G-L}, kPa

ΔP_{G-L}, ins water

NITROGEN FLOW RATE, g/s

Figure 4.51 Pressure differential between atomizing gas and liquid [67].

ing through the injector orifice and the smaller the drops produced in atomization. When the gas bubbles emerge from the nozzle they "explode," in much the same manner as in flashing injection, as described by Solomon et al. [66]. This rapid expansion of multitudinous small bubbles in the nozzle efflux shatters the surrounding liquid shreds and ligaments leaving the final orifice into small drops. Some of the results obtained with this type of atomizer are presented in Chapter 6.

The advantages offered by effervescent atomization are the following:

1. Atomization is very good even at very low injection pressures and low gas flow rates. Mean drop sizes are comparable to those obtained with air-assist atomizers for the same gas/liquid ratio.
2. The system has large holes and passages so that problems of "plugging" are greatly reduced. This could be an important advantage for combustion devices that burn residual fuels, slurry fuels, or any type of fuel where atomization is impeded by the necessity of using large hole and passage sizes to avoid plugging of the nozzle.
3. For combustion applications, the aeration of the spray created by the presence of the air bubbles could prove very beneficial in alleviating soot formation and exhaust smoke.
4. The basic simplicity of the device lends itself to good reliability, easy maintenance, and low cost.

The only apparent drawback of this method is the need for a separate supply of atomizing air or gas, which must be provided at essentially the same pressure as that of the liquid.

More work is needed to study the effects of nozzle size, geometry, liquid properties, and nozzle operating conditions on the atomizing performance of this type of injector.

ELECTROSTATIC ATOMIZERS

A basic requirement for atomizing any liquid is to make some area of its surface unstable. The surface will then rupture into ligaments, which subsequently disintegrate into drops. In electrical atomization, the energy causing the surface to disrupt comes from the mutual repulsion of like charges that have accumulated on the surface. An electrical pressure is created that tends to expand the surface area. This pressure is opposed by surface tension forces, which tend to contract or minimize surface area. When the electrical pressure exceeds the surface tension forces, the surface becomes unstable and droplet formation begins. If the electrical pressure is maintained above the critical value consistent with the liquid flow rate, then atomization is continuous.

The electrical pressure, P_e, has been derived by Graf [69] as

$$P_e = \frac{FV^2}{2\pi D^2} \tag{4.2}$$

where V is the applied voltage, D is the drop diameter, and F is a charging factor that represents the fraction of the applied potential attained on the drop surface. It is found that F decreases with increasing liquid conductivity and increasing electrode spacing.

Many configurations of atomizing electrode have been tested, including hypodermic needles, sintered bronze filters, and cones. Both direct- and alternating-current electrical systems have been used to generate the high voltage needed for fine atomization. A typical dc circuit, of the type used successfully by Luther [70], is illustrated in Fig. 4.52.

The results of many experimental studies on electrostatic atomization [71–78] show that, in general, the size of drops produced depends on the applied voltage, surface tension, electrode size and configuration, liquid flow rate, and electrical properties of the liquid, such as dielectric constant and electrical conductivity. The smallest drops are produced by the highest voltage that can be used without encountering excessive corona losses.

The very low liquid flow rates that are generally associated with electrostatic atomization have tended to restrict its practical application to electrostatic painting and nonimpact printing. However, an invention by Kelly [77] called the spray triode, shown schematically in Fig. 4.53, appears to have great promise for the development of electrostatic atomizers capable of handling the high fuel flow rates required by most practical combustion devices. This figure shows that a voltage differential is impressed between a central submerged emitter electrode and a blunt orifice electrode. The two electrodes form a submerged electron gun and serve to charge the fluid that flows around the emitter electrode and exits via the orifice. Once free of the confines of the interelectrode region, the charged fluid undergoes disruption and spray formation. Charge is returned to the circuit by a collector electrode, which in a combustion system would be the flame front and combustion chamber wall. The resistor limits electrode current in the event of an internal breakdown in the fluid. It may be noted that the active electrodes are in an ideal location; they are submerged in an insulating fluid. This arrangement precludes the possibility of external corona breakdown and permits operation in ionizing combustion gases.

A key feature of Kelly's atomizer is the special emitter electrode, which comprises a multiplicity of very small tungsten fibers embedded in a refractory ma-

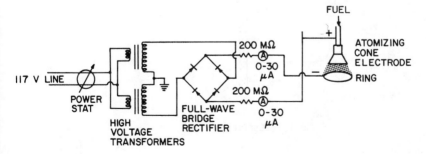

Figure 4.52 Cone and ring electrode system for electrostatic atomization *(Luther [70])*.

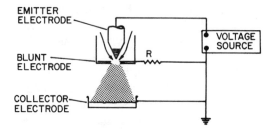

Figure 4.53 Schematic diagram of spray triode atomizer *(Kelly [77]).*

terial, as illustrated in Fig. 4.54. It is claimed that submicron diameter, continuous monocrystalline fibers can be developed at densities up to $10^7/cm^2$ [77]. As each fiber is capable of handling flow rates up to 1 ml/s, this implies that even a 1-mm-diameter electrode could effectively handle fuel flow rates up to 30 kg/h. This is clearly very satisfactory from the standpoint of its potential application to practical combustion systems.

ULTRASONIC ATOMIZERS

When a liquid is introduced onto a rapidly vibrating solid surface, a checkerboard-like wave pattern appears in the film that forms as the liquid spreads over the surface. As the amplitude of the surface vibration is increased, the wave crest height in the film also increases. It was demonstrated by Lang [79] that atomi-

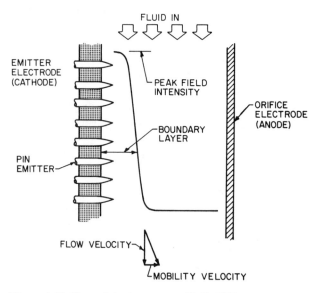

Figure 4.54 Charge injection process *(Kelly [77]).*

zation occurs when the amplitude of the vibrating surface increases to the point where the wave crests in the film become unstable and collapse. This causes a mist of small drops to be ejected from the surface.

An early practical application of ultrasonic nozzles was to small boilers for domestic heating. This took place in the early 1960s, and for the next 20 years or so the emphasis remained in the combustion area. However, in recent years ultrasonic nozzle technology has found a variety of industrial and laboratory applications and is now used in fields as diverse as semiconductor processing, humidification, pharmaceutical coatings, and vaporization of volatile anesthetic agents for general anesthesia [80].

One class of widely applicable nozzles has been described by Berger [81]. As shown in Fig. 4.55, the nozzle is an acoustically resonant device consisting of a pair of piezoelectric transducer elements sandwiched between a pair of titanium horn sections. The surface on which atomization takes place is at the end of the small-diameter stem. Liquid is delivered to this surface by means of an axial tube running the length of the device.

The common contact plane between the two piezoelectric transducer disks forms one of the two electrical input terminals for the high-frequency electrical signal used to energize the device. The other terminal is the metal body. With this transducer orientation both disks will, depending on the polarity of the input signal, either expand or contract simultaneously by equal amounts. This cyclic expansion and contraction results in the propagation of traveling pressure waves longitudinally outward in both directions along the nozzle, with a frequency equal to that of the input signal. However, the nozzle is designed such that its total length is equivalent to one wavelength of the pressure wave. This results in reflections of the wave at each end and the establishment of a standing wave pattern, which is the necessary condition for resonant operation. Figure 4.56 depicts the amplitude envelope of such a standing wave. The interface between transducer disks is fixed as a nodal plane, since the equal and opposite forces developed there preclude any net motion. At each free end the boundary conditions dictate the existence of antinodes, that is, planes where the amplitude of the standing wave is a maximum. The much greater amplitude at the atomizing end results from the placement of a diameter step transition at the second nodal plane. The use of stepped horns is a well-known method for increasing vibrational amplitude.

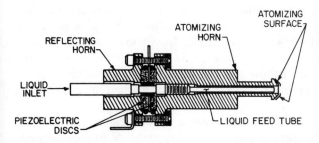

Figure 4.55 Ultrasonic nozzle assembly *(Berger [81]).*

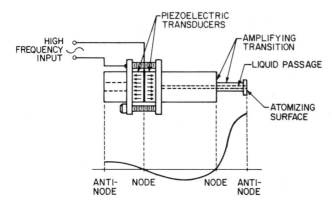

Figure 4.56 Standing wave pattern *(Berger [81])*.

The amplitude change achieved is equal to the ratio of the larger to the smaller cross-sectional area. This amplification is necessary to realize the vibrational levels necessary to effect atomization. Minimum vibrational amplitudes are typically on the order of several micrometers. To achieve these levels within the input power-handling capabilities of these nozzles, minimum amplitude gain ratios between 6 to 8 are required.

A most useful attribute of the ultrasonic atomizer is its low spray velocity. This makes it an easy matter to entrain the spray in a moving stream and convey the drops in a controlled manner as a uniform mist. This is especially important in coating applications and in such processes as humidification, product moisturizing, and spray drying.

Another asset of the ultrasonic nozzle is its ability to provide very fine atomization at the extremely low flow rates required for certain pharmaceutical and lubrication processes. Nebulizers, as they are termed in medicine, produce very small drops in the size range from 1 to 5 μm, which, when inhaled, can penetrate to the extreme air passages of the lungs for the purpose of medication or humidification [82]. However, this type of atomizer is much less successful in handling the high flow rates required for most engine and furnace applications. Typical maximum flow rates are around 7 liters/h, at a frequency of 55 kHz. This is due primarily to the low amplitude of oscillations that ultrasonic transducers generate. One method of increasing the flow rate is to combine the principle of ultrasonic atomization with that of whistle-type atomization. In this manner the amplitude of the signals produced by the ultrasonic vibrator is further amplified by the resonant effects created within the hollow space of the horn [83]. However, a drawback of this approach is that the operating frequency of the transducer is limited to one value.

Some of the ultrasonic atomizers manufactured by the Delavan company are illustrated in Fig. 4.57. The three nozzles shown are designed to produce three different spray angles: narrow, standard, and wide. They operate at a frequency

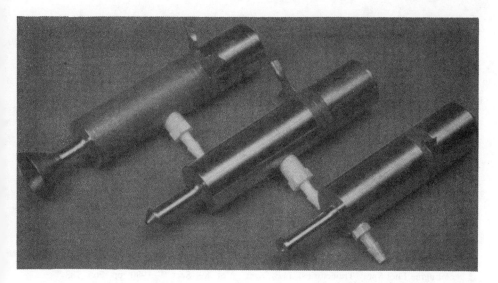

Figure 4.57 Three types of Delavan ultrasonic nozzles: narrow, standard, and wide spray angles.

that is nominally 50 kHz, and drop sizes are generated in the size range from 30 to 60 μm. In addition to combustion applications, these atomizers are used to spray acid or alkali solutions to etch extremely small parts such as computer chips and also to spray dry chemicals or pharmaceuticals in laboratory driers.

WHISTLE ATOMIZERS

As in an ultrasonic atomizer using a transducer, liquid can also be disintegrated into drops by directing high-pressure gas into the center of a liquid jet, as shown in Fig. 4.58. Due to the strong sound waves created inside the nozzle by the focusing airflow, this atomizer is frequently called a whistle or stem-cavity type. It generally operates at a sound frequency of about 10 kHz and produces droplets around 50 μm in diameter at flow rates up to 1.25 liters/s.

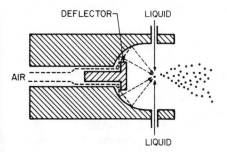

Figure 4.58 Schematic diagram of whistle-type atomizer *(Lee et al. [83])*.

A drawback of whistle atomizers is that the drop size cannot easily be controlled unless the nozzle dimension is changed. Wilcox and Tate [84] studied this type of nozzle systematically and concluded that the sound field is not an important variable in the atomizing process. This led Topp and Eisenklam [82] to suspect that all whistle atomizers operate simply as air-assist types; i.e., the liquid is disintegrated primarily by the aerodynamic interactions between the gas and the liquid. No reliable or proven theoretical analysis on the performance of whistle atomizers seems to be available [83].

REFERENCES

1. Elliott, M. A., Combustion of Diesel Fuel, *SAE Trans.*, Vol. 3, No. 3, 1949, pp. 490–497.
2. Rosselli, A., and Badgley, P., Simulation of the Cummins Diesel Injection System, *SAE Trans.*, Vol. 80, Pt. 3, 1971, pp. 1870–1880.
3. Sinnamon, J. F., Lancaster, D. R., and Steiner, J. C., An Experimental and Analytical Study of Engine Fuel Spray Trajectories, *SAE Trans.*, Vol. 89, Sect. 1, 1980, pp. 765–783.
4. Elkotb, M. M., Fuel Atomization for Spray Modeling, *Prog. Energy Combust. Sci.*, Vol. 8, 1982, pp. 61–91.
5. Borman, G. L., and Johnson, J. H., Unsteady Vaporization Histories and Trajectories of Fuel Drops Injected into Swirling Air, SAE Paper 598C, 1962.
6. Adler, D., and Lyn, W. T., The Evaporation and Mixing of a Fuel Spray in a Diesel Air Swirl, *Proc. Inst. Mech. Eng.*, Vol. 184, Pt. 3J, Paper 16, 1969.
7. Arai, M., Tabata, M., Hiroyasu, H., and Shimizu, M., Disintegrating Process and Spray Characterization of Fuel Jet Injected by a Diesel Nozzle, *SAE Trans.*, Vol. 93, Sect. 2, 1984, pp. 2358–2371.
8. Hiroyasu, H., Diesel Engine Combustion and Its Modelling, International JSME Symposium on Diagnostics and Modeling on Combustion in Reciprocating Engines, Tokyo, September 4–6, 1985.
9. Schweitzer, P. H., Penetration of Oil Sprays, Pennsylvania State Engineering Experimental Station Bulletin No. 46, 1937.
10. Lyn, W. T., and Valdamanis, E., The Application of High-Speed Schlieren Photography to Diesel Combustion Research, *J. Photogr. Sci.*, Vol. 10, 1962, pp. 74–82.
11. Huber, E. W., Stock, D., and Pischinger, F., Investigation of Mixture Formation and Combustion in Diesel Engine with the Aid of Schlieren Method, 9th International Congress on Combustion Engines (CIMAC), Stockholm, Sweden, 1971.
12. Rife, J., and Heywood, J. B., Photograhpic and Performance Studies of Diesel Combustion with a Rapid Compression Machine, *SAE Trans.*, Vol. 83, 1974, pp. 2942–2961.
13. Hiroyasu, H., and Arai, M., Fuel Spray Penetration and Spray Angle in Diesel Engines, *Trans. JSAE*, Vol. 21, 1980, pp. 5–11.
14. Reitz, R. D., and Bracco, F. V., On the Dependence of Spray Angle and Other Spray Parameters on Nozzle Design and Operating Conditions, *SAE Tech. Pap. Ser.* 790494, 1979.
15. Chiu, S., Shahed, S. M., and Lyn, W. T., A Transient Spray Mixing Model for Diesel Combustion, *SAE Trans.*, Vol. 85, Sect. 1, 1976, pp. 502–512.
16. Dent, J. C., A Basis for Comparison of Various Experimental Methods for Studying Spray Penetration, *SAE Trans.*, Vol. 80, Pt. 3, 1971, pp. 1881–1884.
17. Hiroyasu, H., and Kadota, T., Models for Combustion and Formation of Nitric Oxide and Soot in Direct-Injection Diesel Engines, *SAE Trans.*, Vol. 85, 1976, pp. 513–526.
18. Haselman, L. C., and Westbrook, C. K., A Theoretical Model for Two-Phase Fuel Injection in Stratified Charge Engines, SAE Paper 780318, 1978.

19. Pirouz-Panah, V., and Williams, T. J., Influence of Droplets on the Properties of Liquid Fuel Jets, *Proc. Inst. Mech. Eng.*, Vol. 191, No. 28, 1977, pp. 299–306.
20. Melton, R. B., Diesel Fuel Injection Viewed as a Jet Phenomenon, SAE Paper 710132, 1971.
21. Lewis, J. D., Studies of Atomization and Injection Processes in the Liquid Propellant Rocket Engine, Combustion and Propulsion, Fifth AGARD Colloquium on High Temperature Phenomena, Pergamon, London, 1963, pp. 141–174.
22. Delavan Industrial Nozzles and Accessories, brochure, Delavan Ltd., Gorsey Lane, Widnes, Cheshire, WA80RJ, England.
23. Tate, R., Sprays, *Kirk-Othmer Encyclopedia of Chemical Technology*, Vol. 18, 2nd ed., Interscience Publishers, New York, 1969, pp. 634–654.
24. Joyce, J. R., The Atomization of Liquid Fuels for Combustion, *J. Inst. Fuel*, Vol. 22, No. 124, 1949, pp. 150–156.
25. Jones, A. R., Design Optimization of a Large Pressure Jet Atomizer for Power Plant, *Proceedings of the 2nd International Conference on Liquid Atomization and Spray Systems*, Madison, Wis., 1982, pp. 181–185.
26. Joyce, J. R., Report ICT 15, Shell Research Ltd., London, 1947.
27. Mock, F. C., and Ganger, D. R., Practical Conclusions on Gas Turbine Spray Nozzles, *SAE Q. Trans.*, Vol. 4, No. 3, July 1950, pp. 357–367.
28. Carey, F. H., The Development of the Spill Flow Burner and Its Control System for Gas Turbine Engines, *J. R. Aeronaut. Soc.*, Vol. 58, No. 527, November 1954, pp. 737–753.
29. Radcliffe, A., The Performance of a Type of Swirl Atomizer, *Proc. Inst. Mech. Eng.*, Vol. 169, 1955, pp. 93–106.
30. Radcliffe, A., Fuel Injection, *High Speed Aerodynamics and Jet Propulsion*, Section D, Vol. XI, Princeton University Press, Princeton, N.J., 1960.
31. Tipler, W., and Wilson, A. W., Combustion in Gas Turbines, Paper No. B9, *Proceedings of the Congress International des Machines a Combustion (CIMAC)*, Paris, 1959, pp. 897–927.
32. Dombrowski, N., and Munday, G., Spray Drying, *Biochemical and Biological Engineering Science*, Vol. 2, Chapter 16, Academic Press, New York, 1968, pp. 209–320.
33. Pilcher, J. M., and Miesse, C. C., Methods of Atomization, *Injection and Combustion of Liquid Fuels*, Chapter 2, WADC-TR-56-3 AD 118142, Battelle Memorial Institute, Columbus, Ohio, March 1957.
34. Tyler, S. R., and Turner, H. G., Fuel Systems and High Speed Flight, *Shell Aviat. News*, No. 233, 1957.
35. Clare, H., Gardiner, J. A., and Neale, M. C., Study of Fuel Injection in Air Breathing Combustion Chambers, *Experimental Methods in Combustion Research*, Pergamon, London, 1964, pp. 5–20.
36. Snyder, H., Senser, D. W., Lefebvre, A. H., and Coutinho, R. S., paper presented at the 1987 IEEE Industry Applications Society 22nd Annual Meeting, Atlanta, Ga., October 1987.
37. Fraser, R. P., and Eisenklam, P., Liquid Atomization and the Drop Size of Sprays, *Trans. Inst. Chem. Eng.*, Vol. 34, 1956, pp. 295–319.
38. Dombrowski, N., and Fraser, R. P., A Photographic Investigation into the Disintegration of Liquid Sheets, *Philos. Trans. R. Soc. London Ser. A*, Vol. 247, No. 924, September 1954, pp. 101–130.
39. Dombrowski, N., Hasson, D., and Ward, D. E., Some Aspects of Liquid Flow Through Fan Spray Nozzles, *Chem. Eng. Sci.*, Vol. 12, 1960, pp. 35–50.
40. Carr, E., The Combustion of a Range of Distillate Fuels in Small Gas Turbine Engines, ASME Paper 79-GT-175, 1979.
41. Carr, E., Further Applications of the Lucas Fan Spray Fuel Injection System, ASME Paper 85-IGT-116, 1985.
42. Watkins, S. C., Simplified Flat Spray Fuel Nozzle, U.S. Patent No. 3,759,448, 1972.
43. Masters, K., *Spray Drying*, 2nd ed., John Wiley and Sons, New York, 1976.
44. Hinze, J. O., and Milborn, H., Atomization of Liquids by Means of a Rotating Cup, *ASME J. Appl. Mech.*, Vol. 17, No. 2, 1950, pp. 145–153.

45. Kayano, A., and Kamiya, T., Calculation of the Mean Size of the Droplets Purged from the Rotating Disk, *Proceedings of the 1st International Conference on Liquid Atomization and Spray Systems,* Tokyo, 1978, pp. 133–143.

46. Tanasawa, Y., Miyasaka, Y., and Umehara, M., Effect of Shape of Rotating Disks and Cups on Liquid Atomization, *Proceedings of the 1st International Conference on Liquid Atomization and Spray Systems,* Tokyo, 1978, pp. 165–172.

47. Matsumoto, S., Crosby, E. J., and Belcher, D. W., Rotary Atomizers; Performance Understanding and Prediction, *Proceedings of the 3rd International Conference on Liquid Atomization and Spray Systems,* London, July 1985, pp. 1A/1/1–20.

48. Karim, G. A., and Kumar, R., The Atomization of Liquids at Low Ambient Pressure Conditions, *Proceedings of the 1st International Conference on Liquid Atomization and Spray Systems,* Tokyo, August 1978, pp. 151–155.

49. Masters, K., Rotary Atomizers, *Proceedings of the 1st International Conference on Liquid Atomization and Spray Systems,* Tokyo, August 1978, p. 456.

50. Wehner, H., Combustion Chambers for Turbine Power Plants, *Interavia,* Vol. 7, No. 7, 1952, pp. 395–400.

51. Maskey, H. C., and Marsh, F. X., The Annular Combustion Chamber with Centrifugal Fuel Injection, SAE Preprint No. 444C, 1962.

52. Nichol, I. W., The T65 and T72 Shaft Turbines, SAE Preprint No. 624A, 1963.

53. Burgher, M. W., Interrelated Parameters of Gas Turbine Engine Design and Electrical Ignition Systems, SAE Preprint No. 682C, 1963.

54. Romp, H. A., *Oil Burning,* Martinus Nijhoff, The Hague; Stechert & Company, New York, 1937.

55. Gretzinger, J., and Marshall, W. R., Jr., Characteristics of Pneumatic Atomization, *J. Am. Inst. Chem. Eng.,* Vol. 7, No. 2, June 1961, pp. 312–318.

56. Mullinger, P. J., and Chigier, N. A., The Design and Performance of Internal Mixing Multi-Jet Twin-Fluid Atomizers, *J. Inst. Fuel,* Vol. 47, 1974, pp. 251–261.

57. Bryce, W. B., Cox, N. W., and Joyce, W. I., Oil Droplet Production and Size Measurement from a Twin-Fluid Atomizer using Real Fluids, *Proceedings of the 1st International Conference on Liquid Atomization and Spray Systems,* Tokyo, 1978, pp. 259–263.

58. Sargeant, M., Blast Atomizer Developments in the Central Electricity Generating Board, *Proceedings of the 2nd International Conference on Liquid Atomization and Spray Systems,* Madison, Wis., 1982, pp. 131–135.

59. Hurley, J. F., and Doyle, B. W., Design of Two-Phase Atomizers for Use in Combustion Furnaces, *Proceedings of the 3rd International Conference on Liquid Atomization and Spray Systems,* London, July 1985, pp. 1A/3/1–13.

60. Rosfjord, T. J., Atomization of Coal Water Mixtures: Evaluation of Fuel Nozzles and a Cellulose Gum Simulant, ASME Paper 85-GT-38, 1985.

61. Wigg, L. D., The Effect of Scale on Fine Sprays Produced by Large Airblast Atomizers, National Gas Turbine Establishment Report No. 236, 1959.

62. Stambuleanu, A., *Flame Combustion Processes in Industry,* Abacus Press, Turnbridge Wells, Kent, 1976.

63. Jasuja, A. K., Atomization of Crude and Residual Fuel Oils, *ASME J. Eng. Power,* Vol. 101, No. 2, 1979, pp. 250–258.

64. Sher, E., and Elata, D., *Ind. Eng. Chem. Process Des. Dev.,* Vol. 16, 1977, pp. 237–242.

65. Marek, C. J., and Cooper, L. P., U.S. Patent No. 4,189,914, 1980.

66. Solomon, A. S. P., Rupprecht, S. D., Chen, L. D., and Faeth, G. M., Flow and Atomization in Flashing Injectors, *Atomization Spray Technol.,* Vol. 1, No. 1, 1985, pp. 53–76.

67. Lefebvre, A. H., Wang, X. F., and Martin, C. A., Spray Characteristics of Aerated Liquid Pressure Atomizers, *AIAA J. Propul. Power,* in press.

68. Lefebvre, A. H., *Gas Turbine Combustion,* Hemisphere, Washington, D.C., 1983.

69. Graf, P. E., Breakup of Small Liquid Volume by Electrical Charging, *Proceedings of API Research Conference on Distillate Fuel Combustion,* API Publication 1701, Paper CP62-4, 1962.

70. Luther, F. E., Electrostatic Atomization of No. 2 Heating Oil, *Proceedings of API Research Conference on Distillate Fuel Combustion,* API Publication 1701, Paper CP62-3, 1962.

71. Peskin, R. L., Raco, R. J., and Morehouse, J., A Study of Parameters Governing Electrostatic Atomization of Fuel Oil, *Proceedings of API Research Conference on Distillate Fuel Combustion,* API Publication 704, 1965.

72. Bollini, R., Sample, B., Seigal, S. D., and Boarman, J. W., Production of Monodisperse Charged Metal Particles by Harmonic Electrical Spraying, *J. Interface Sci.,* Vol. 51, No. 2, 1975, pp. 272–277.

73. Drozin, V. G., The Electrical Dispersion of Liquids as Aerosols, *J. Colloid Sci.,* Vol. 10, No. 2, 1955, pp. 158–164.

74. Macky, W. A., Some Investigations on the Deformation and Breaking of Water Drops in Strong Electric Fields, *Proc. R. Soc., London Ser. A,* Vol. 133, 1931, pp. 565–587.

75. Nawab, M. A., and Mason, S. C., The Preparation of Uniform Emulsions by Electrical Dispersion, *J. Colloid Sci.,* Vol. 13, 1958, pp. 179–187.

76. Vonnegut, B., and Neubauer, R. L., Production of Monodisperse Liquid Particles by Electrical Atomization, *J. Colloid Sci.,* Vol. 7, 1952, pp. 616–622.

77. Kelly, A. J., The Electrostatic Atomization of Hydrocarbons, *Proceedings of the 2nd International Conference on Liquid Atomization and Spray Systems,* Madison, Wis., 1982, pp. 57–65.

78. Bailey, A. G., The Theory and Practice of Electrostatic Spraying, *Atomization Spray Technol.,* Vol. 2, 1986, pp. 95–134.

79. Lang, R. J., Ultrasonic Atomization of Liquids, *J. Acoust. Soc. Am.,* Vol. 34, No. 1, 1962, pp. 6–8.

80. Cabler, P., Geddes, L. A., and Rosborough, J., The Use of Ultrasonic Energy to Vaporize Anaesthetic Liquids, *Br. J. Aneasth.,* Vol. 47, 1975, pp. 541–545.

81. Berger, H. L., Characterization of a Class of Widely Applicable Ultrasonic Nozzles, *Proceedings of the 3rd International Conference on Liquid Atomization and Spray Systems,* London, July 1985, pp. 1A/2/1–13.

82. Topp, M. N., and Eisenklam, P., Industrial and Medical Use of Ultrasonic Atomizers, *Ultrasonics,* Vol. 10, No. 3, 1972, pp. 127–133.

83. Lee, K. W., Putnam, A. A., Gieseke, J. A., Golovin, M. N., and Hale, J. A., Spray Nozzle Designs for Agricultural Aviation Applications, NASA CR 159702, 1979.

84. Wilcox, R. L., and Tate, R. W., Liquid Atomization in a High Intensity Sound Field, *J. Am. Inst. Chem. Eng.,* Vol. 11, No. 1, 1965, pp. 69–72.

FLOW IN ATOMIZERS

INTRODUCTION

In twin-fluid atomizers of the airblast and air-assist types, atomization and spray dispersion tend to be dominated by air momentum forces, with hydrodynamic processes playing only a secondary role. With pressure-swirl nozzles, however, the internal flow characteristics are of primary importance, because they govern the thickness and uniformity of the annular liquid film formed in the final discharge orifice as well as the relative magnitude of the axial and tangential components of velocity of this film. It is therefore of great practical interest to examine the interrelationships that exist between internal flow characteristics, nozzle design variables, and important spray features such as cone angle and mean drop size. The various equations that have been derived for nozzle discharge coefficient are discussed at some length, because this coefficient not only affects the flow rate of any given nozzle but also can be used to calculate its velocity coefficient and spray cone angle.

Consideration is also given to the complex flow situations that arise on the surface of a rotating cup or disk. These flow characteristics are of basic importance to the successful operation of these atomizers, because they exercise a controlling influence on the nature of the atomization process, the quality of atomization, and distribution of drop sizes in the spray.

FLOW NUMBER

The effective flow area of a pressure atomizer is usually described in terms of a flow number, which is expressed as the ratio of the nozzle throughput to the square root of the fuel-injection pressure differential. Two definitions of flow number are in general use: a British version, based on the volume flow rate, and the other is an American version, based on the mass flow rate. They are

$$FN_{UK} = \frac{\text{flow rate, UK gal/h}}{(\text{injection pressure differential, psid})^{0.5}} \quad (5.1)$$

$$FN_{US} = \frac{\text{flow rate, lb/h}}{(\text{injection pressure differential, psid})^{0.5}} \quad (5.2)$$

Note that 1 UK gallon = 1.2 US gallons.

Equations (5.1) and (5.2) have the advantage of being expressed in units that are in general use. Unfortunately, they are basically unsound. For example, they do not allow a fixed and constant value of flow number to be assigned to any given nozzle. Thus, although it is customary to stamp or engrave a value of flow number on the body of a simplex atomizer, this value is correct only when the nozzle is flowing a standard calibrating fluid of density 765 kg/m^3. In the past this has posed no problems with aircraft gas turbines because 765 kg/m^3 roughly corresponds to the density of aviation gasoline. However, for liquids of other densities, these two definitions of flow number could lead to appreciable errors when used to calculate liquid flow rates or injection pressures.

The basic deficiency in Eqs. (5.1) and (5.2) is the omission of liquid density. Inclusion of this property would not only allow these equations to be rewritten in dimensionally correct form but also enable the flow number to be defined in a much more positive and useful manner than at present, namely as the effective flow area of the nozzle. Thus the flow number of any given nozzle would have a fixed and constant value for all liquids.

By including liquid density, the flow number in square meters is obtained as

$$FN = \frac{\text{flow rate, kg/s}}{(\text{pressure differential, Pa})^{0.5}(\text{liquid density, kg/m}^3)^{0.5}} \quad (5.3)$$

The standard UK and US flow numbers may be calculated from Eq. (5.3) using the formulas:

$$FN_{UK} = 0.66 \times 10^8 \times \rho_L^{-0.5} \times FN \quad (5.4)$$

$$FN_{US} = 0.66 \times 10^6 \times \rho_L^{0.5} \times FN \quad (5.5)$$

PLAIN–ORIFICE ATOMIZER

It is generally considered that the flow in a plain-orifice atomizer is similar to that in a pipe, and consequently the nature of the flow is related to the Reynolds number. Measurements of discharge coefficient carried out by Bird [1] and Gellales [2] showed great differences at high as compared with low Reynolds number. Schweitzer [3] noted that the spray was more rapidly dispersed when the Reynolds number exceeded a certain value, and he attributed this to the transition from laminar to turbulent flow, which resulted in more rapid disruption of the jet.

The critical value of Reynolds number, that is, the value at which the nature of flow changes from laminar to turbulent, usually lies between 2000 and 3000. For values below the critical Reynolds number the flow tends to be laminar, while at higher values it tends to be turbulent. One reason why the critical value of Reynolds number cannot be precisely defined is that it depends on the geometry of the atomizer and on the properties of the liquid [3, 4]. In general, laminar flow is promoted by

1. A rounded entrance to the orifice
2. Smooth passage walls
3. Absence of bends
4. High liquid viscosity
5. Low liquid velocity

Turbulent flow is promoted by

1. Large passage diameters
2. Changes in flow velocity and direction
3. Abrupt changes in cross-sectional area
4. Surface roughness
5. Imperfections in atomizer geometry
6. Mechanical vibrations
7. Low liquid viscosity
8. High liquid velocity

Turbulence is conducive to good atomization but usually at the expense of some increase in pressure loss.

For low-viscosity liquids, such as water, kerosine, and light diesel oil, the flow in the atomizer is normally turbulent. However, in intermittent injection systems there is a period at the beginning of each injection when the velocity is rising from zero and a corresponding period at the end when it falls again to zero, during which the flow is either laminar or semiturbulent.

Discharge Coefficient

The discharge coefficient of a plain-orifice atomizer is governed partly by the pressure losses incurred in the nozzle flow passages and also by the extent to which the liquid flowing through the final discharge orifice makes full use of the available flow area. Discharge coefficient is related to nozzle flow rate by the equation

$$\dot{m}_L = C_D A_o (2\rho_L \, \Delta P_L)^{0.5} \tag{5.6}$$

$$= 1.111 C_D d_o^2 (\rho_L \, \Delta P_L)^{0.5} \tag{5.7}$$

$$= 35.12 C_D d_o^2 (SG \, \Delta P_L)^{0.5} \tag{5.8}$$

Factors influencing discharge coefficient. Measurements of discharge coefficient carried out on various orifice configurations over wide ranges of operating conditions indicate that the most important parameters are Reynolds number, length/diameter ratio, injection pressure differential, ambient gas pressure, inlet chamfer, and cavitation.

Reynolds number. The influence of Reynolds number on discharge coefficient has been investigated by Bird [1], Gellales [2], Bergwerk [5], Spikes and Pennington [6], Lichtarowicz et al. [7], Arai et al. [8], and others. The manner in which Reynolds number influences the discharge coefficient of a plate orifice is illustrated in Fig. 5.1. This figure exhibits three distinct sections. In the first stage, corresponding to laminar flow, C_D increases almost linearly with $\sqrt{Re}$. During the second stage, corresponding to semiturbulent flow, C_D at first increases with Re up to a maximum value, beyond which further increase in Re causes C_D to decline. In the fully turbulent stage, which is of most practical interest, C_D remains sensibly constant.

Length/diameter ratio. The characteristic peak in the curve of C_D against Re for a sharp-edged orifice diminishes rapidly as l_o/d_o is increased from 0.5 to unity, as illustrated in Fig. 5.2. This figure is based on experimental data compiled from several sources by Lichtarowicz et al. [7]. It shows that for $l_o/d_o = 0.5$ the passage is similar to a simple plate orifice, as illustrated in Fig. 5.3. The discharge coefficient is low because the liquid jet forms a vena contracta, which, in the short length available, has no time to reexpand and fill the nozzle. With increase in l_o/d_o the jet expands in the passage and C_D increases, reaching a maximum at a

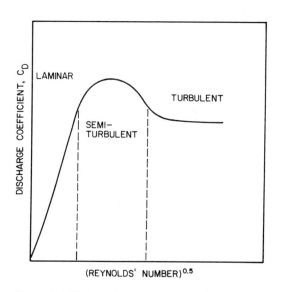

Figure 5.1 Variation in discharge coefficient with Reynolds number *(Giffen and Muraszew [4])*.

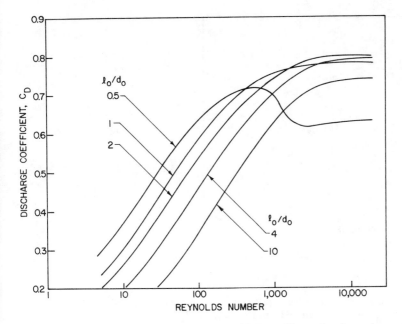

Figure 5.2 Variation of discharge coefficient with Reynolds number for various orifice length/diameter ratios. Based on experimental data compiled by Lichtarowicz et al. [7].

value of l_o/d_o of around 2. Further increase in l_o/d_o reduces C_D due to increase in frictional losses.

Figure 5.2 shows that discharge coefficients generally increase with increase in Reynolds numbers until a maximum value is attained at a Reynolds number of around 10,000. Beyond this point, the value of C_D remains sensibly constant and independent of Reynolds number. The maximun values of C_D are shown plotted against l_o/d_o in Fig. 5.4. The experimental data on which this figure is based were drawn from reference [7], but actual data points have been omitted for clarity. The figure shows $C_{D_{max}}$ rising steeply from about 0.61 to a maximum value of about 0.81 as l_o/d_o increases from 0 to 2. Further increase in l_o/d_o causes $C_{D_{max}}$ to decline slowly in a nearly linear fashion to about 0.74 at $l_o/d_o = 10$. Figure

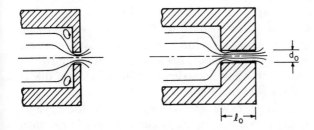

Figure 5.3 Influence of final orifice length/diameter ratio on flow pattern.

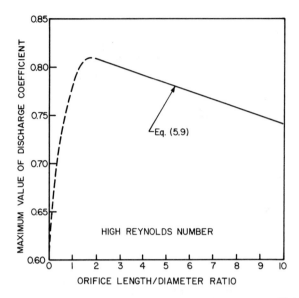

Figure 5.4 Variation of maximum value of discharge coefficient with orifice length/diameter ratio. Based on experimental data compiled by Lichtarowicz et al. [7].

5.4 indicates a linear relationship between $C_{D_{max}}$ and l_o/d_o for the range of l_o/d_o between 2 and 10. For this region, Lichtarowicz et al. [7] proposed the following expression, which is claimed to fit the experimental data to within about 1%:

$$C_{D_{max}} = 0.827 - 0.0085 \frac{l_o}{d_o} \tag{5.9}$$

Injection pressure. The influence of liquid injection pressure differential on C_D is quite small. For example, Gellales [2] found for diesel oil an increase in C_D from 0.91 to 0.93 for a five-fold increase in injection pressure. The result was obtained at an l_o/d_o ratio of 3. For higher values of l_o/d_o the discharge coefficient decreases with an increase in injection pressure. This is due mainly to higher frictional losses, which increase with the square of the velocity [4].

Ambient pressure. The effect of ambient air or gas pressure on discharge coefficient has been investigated by Arai et al. [8]. Their results, with data points omitted, are shown in Fig. 5.5. They were obtained when flowing water through a circular nozzle 1.2 mm long and 0.3 mm in diameter. It is of interest to observe in this figure that C_D changes not only with Re but also with ambient gas pressure. At normal atmospheric pressure, C_D reaches a maximum value of around 0.8 at Re = 3000. For Reynolds numbers between 3000 and 15,000 it has two values. The higher value corresponds to Re increasing from 3000 and the lower value corresponds to Re decreasing from 15,000. For Reynolds numbers larger than 15,000, C_D remains single-valued at around 0.7.

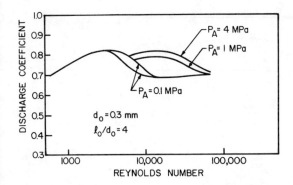

Figure 5.5 Influence of ambient air pressure on discharge coefficient. Based on experimental data of Arai et al. [8].

Double values of C_D were not observed at higher ambient pressures. Instead, C_D remained fairly constant and independent of ambient pressure at a value of about 0.8 over a range of Re from 2000 to 20,000. At higher values of Re, C_D slowly declines from 0.8 at Re = 20,000 to around 0.72 at Re = 50,000.

Inlet chamfer. During tests on diesel injection nozzles, Bergwerk [5] found that an inlet chamfer on the nozzle could raise the discharge coefficient. Similar results were obtained by Zucrow [9], who showed that the discharge coefficient reaches a maximum for submerged orifices with low pressure drop when the included angle of the chamfer is between 20° and 60°.

Detailed information on the influence of inlet chamfer on discharge coefficient has been provided by Spikes and Pennington [6]. These workers carried out a comprehensive test program on submerged orifices to determine the optimum angle and depth of chamfer for maintaining a constant discharge coefficient over the greatest possible range of operating conditions. In a first series of tests they varied the chamfer angle. An orifice 1.57 mm in diameter, having a fixed parallel throat length of 0.51 mm, was chosen for these tests and a chamfer 0.51 mm deep was cut on the upstream edge. The main feature of the results obtained using chamfered orifices is the increase in discharge coefficient given by the chamfer, as illustrated in Fig. 5.6. This figure indicates an optimum chamfer angle of about 50°, which is consistent with Zucrow's [9] findings.

Further tests were made to examine the effects of chamfer depth. Figure 5.7 shows the variation in discharge coefficient with chamfer depth for an orifice with a chamfer having an included angle of 50°. This figure demonstrates a strong dependence of discharge coefficient on depth of chamfer and shows that even slight variations in chamfer depth can have a marked effect on orifice flow rate.

Cavitation. In flow regions of low static pressure, gas or vapor may be released from the liquid to form bubbles, which can have a pronounced influence on discharge coefficient. The subsequent explosion or collapse of these bubbles can also

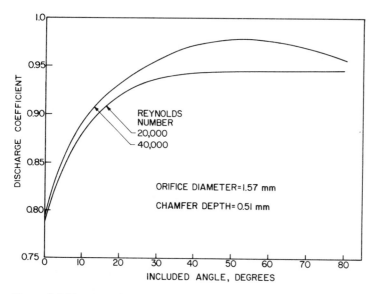

Figure 5.6 Variation of discharge coefficient with angle of upstream chamfer. Based on experimental data of Spikes and Pennington [6].

accelerate jet breakup. In addition to its adverse effect on discharge coefficient, cavitation may cause severe erosion of nozzle passages. The normal working range of diesel nozzles is such that both cavitating and noncavitating flow may occur under certain conditions.

The results of tests carried out by Spikes and Pennington [6] using square-edged orifices are shown in Fig. 5.8. In these curves the discharge coefficient is shown as a function of Reynolds number and cavitation number C, the latter being defined as

$$C = \frac{p_1 - p_2}{p_2 - p_v} \qquad (5.10)$$

where p_1 is the upstream pressure, p_2 is the downstream pressure, and p_v is the vapor pressure. Cavitation occurs if the local pressure becomes equal to the vapor pressure of the liquid. The fluid used in Spikes and Pennington's tests was aviation kerosine. The vapor pressure of this liquid is sufficiently small to be neglected. Under such conditions the cavitation number may be expressed more simply as

$$C = \frac{p_1 - p_2}{p_2} = \frac{\text{pressure drop}}{\text{downstream pressure}} \qquad (5.11)$$

Figure 5.8 shows that the effects of cavitation can be high and can cause changes of discharge coefficient greater than those associated with Reynolds number when the latter is high.

Many orifices show instability around the point of changeover from cavitating to noncavitating flow. With long holes this is due to the tendency of the liquid

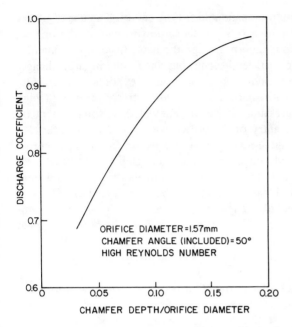

Figure 5.7 Variation of discharge coefficient with depth of upstream chamfer. Based on experimental data of Spikes and Pennington [6].

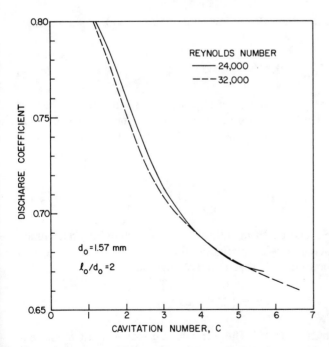

Figure 5.8 Influence of cavitation on discharge coefficient for a square-edged round orifice. Based on experimental data of Spikes and Pennington [6].

downstream of the vena contracta to expand and fill the cavity. The use of a very short length places the cavitation zone some distance downstream of the orifice, so that it cannot influence the flow pattern through the hole. Instabilities can also be eliminated by making the hole sufficiently long, but this results in large changes in discharge coefficient with both cavitation number and Reynolds number [6].

As cavitation is associated with sharp corners, which create regions of high liquid velocity and low static pressure, it can be alleviated to some extent by replacing the sharp corner at the entry of the orifice with a lead-in chamfer. An ideal chamfer would have a profile similar to that of a jet stream during contraction downstream of a sharp-edged entry, and the orifice itself would have a diameter equal to that of the vena contracta. The experiments of Spikes and Pennington [6] have shown that this ideal condition can be approached closely with a straight-sided chamfer if the included angle of the chamfer is 50° and the chamfer depth is 0.30 times the diameter of the orifice. However, it should be noted that although this chamfer will suppress cavitation over a wide range, Reynolds number effects can still cause appreciable changes in the discharge coefficient [6].

Empirical expressions. The lack of any quantitative theory for flow in orifices over wide ranges of Reynolds number has led to a number of empirical correlations for noncavitating flows. According to Nakayama [10]

$$C_D = \frac{Re^{5/6}}{17.11 l_o/d_o + 1.65 Re^{0.8}} \tag{5.12}$$

for l_o/d_o in the range 1.5 to 17 and Re in the range 550 to 7000, to an accuracy of within 2.8%. Nakayama also suggests

$$C_{D_{max}} = 0.868 - 0.0425 \left(\frac{l_o}{d_o}\right)^{0.5} \tag{5.13}$$

as an expression for discharge coefficient at high Reynolds numbers, valid over the same range of l_o/d_o. This formula gives lower values than Eq. (5.9), from Lichtarowicz et al. [7], since in Nakayama's tests the range of Re was restricted at the upper end.

Asihmin et al. [11] suggest

$$C_D = \left[1.23 + \frac{58(l_o/d_o)}{Re} \right]^{-1} \tag{5.14}$$

for l_o/d_o in the range 2 to 5 and Re from 100 to 1.5×10^5, claiming an accuracy within 1.5%. To accommodate a wider variation in l_o/d_o, Lichtarowicz et al. [7] proposed the following modification of Asihmin's equation:

$$\frac{1}{C_D} = \frac{1}{C_{D_{max}}} + \frac{20}{Re}\left(1 + 2.25\frac{l_o}{d_o}\right) \tag{5.15}$$

in which $C_{D_{max}}$ is given by Eq. (5.9). This provides an excellent fit to the experimental data [7] in the range of l_o/d_o from 2 to 10 and for Re in the range from 10 to 20,000.

It is important to note that these empirical equations for C_D do not apply under conditions where cavitation is present. Tests carried out by Ruiz and Chigier [12] on a diesel electromagnetic injector showed the occurrence of cavitation in this type of injector over most of its operating range. For Reynolds numbers up to about 40,000, C_D remained sensibly constant at around 0.7, but cavitation effects caused C_D to fall to around 0.6 with increase in Reynolds number from 40,000 to 50,000.

Another complication with diesel injectors is that the value of Reynolds number varies during the injection period. Varde and Popa [13] quote values of C_D varying from 0.5 to 0.8 depending on the average Reynolds number.

PRESSURE–SWIRL ATOMIZER

In this type of nozzle the liquid is injected through tangential or helical passages into a swirl chamber, from which it emerges with both tangential and axial velocity components to form a thin conical sheet at the nozzle exit. This sheet rapidly attenuates, finally disintegrating into ligaments and then drops.

Despite the geometric simplicity of the simplex swirl atomizer, the hydrodynamic processes occurring within the nozzle are highly complex. Nevertheless, the early theories based on the assumption of frictionless flow soon led to the formulation of quantitative relationships between the main atomizer dimensions and various flow parameters, such as discharge coefficient, and initial spray cone angle [14, 15]. Subsequently, Taylor [16] showed that although the motion in the bulk of the fluid can be considered irrotational, the viscous effect of the retarded boundary layer is far from negligible. The liquid in contact with the swirl chamber end walls cannot rotate at a sufficient rate to hold it in a circular path against the radial pressure gradient, and hence a current directed toward the orifice is set up through the surface layer. Other workers [17, 18] have shown that for real fluids an outward flow may also occur through a boundary layer around the air core. However, as Dombrowski and Hassan [19] and Jones [20] have emphasized, for low-viscosity liquids simple inviscid analysis still provides a basic understanding of the flow characteristics of pressure-swirl atomizers and gives a reasonable guide to discharge coefficients and cone angle.

Discharge Coefficient

The discharge coefficient of a swirl atomizer is inevitably low, owing to the presence of the air core, which effectively blocks off the central portion of the orifice. Radcliffe [21] studied the performance of a family of injectors based on common design rules, using fluids that covered wide ranges of density and viscosity. He demonstrated the existence of a unique relationship between the discharge coefficient and the Reynolds number based on orifice diameter. At low Reynolds numbers, the effect of viscosity is to thicken the fluid film in the final orifice and thereby increase in discharge coefficient. With nozzles of small flow number this effect can be significant at low flow rates. However, for Reynolds numbers larger than 3000, that is, over most of the normal working range, the discharge coefficient is practically independent of Reynolds number. Thus, for liquids of low viscosity, the convention is to disregard conditions at low Reynolds number and assume that any given atomizer has a constant discharge coefficient.

The following analysis, from Giffen and Muraszew [4], refers to a simplex atomizer, but the results obtained can be applied to other types of pressure-swirl nozzles, such as duplex, dual-orifice, and spill-return. The liquid flow pattern is produced by the imposition of a spiral motion on a free vortex, as illustrated in Fig. 5.9. Conservation of angular momentum provides the following relationship between tangential velocity v and radius r:

$$vr = v_i R_s \qquad (5.16)$$

where v_i is the inlet velocity to the swirl chamber and R_s is its radius. Also, we have

$$v_i = \frac{\dot{m}_L}{\rho_L A_p} \qquad (5.17)$$

where A_p is the total cross-sectional area of the inlet ports. This equation implies the existence of an air core at the center of the swirl chamber, which is always observed in practice, since with $r = 0$ the velocity v would otherwise be infinite.

With the assumption that no losses occur within the atomizer, the total head can be considered constant throughout the swirl chamber and equal to the injection pressure P. The total pressure at any point in the liquid flowing through the orifice is then given by Bernoulli's equation as

$$P = p + 0.5\rho_L u^2 + 0.5\rho_L v^2 = \text{constant} \qquad (5.18)$$

where p is the static pressure at any point in the liquid and u is the axial velocity in the orifice. But for a free vortex alone it can be shown that $p + 0.5\rho_L v^2 = $ constant, with the result that, for the steady flow of a nonviscous fluid, the axial velocity u is uniform and constant for all values of r in the liquid annulus around the air core in the orifice. At the air core the static pressure is the back pressure of the ambient atmosphere, i.e., $p = 0$, and so

$$P = 0.5\rho_L(u_{r_a}^2 + v_{r_a}^2) \qquad (5.19)$$

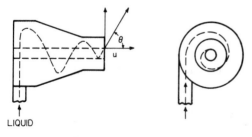

LIQUID

Figure 5.9 Illustration of flow path in simplex atomizer.

Since u remains constant, we have $u_{r_a} = u$. Substitution of u for u_{r_a} in Eq. (5.19) gives

$$P = 0.5\rho_L(v_{r_a}^2 + u^2) \tag{5.20}$$

The axial component of velocity in the orifice is given by

$$u = \frac{\dot{m}_L}{\rho_L(A_o - A_a)} \tag{5.21}$$

where A_o is the orifice area and A_a is the air core area.

From Eqs. (5.16) and (5.17) we have

$$v_{r_a} = \frac{\dot{m}_L R_s}{\rho_L A_p r_a} \tag{5.22}$$

Substituting into Eq. (5.20) u and v_{r_a} from Eqs. (5.21) and (5.22), respectively, gives

$$P = 0.5\rho_L\left[\left(\frac{\dot{m}_L R_s}{\rho_L A_p r_a}\right)^2 + \left(\frac{\dot{m}_L}{\rho_L(A_o - A_a)}\right)^2\right] \tag{5.23}$$

Now

$$\dot{m}_L = C_D A_o(2\rho_L P)^{0.5} \tag{5.6}$$

Substitution of $\dot{m}_L$ from Eq. (5.6) into Eq. (5.23) gives

$$\frac{1}{C_D^2} = \frac{1}{K_1^2 X} + \frac{1}{(1 - X)^2} \tag{5.24}$$

where $X = A_a/A_o$ and $K_1 = A_p/\pi r_o R_s$.

Equation (5.24) provides a relationship between the atomizer dimensions, the size of the air core, and the discharge coefficient of the nozzle. To eliminate one of these variables we may apply the condition that, for any given value of K_1, the size of the air core will always be such as to give maximum flow; i.e., the value of C_D, expressed as a function of X, is a maximum [4].

Putting $d(1/C_D^2)/dX = 0$ leads to

$$2K_1^2 X^2 = (1 - X)^3 \tag{5.25}$$

Substituting in Eq. (5.24) the value of K_1 from Eq. (5.25) gives

$$C_D = \left[\frac{(1 - X)^3}{1 + X}\right]^{0.5} \tag{5.26}$$

Since, from Eq. (5.25) X is a unique function of K_1, it follows from Eq. (5.26) that C_D also depends solely on K and is independent of injection pressure. Figure 5.10 illustrates the relationship of C_D with the atomizer constant K, where $K = A_p/D_s d_o = \pi K_1/4$.

Giffen and Muraszew [4] observed that Eq. (5.26) gave values of C_D that

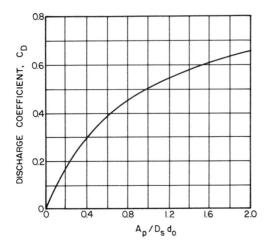

Figure 5.10 Theoretical relationship between discharge coefficient and atomizer dimensions.

were too low in comparison with the experimental data. To accommodate this they introduced a constant into Eq. (5.26), which then becomes

$$C_D = 1.17 \left[\frac{(1 - X)^3}{1 + X} \right]^{0.5}$$ (5.27)

Several other equations for discharge coefficient have been derived. According to Taylor [16], for inviscid flow in a swirl atomizer the discharge coefficient is given by

$$C_D^2 = 0.225 \frac{A_p}{D_s d_o}$$ (5.28)

This equation requires modification in the light of Carlisle's [22] evidence on the effect of D_s/d_o and L_s/D_s on discharge coefficient. The relationship between C_D and D_s/d_o is indicated by the curve drawn through the data points in Fig. 5.11, which corresponds to the relationship

$$\frac{C_{D,\text{meas}}}{C_{D,\text{theor}}} = 0.55 \left(\frac{D_s}{d_o} \right)^{0.5}$$ (5.29)

Thus the influence of D_s/d_o can readily be accommodated using the correction factor $0.55(D_s/d_o)^{0.5}$. With regard to L_s/D_s, inspection of Fig. 5.12 reveals that over the range of most interest (i.e., from 0.5 to 1.0) the appropriate correction factor is constant at about 0.95. Incorporating these two terms into Eq. (5.28) gives the relationship

$$C_D^2 = 0.0616 \frac{D_s}{d_o} \frac{A_p}{D_s d_o}$$ (5.30)

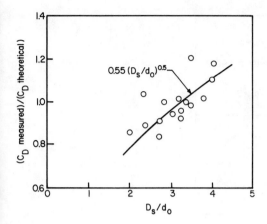

Figure 5.11 Influence of swirl chamber dimensions on discharge coefficient *(Carlisle [22])*.

This equation is shown plotted in Fig. 5.13 for two values of D_s/d_o, namely 3.5 and 5.0.

Eisenklam [23] and Dombrowski and Hassan [19] used the following dimensionally correct group to correlate their experimental data on discharge coefficients:

$$\frac{A_p}{D_s d_o} \left(\frac{D_s}{d_o}\right)^{1-n} \tag{5.31}$$

Eisenklam states that the value of n varies from 0.1 to 0.5, but Dombrowski and Hassan claim that a unique correlation is obtained for $n = 0.5$.

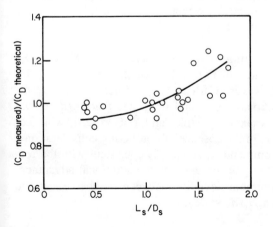

Figure 5.12 Influence of swirl chamber dimensions on discharge coefficient *(Carlisle [22])*.

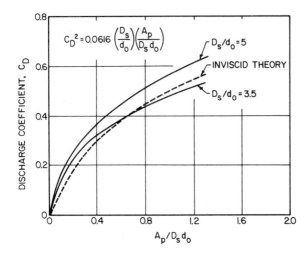

Figure 5.13 Practical relationship between discharge coefficient and atomizer dimensions.

Rizk and Lefebvre [24] derived the following relationship for C_D:

$$C_D = 0.35 \left(\frac{A_p}{D_s d_o}\right)^{0.5} \left(\frac{D_s}{d_o}\right)^{0.25} \tag{5.32}$$

This relationship is illustrated in Fig. 5.14.

Design considerations. Frictional losses are a major consideration in the design of swirl atomizers. They have two opposing effects on discharge coefficient. First, frictional losses represent a wasteful dissipation of atomization energy, which reduces the effective pressure drop across the atomizer and also the discharge coefficient. Second, by impeding the rotating flow in the swirl chamber, friction reduces the diameter of the air core and thereby increases the discharge coefficient. The relative importance of these two opposing effects depends mainly on various geometric features, which are discussed below.

Ratio of swirl chamber diameter to final orifice diameter. The effect of D_s/d_o on discharge coefficient is illustrated in Fig. 5.11, which shows that C_D increases with increase in D_s/d_o. However, Carlisle [22] pointed out that D_s/d_o should be kept small to reduce frictional losses and suggested that D_s/d_o should not exceed 5.0. A similar view was taken by Tipler and Wilson [25], who recommended a value of 2.5. However, in the absence of conflicting considerations, a D_s/d_o ratio of 3.3 has much to commend it, since it is consistent with the recommendations of Carlisle, Tipler, and Wilson and has the additional advantage, as shown in Fig. 5.11, that the minimum deviation between the theoretical and measured values of C_D occurs when $D_s/d_o = 3.3$.

Length/diameter ratio of swirl chamber. This should be kept short to minimize frictional losses. However, sufficient length must be provided for the separate jets

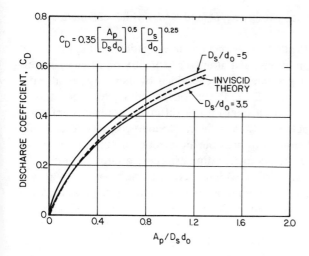

Figure 5.14 Practical relationship between discharge coefficient and atomizer dimensions *(Rizk and Lefebvre [24])*.

issuing from the swirl ports to coalesce into a uniform vortex sheet. In most current designs the L_s/D_s ratio lies between 0.5 and 1.0, although it has been suggested that higher values of L_s/D_s, up to a maximum of 2.75, would result in improved atomization [26].

Length/diameter ratio of final orifice. The high frictional losses incurred in this passage call for the shortest possible length. On large atomizers l_o/d_o can be made as small as 0.2, but with small atomizers the difficulty of manufacturing small-scale components to the required degree of accuracy usually dictates minimum values for l_o/d_o of around 0.5.

Length/diameter ratio of swirl ports. Tipler and Wilson [25] recommended that this ratio should not be less than 1.3, as short slots discharge the liquid in a diffused manner and may result in an uneven spray.

Manufacture. Good atomization and uniform drop distribution can only be obtained with an atomizer that is accurately made and well finished. The various faults that can arise in manufacture and impair atomizer performance have been fully described by Joyce [27]. It is also important to ensure that nozzle orifices do not get damaged during installation.

Comprehensive equation for discharge coefficient. From the foregoing discussion it is clear that the discharge coefficient is not solely dependent on the dimensionless parameters $A_p/D_s d_o$ and D_s/d_o. Although the simply derived equations for C_D based on these terms hold reasonably well in practice, if more precise values are needed then consideration must also be given to some of the other geometric ratios discussed above, namely L_s/D_s and l_o/d_o, which can have a small but significant effect on C_D.

Jones [20] conducted a detailed experimental investigation on the effects of liquid properties, operating conditions, and atomizer geometry on discharge coefficient. The effect of atomizer geometry was examined by using a specially designed, three-piece, large atomizer. Several versions of these three constituent pieces were manufactured, allowing a total of 159 different atomizer configurations to be constructed. In practice, however, it was found necessary to use only a small proportion of the total number of possible atomizer configurations to cover a satisfactory range of each of the geometric dimensionless groups.

Table 5.1 illustrates the range over which each nondimensional group was investigated and compares the values with those typical of the large pressure-swirl atomizers employed in power generation by the Central Electricity Generating Board in the United Kingdom.

From analysis of the experimental data, Jones [20] obtained the following empirical equation for discharge coefficient:

$$C_D = 0.45 \left(\frac{d_o \rho_L U}{\mu_L}\right)^{-0.02} \left(\frac{l_o}{d_o}\right)^{-0.03} \left(\frac{L_s}{D_s}\right)^{0.05} \left(\frac{A_p}{D_s d_o}\right)^{0.52} \left(\frac{D_s}{d_o}\right)^{0.23} \quad (5.33)$$

In this equation it is of interest to note that the exponents for the two main

Table 5.1 Range of values of nondimensional groups covered by Jones [20]

Dimensionless group	Range covered	Typical value	
$\dfrac{l_o}{d_o}$	0.1–0.9	0.15	
$\dfrac{L_s}{D_s}$	0.31–1.26	0.7	
$\dfrac{L_p}{D_p}$	0.79–3.02	1.2	
$\dfrac{A_p}{d_o D_s}$	0.19–1.21	0.52	
$\dfrac{D_s}{d_o}$	1.41–8.13	2.7	
$\dfrac{d_o \rho_L U^2}{\sigma}$	11.5×10^3–3.55×10^5	Low pressure 2.4 MPa (350 psi) 1.08×10^5	High pressure 6.3 MPA (900 psi) 3.88×10^5
$\dfrac{d_o \rho_L U}{\mu_L}$	1.913×10^3–21.14×10^3	6.45×10^3	23.64×10^3
$\dfrac{\mu_L}{\mu_A}$	279–2235	750	
$\dfrac{\rho_L}{\rho_A}$	694–964	700	

terms, $A_p/D_s d_o$ and D_s/d_o, are very close to those in Eq. (5.32). In Eq. (5.33) these exponents are 0.52 and 0.23, as compared with 0.5 and 0.25 in Eq. (5.32).

Babu et al. [28] followed Eisenklam [23] and Dombrowski and Hassan [19] in assuming a vortex flow pattern within the nozzle of the form

$$vr^n = \text{constant} \qquad (5.34)$$

Their theoretical study, supplemented by analysis of experimental data, led to the following empirical expression for discharge coefficient:

$$C_D = \frac{K_{cd}}{[1/(1 - X)^2 + (\pi/4B)^2/X^n]^{0.5}} \qquad (5.35)$$

where

$$K_{cd} = 7.3423 \frac{A_o^{0.13735} A_s^{0.07782}}{A_p^{0.041066}}$$

$$B = \frac{A_p}{D_m d_o} \left(\frac{D_m}{d_o}\right)^{1-n}$$

$$n = 17.57 \frac{A_o^{0.1396} A_p^{0.2336}}{A_s^{0.1775}} \qquad \text{for } \Delta P_L > 2.76 \text{ MPa (400 psi)}$$

$$n = 28 \frac{A_o^{0.14176} A_p^{0.27033}}{A_s^{0.17634}} \qquad \text{for } \Delta P_L = 0.69 \text{ MPa (100 psi)}$$

The main advantage of Eq. (5.35) is that it is supported by a considerable body of experimental data obtained with several different atomizers. However, it takes no account of variations in the geometric parameters L_s/D_s and l_o/d_o. Furthermore, it demands a knowledge of the air core area A_a to calculate the value of X for insertion into Eq. (5.35). Such information is not always readily available.

Film Thickness

In pressure-swirl atomizers the liquid emerges from the nozzle as a thin conical sheet that rapidly attenuates as it spreads radially outward, finally disintegrating into ligaments and then drops. In the prefilming airblast atomizer the liquid is also spread out into a thin continuous sheet before being exposed to the high-velocity air. It is of interest, therefore, to examine the factors that govern the thickness of this liquid film.

Theory. For both pressure-swirl and airblast types of atomizers, it has long been recognized that the thickness of the annular liquid film produced at the nozzle exit has a strong influence on the mean drop size of the spray [29]. In pressure-swirl atomizers the thickness of the liquid film in the final orifice is directly related to the area of the air core. Giffen and Muraszew's analysis of the flow conditions within a simplex nozzle, assuming a nonviscous fluid, led to the relationship between atomizer dimensions and the size of the air core expressed in Eq. (5.25). Substituting in this equation for $K_1 = 4A_p/\pi D_s d_o$ gives

$$\left(\frac{A_p}{D_s d_o}\right)^2 = \frac{\pi^2}{32} \frac{(1-X)^3}{X^2} \tag{5.36}$$

where X is the ratio of the area of the air core to the area of the final discharge orifice.

After calculating X from Eq. (5.36), the corresponding value of liquid film thickness t can readily be derived, since from geometric considerations

$$X = \frac{(d_o - 2t)^2}{d_o^2} \tag{5.37}$$

In Eq. (5.27) the nozzle discharge coefficient is expressed as a unique function of X. In Eq. (5.32) the discharge coefficient is expressed in terms of nozzle dimensions. Combining these two equations yields the following expression, from which X and hence t can be calculated from the nozzle dimensions:

$$\frac{(1-X)^3}{1+X} = 0.09 \left(\frac{A_p}{D_s d_o}\right)\left(\frac{D_s}{d_o}\right)^{0.5} \tag{5.38}$$

Simmons and Harding [30] derived the following expression for t:

$$t = \frac{0.48FN}{d_o \cos \theta} \tag{5.39}$$

In Eq. (5.39) it should be noted that film thickness is expressed in micrometers, FN is the nozzle flow number in $(\text{lb/h})/(\text{psid})^{0.5}$ for standard calibrating fluid (MIL-C-7024II), d_o is the discharge orifice diameter in inches, and θ is half the included spray angle in degrees.

In SI units Eq. (5.39) becomes

$$t = \frac{0.00805 \sqrt{\rho_L}FN}{d_o \cos \theta} \tag{5.40}$$

The implication of Eqs. (5.36), (5.38), and (5.40) is that film thickness is independent of liquid viscosity and liquid injection pressure.

Rizk and Lefebvre [24] used a theoretical approach to investigate the internal flow characteristics of pressure-swirl atomizers. In particular, they examined the effects of atomizer dimensions and operating conditions on spray cone angle, velocity coefficient, and the thickness of the annular liquid film formed at the discharge orifice. A general expression for film thickness was derived in terms of atomizer dimensions, liquid properties, and liquid injection pressure as

$$t^2 = \frac{1560\dot{m}_L\mu_L}{\rho_L d_o \Delta P_L} \frac{1+X}{(1-X)^2} \tag{5.41}$$

or, by substituting $FN = \dot{m}_L/(\Delta P_L \rho_L)^{0.5}$

$$t^2 = \frac{1560 FN \mu_L}{\rho_L^{0.5} d_o \, \Delta P_L^{0.5}} \frac{(1 + X)}{(1 - X)^2} \tag{5.42}$$

Since X is dependent on t—in fact, $X = A_a/A_o = (d_o - 2t)^2/d_o^2$—some trial-and-error procedures are involved in the solution of Eqs. (5.41) and (5.42).

Rizk and Lefebvre [24] used Eq. (5.41) to calculate film thicknesses for different nozzle dimensions and operating conditions. Their results are shown in Figs. 5.15 to 5.19 as plots of film thickness against injection pressure differential. Also shown in these figures are the corresponding measured values of Kutty et al. [31, 32]. Theory and experiment both indicate that a higher nozzle pressure drop produces a thinner liquid sheet. Thus the improvement in atomization quality that always accompanies an increase in nozzle pressure drop, which is usually attributed to increase in liquid discharge velocity, may also be due in part to the decrease in film thickness caused by the increase in nozzle pressure drop, as shown in Figs. 5.15 to 5.19.

Increase in final orifice diameter leads to a thicker film, as shown in Fig. 5.15. This is because increase in d_o lowers C_D [see, for example, Eqs. (5.28) to (5.33)]. The variation of film thickness with liquid ports area is shown in Fig. 5.16. Increasing the inlet area raises the flow rate through the nozzle, which re-

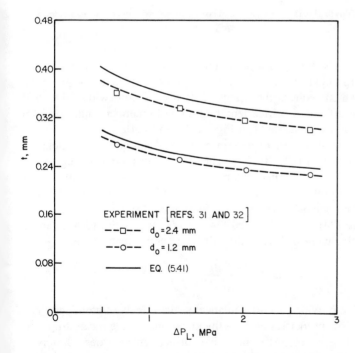

Figure 5.15 Variation of film thickness with injection pressure for different orifice diameters *(Rizk and Lefebvre [24])*.

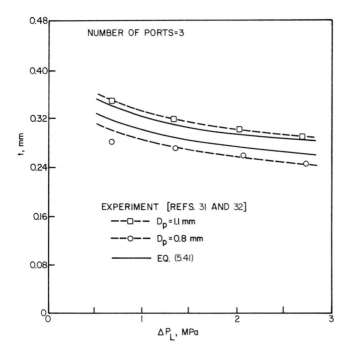

Figure 5.16 Variation of film thickness with injection pressure for different inlet port diameters *(Rizk and Lefebvre [24])*.

sults in a thicker film. The effect of a reduction in swirl chamber diameter is to increase the liquid film thickness, as shown in Fig. 5.17. This is attributed to the lower swirl action, which reduces the diameter of the air core within the final discharge orifice. The effects of orifice length and swirl chamber length on film thickness are quite small, as illustrated in Figs. 5.18 and 5.19.

Equation (5.41) was also used to determine values of liquid film thickness for the SMD data reported in [24, 30, 33]. The results are plotted in Fig. 5.20; they indicate that SMD $\propto t^{0.39}$. This exponent 0.39 is almost identical to that obtained by Rizk and Lefebvre [34] for prefilming airblast atomizers and also agrees closely with the conclusion reached by Simmons [35].

Rizk and Lefebvre [24] have suggested that for $t/d_o \ll 1$ Eq. (5.42) can be written more succinctly, while still retaining all its essential features, as

$$t = 3.66 \left[\frac{d_o F N \mu_L}{(\Delta P_L \rho_L)^{0.5}} \right]^{0.25} \tag{5.43}$$

The above equation provides useful guidance on the effects of atomizer characteristics and liquid flow properties on film thickness and on the mean drop size of the ensuing spray. It is of interest to note that surface tension does not appear in the above expression for film thickness although, of course, it does play a major role in the subsequent breakup of the liquid sheet into ligaments and drops. The

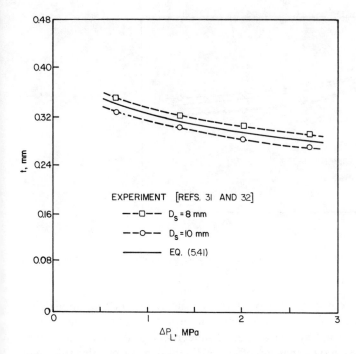

Figure 5.17 Variation of film thickness with injection pressure for different swirl chamber diameters *(Rizk and Lefebvre [24])*.

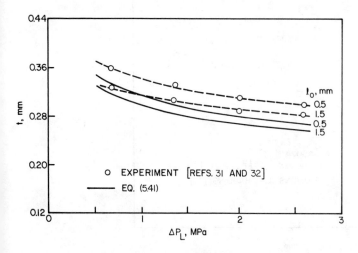

Figure 5.18 Variation of film thickness with injection pressure for different orifice lengths *(Rizk and Lefebvre [24])*.

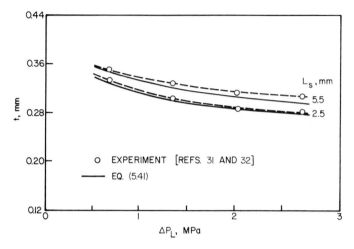

Figure 5.19 Variation of film thickness with injection pressure for different swirl chamber lengths *(Rizk and Lefebvre [24])*.

influence of liquid viscosity is clearly of major importance in the atomization process, because viscous forces impede atomization in two ways, by increasing the initial film thickness and by resisting the disintegration of the liquid sheet into drops. Equation (5.43) shows that the effect of liquid density on film thickness is quite small ($t \propto \rho_L^{-0.125}$). Thus its influence on atomization quality should also be small, and this is confirmed by the results of measurements of mean drop size for pressure-swirl atomizers [36].

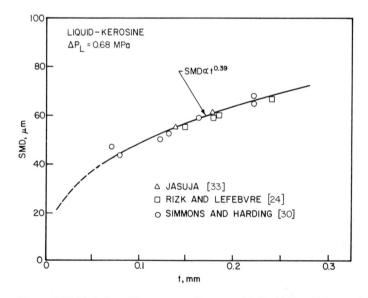

Figure 5.20 Variation of Sauter mean diameter with liquid film thickness *(Rizk and Lefebvre [24])*.

By making the substitution

$$FN = \frac{\dot{m}_L}{(\Delta P_L \rho_L)^{0.5}}$$

Eq. (5.43) becomes

$$t = 3.66 \left(\frac{d_o \dot{m}_L \mu_L}{\rho_L \Delta P_L} \right)^{0.25} \tag{5.44}$$

This equation indicates that liquid film thickness increases with increases in nozzle size, liquid flow rate, and liquid viscosity and diminishes with increase in liquid density and/or liquid injection pressure.

Experiment. Several attempts have been made to measure liquid film thickness in the final orifice of a pressure-swirl atomizer. In some experiments the film thickness was measured directly; in others it was obtained as half the difference between the measured air core diameter and the orifice diameter.

Kutty et al. [31] used a photographic technique to investigate the influence of liquid pressure differential on air core size. Photographs were taken with the camera pointing upstream through the nozzle outlet orifice. Illumination was achieved by fitting a transparent window at the rear of the swirl chamber. Air core diameters were measured from the negatives by enlarging them in a microfilm reader having a total magnification of 100.

Suyari and Lefebvre [37] used a method based on measurement of electrical conductance. The procedure is to flow water through the atomizer and measure the electrical conductance between two electrodes located in the discharge orifice. As the electrical conductivity of water is known, this measurement provides a direct indication of the average liquid film thickness in the flow path between the two electrodes.

The system is calibrated by flowing water through the nozzle and measuring electrical conductance with a plastic rod of low electrical conductivity inserted along the axis of the nozzle discharge orifice. By repeating this measurement with rods of different diameter, a calibration curve can be drawn to relate voltmeter reading to film thickness.

To examine the effect of variations in nozzle geometry on liquid film thickness, a number of inserts were produced to provide four different values of inlet port diameter, namely 0.749, 1.143, 1.245, and 1.346 mm. The results for liquid film thickness obtained with these different nozzle geometries are shown in Fig. 5.21, in which film thickness is plotted against water injection pressure differential for an ambient air pressure of 1 atm.

Comparison of theory and experiment. It is clearly of interest to compare the measured values of liquid film thickness with the corresponding predicted values from Eqs. (5.36), (5.38), (5.39), and (5.43). This is done in Fig. 5.22 for conditions of normal atmospheric air pressure and a liquid pressure differential of 0.69 MPA (100 psi).

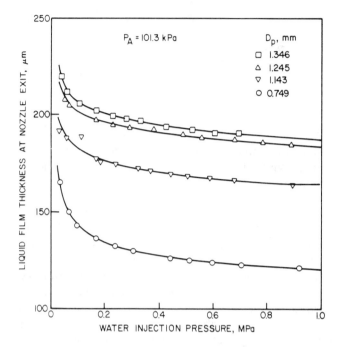

Figure 5.21 Influence of water injection pressure and inlet port diameter on liquid film thickness in final orifice *(Suyari and Lefebvre [37])*.

Inspection of Fig. 5.22 shows good agreement between the experimental data and values of t calculated using the simple inviscid relationship between atomizer dimensions and air core size provided by Giffen and Muraszew [4] [see Eq. (5.36)]. However, Eq. (5.38) provides the best fit to the experimental data. The equation proposed by Simmons and Harding [Eq. (5.39)] also demonstrates good agreement with the measured values.

The predictive capability of Eq. (5.43), due to Rizk and Lefebvre [24], is less satisfactory. This may be because the constant 3.66 in this equation was chosen to provide the best fit to the data of Kutty et al. [31]. Their photographic method measures the *maximum* film thickness to be found anywhere in the final orifice. The presence of ripples or undulations on the liquid surface could cause their measured values of t to be higher than those obtained by the conductance method, which measures the *average* film thickness in the final orifice. If the average values of t as determined by Suyari and Lefebvre [37] are used to obtain the constant in Eq. (5.43), it becomes

$$t = 2.7 \left[\frac{d_o F N \mu_L}{(\Delta P_L \rho_L)^{0.5}} \right] \quad (5.45)$$

This equation demonstrates a marked improvement in predictive capability, as indicated by the dashed line in Fig. 5.22, but the level of agreement with experimental data is still not as good as that obtained with Eq. (5.38).

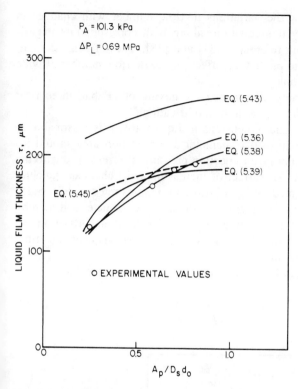

Figure 5.22 Comparison of measured values of film thickness with theoretical predictions *(Suyari and Lefebvre [37])*.

Flow Number

The flow number of an atomizer is easily determined. The liquid flow rate $\dot{m}_L$ is obtained by measuring the time required to pass a given volume (or mass) of liquid while maintaining the pressure differential across the nozzle at a constant value. Usually, the pressure differential is 0.69 MPa (100 psi) and the liquid is standard calibrating fluid (MIL-C-7024II), which has a density of 765 kg/m³ at room temperature. These values of $\dot{m}_L$, ΔP_L, and ρ_L are then substituted into Eq. (5.3) to obtain the flow number.

The influence of nozzle dimensions on flow number was studied experimentally by Kutty et al. [31, 38]. They manufactured a large number of simplex nozzles, all having three equispaced inlet ports, in a manner designed to elucidate the effect of variations in each of the key dimensions on nozzle flow characteristics. Rizk and Lefebvre's [24] analysis of their data yielded the following empirical expression for nozzle flow number:

$$FN = 0.0308\left(\frac{A_p^{0.5}d_o}{D_s^{0.45}}\right)$$

(5.46)

Figures 5.23 to 5.27 show the variations in flow number with changes in nozzle dimensions as given by this equation, along with corresponding experimental results for kerosine from references [31] and [38]. These figures demonstrate the ability of Eq. (5.46) to predict the effect on nozzle flow number of wide variations in atomizer dimensions.

Equation (5.46) should apply also to nozzles having other than three inlet ports, but the constant (0.0308) may require modification.

Considering the various nozzle dimensions in Eq. (5.46), it is reasonable to expect that higher values of final orifice diameter and inlet port area should increase the liquid flow rate for any given injection pressure differential, due to the increase in available flow area. The effect of increasing swirl chamber diameter in reducing flow rate, as indicated in Eq. (5.46) and demonstrated in Fig. 5.25, is due to the higher swirl action, which enlarges the air core diameter and thereby reduces the effective flow area of the final orifice. The frictional losses in the nozzle are relatively small, so that the lengths of the swirl chamber and final orifice have little effect on flow number.

Equation (5.46) can be modified to a dimensionally correct form as

$$FN = 0.395 \left(\frac{A_p^{0.5} d_o^{1.25}}{D_s^{0.25}} \right) \tag{5.47}$$

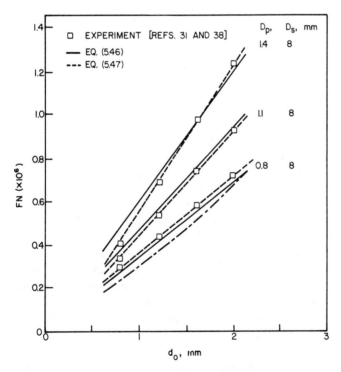

Figure 5.23 Influence of discharge orifice diameter on nozzle flow number *(Rizk and Lefebvre [24]).*

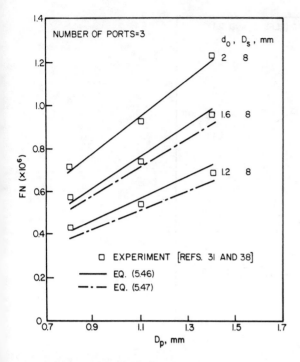

Figure 5.24 Influence of inlet port diameter on nozzle flow number *(Rizk and Lefebvre [24]).*

Although this equation does not predict the measured flow numbers to quite the same degree of accuracy as Eq. (5.46), the level of agreement between theory and experiment is still fairly close, as illustrated in Figs. 5.23 to 5.27.

Velocity Coefficient

This is defined as the ratio of the actual discharge velocity to the theoretical velocity corresponding to the total pressure differential across the nozzle, i.e.,

$$K_V = \frac{U}{(2\Delta P_L / \rho_L)^{0.5}} \tag{5.48}$$

Accurate values of K_V are essential for calculating the true velocity at which the liquid is discharged from the nozzle. This velocity is of prime importance to atomization, since it is the relative velocity between the initial liquid sheet and the surrounding air or gas that largely determines the mean drop size of the spray.

It has been shown [39] that K_V can be related to nozzle dimensions and spray cone angle by the expression

$$K_V = \frac{C_D}{(1 - X) \cos \theta} \tag{5.49}$$

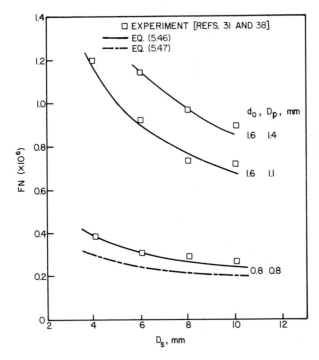

Figure 5.25 Influence of swirl chamber diameter on nozzle flow number *(Rizk and Lefebvre [24])*.

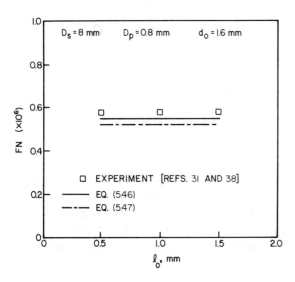

Figure 5.26 Influence of discharge orifice length on nozzle flow number *(Rizk and Lefebvre [24])*.

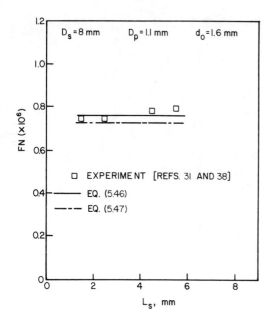

Figure 5.27 Influence of swirl chamber length on nozzle flow number *(Rizk and Lefebvre [24])*.

Rizk and Lefebvre [39] used the available experimental data on cone angle [31], air core size [32], and liquid flow rate [38] to calculate the velocity of the liquid discharging from the nozzle for different atomizer dimensions and different liquid pressure differentials. The ratio of the theoretical pressure drop corresponding to the discharge velocity to the actual measured pressure drop across the nozzle is plotted against final orifice diameter in Fig. 5.28. The figure shows that the ratio of "effective" injection pressure to the actual injection pressure decreases rapidly with increase in orifice diameter. This suggests that for two atomizers of similar size and geometry, differing only in final orifice diameter, the smaller one will produce a higher discharge velocity for the same pressure drop. It is also clear from Fig. 5.28 that with increase in pressure differential the atomizer becomes more efficient in converting the available pressure drop into discharge velocity.

Equation (5.49) was also used to examine the effect on K_V of variations in nozzle dimensions. Calculated values of film thickness, cone angle, and discharge coefficient from Eqs. (5.44), (7.18), and (5.32), respectively, were used in conjunction with measured values of liquid flow rate and nozzle pressure differential to produce the results shown in Figs. 5.29 to 5.33. Figure 5.29 shows that increase in orifice diameter causes a reduction in velocity coefficient. This is because the beneficial effect on K_V of increasing X is outweighed by the corresponding reduction in C_D. The improvement in velocity coefficient at higher levels of ΔP_L is due to the reduction in film thickness.

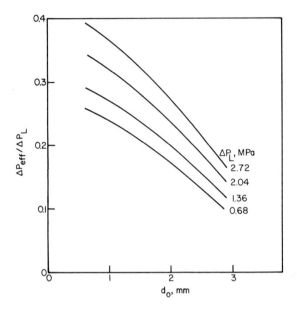

Figure 5.28 Influence of final orifice diameter on effective injection pressure *(Rizk and Lefebvre [39])*.

The higher values of K_V that accompany increases in inlet port diameter, as illustrated in Fig. 5.30, are attributed to higher discharge coefficients, which more than compensate for increases in liquid film thickness. Figure 5.31 illustrates the variation of K_V with swirl chamber diameter. It shows that K_V declines rapidly at first with increase in D_S, but the curves then level off to fairly constant values. These effects can also be explained in the light of Eqs. (5.32) and (5.49).

The variation of velocity coefficient with atomizer constant K is illustrated in Figs. 5.32 and 5.33. The curves drawn in Fig. 5.32 are the result of calculations, while those in Fig. 5.33 are based on actual experimental data. Both figures show that higher values of K and ΔP_L yield higher values of K_V. Comparison of the two figures helps to confirm the validity of Eq. (5.49) for calculating K_V, agreement between theory and experiment being generally satisfactory, especially for low values of K and ΔP_L.

It is well known that viscosity forces impair spray quality by opposing the disintegration of the liquid jets and sheets produced by various types of atomizers, but less attention has been paid to the other adverse effect of viscosity on atomization, which stems from its strong influence on the flow processes occurring within the nozzle itself. Calculated results on the effect of viscosity on K_V [40] show that K_V is markedly reduced by an increase in liquid viscosity, due to the combined effects of increase in film thickness and reduction in cone angle. However, when the film thickness has developed to such an extent that the air core diameter is quite small, changes in X and θ become less significant than increase in discharge coefficient. This causes the decline of K_V to slow down; in fact, K_V

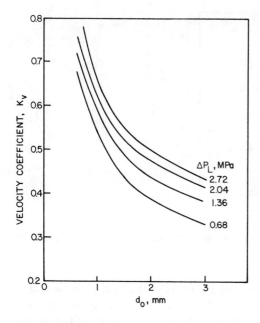

Figure 5.29 Influence of final orifice diameter on velocity coefficient *(Rizk and Lefebvre [39])*.

even rises a little until a level of viscosity is reached at which the discharge orifice is running full. At this condition K_V has a value of around 0.34, regardless of ΔP_L.

This study by Rizk and Lefebvre [40] clearly demonstrates that increase in liquid viscosity impedes atomization, not only by opposing the breakup of the liquid sheet into drops but also by thickening the liquid film in the final orifice

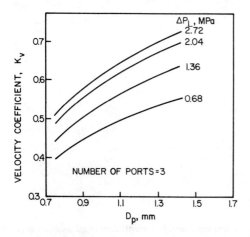

Figure 5.30 Effect of inlet port diameter on velocity coefficient *(Rizk and Lefebvre [39])*.

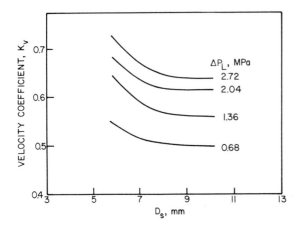

Figure 5.31 Effect of swirl chamber diameter on velocity coefficient *(Rizk and Lefebvre [39])*.

and by reducing the efficiency of the processes whereby the nozzle pressure differential is converted into kinetic energy at the nozzle exit. The results of their analysis show that the main factors governing velocity coefficient are nozzle geometry, nozzle pressure differential, and the liquid properties of density and viscosity. It is found that values of the velocity coefficient, as calculated for wide ranges of atomizer dimensions and liquid properties, conform closely to the empirical expression

$$K_V = 0.00367 K^{0.29}\left(\frac{\Delta P_L \rho_L}{\mu_L}\right)^{0.2} \tag{5.50}$$

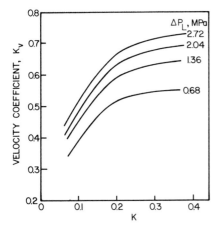

Figure 5.32 Calculated variation of velocity coefficient with atomizer coefficient *(Rizk and Lefebvre [39])*.

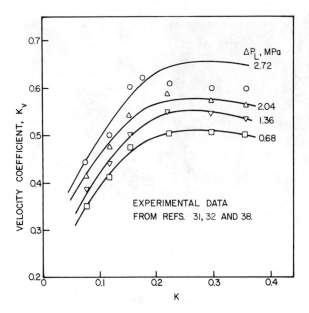

Figure 5.33 Experimental data on variation of velocity coefficient with atomizer constant *(Rizk and Lefebvre [39])*.

ROTARY ATOMIZER

With this device, liquid is supplied to the center of a rotating disk or cup. The friction between the liquid and the cup wall causes the liquid to rotate at roughly the same speed as the cup. This rotary motion creates centrifugal forces within the liquid that induce it to flow radially outward toward the rim of the cup. If the rotational speed of the cup is sufficiently high, the liquid will arrive at the rim in a thin continuous film. The manner in which this film disintegrates into drops will depend on the size and geometry of the cup, its rotational speed, the liquid flow rate, and the physical properties of the liquid [41–44].

It is instructive to consider what happens when the liquid flow rate to a cup rotating at constant speed is gradually increased from zero. At low flow rates the liquid spreads out across the surface and is centrifuged off in the form of drops of sensibly uniform size. This phenomenon is generally known as *direct drop formation*. If the flow rate is increased, ligaments are formed along the periphery and then disintegrate into drops. The number of ligaments increases with increasing flow rate up to a maximum value, beyond which it remains constant, regardless of flow rate. The thickness of the ligaments also increases with flow rate in this mode of atomization [45]. The ligaments themselves are unstable and disintegrate into drops at some distance from the periphery, as illustrated in Fig. 5.34. This process is usually termed *atomization by ligament formation*.

With continuing increase in flow rate a condition is finally reached where the ligaments have attained their maximum number and size and can no longer ac-

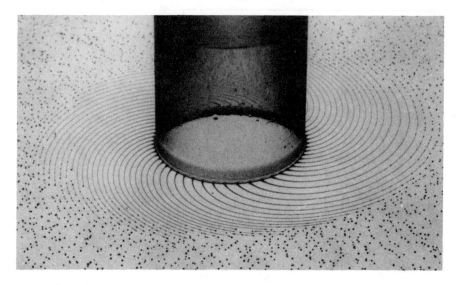

Figure 5.34 Illustration of ligament formation and disintegration into drops *(Courtesy of P. Eisenklam [55]).*

commodate the flow of liquid. A thick film is produced that extends outward to some distance beyond the cup rim. When this film disintegrates into ligaments, it does so in an irregular manner, which results in an appreciable variation in drop size. This process is commonly referred to as *atomization by film formation.*

Critical Flow Rates

Empirical formulas for calculating the critical flow rates corresponding to the transition from one mode of atomization to another have been provided by Tanasawa et al. [46], as follows.

Direct drop formation:

$$q \leq 2.8 \left(\frac{D}{n}\right)^{2/3} \left(\frac{\sigma}{\rho_L}\right) \left[1 + 10\left(\frac{\mu_L}{(\rho_L \sigma D)^{0.5}}\right)^{1/3}\right]^{-1} \tag{5.51}$$

Ligament formation:

$$q = 80 \left(\frac{D}{n}\right)^{2/3} \left(\frac{\sigma}{\rho_L}\right) \left[1 + 10\left(\frac{\mu_L}{(\rho_L \sigma D)^{0.5}}\right)^{1/3}\right]^{-1} \tag{5.52}$$

Film formation:

$$q \geq 5.3 \left(\frac{D}{n}\right)^{2/3} \left(\frac{\sigma}{\rho_L}\right)\left(\frac{\rho_L}{\mu_L}\right)^{1/3} \quad \text{for } \frac{D\rho_L}{\mu_L} < 30 \text{ s/cm} \tag{5.53}$$

$$q \cong 20(D)^{1/2}\left(\frac{1}{n}\right)^{2/3}\left(\frac{\sigma}{\rho_L}\right)^{5/6} \quad \text{for } \frac{D\rho_L}{\mu_L} > 30 \text{ s/cm} \tag{5.54}$$

where q = volume flow rate, cm^3/s
D = disk diameter, cm
n = rotational speed, rpm
s = pitch of ligaments, cm
μ_L = dynamic viscosity of liquid, dyn/cm^2
v_L = kinematic viscosity of liquid, cm^2/s
ρ_L = liquid density, g/cm^3
σ = surface tension, dyn/cm

Film Thickness

Hinze and Milborn [41] analyzed the radial velocity distribution in a thin film on a rotating cup by simply equating the viscous and centrifugal forces. Emslie et al. [47] carried out a similar analysis and included the effects of gravity. Oyama and Endou [48] derived expressions for both radial and tangential velocity distributions. Recently, more exact solutions have been developed [49]. Nikolaev et al. [50] solved the equations of motion for a liquid film on a rotating disk from considerations of the difference between the angular velocity of the film and that of the disk. Bruin [51] obtained simpler expressions for the velocity profiles by using the complex function method to solve the simplified Navier-Stokes equations. Matsumoto et al. [52] predicted the velocity profiles in a film on a rotating disk by numerical integration of nonlinear differential equations deduced from the Navier-Stokes equations.

Matsumoto et al. [49] have compared the solutions by Nikolaev et al. [50], Bruin [51], and Matsumoto et al. [52] for the radial velocity distribution for flow on flat disks. Their results, shown in Fig. 5.35, indicate only small differences between these solutions. Although the lag between the angular velocities of the liquid film and the rotating disk is usually considered quite large, these results suggest that the lag is not more than around 20%.

Theoretical solutions for the film thickness predicted by these results are compared with experimental data in Fig. 5.36. The level of agreement between these different solutions is clearly satisfactory.

Toothed Designs

It has been found that grooves cut into the wall of the cup can help to guide and regulate the flow as it is centrifugally forced to the outside. Additional benefits can be gained by placing teeth on the outer rim of the cup. Such teeth can help to maintain ligament drop formation and avoid film atomization (see Fig. 4.35). The optimum number of teeth depends on cup size, rotational velocity, and liquid

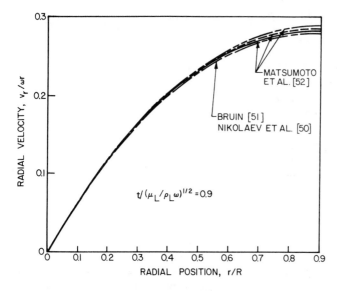

Figure 5.35 Predicted radial velocity distributions from flow on flat disks *(Matsumoto et al. [49])*.

properties. According to Christensen and Steely [45], the optimum number of teeth may be calculated from the expression

$$z = 0.215 \left(\frac{\rho_L \omega^2 D^3}{\sigma} \right)^{5/12} \left(\frac{\rho_L \sigma D}{\mu_L^2} \right)^{1/6} \tag{5.55}$$

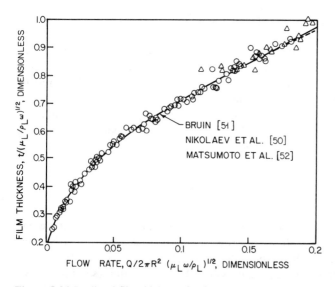

Figure 5.36 Predicted film thickness for flow on flat disks *(Matsumoto et al. [49])*.

This equation shows that the number of teeth needed to maintain ligament formation increases with increases in cup size and rotational velocity and is diminished by increases in surface tension and viscosity. Table 5.2 illustrates the number oı teeth needed as calculated from Eq. (5.55).

Once the liquid reaches the teeth, the various modes of atomization discussed earlier can occur. These modes are illustrated in Fig. 5.37. They include direct drop formation from the teeth, combined drop and ligament formation, ligament formation, and film formation. This figure illustrates that "monodisperse" sprays can only be obtained with direct drop formation or ligament formation.

AIRBLAST ATOMIZER

For plain-jet airblast atomizers the internal hydrodynamics are essentially the same as those of the plain-orifice pressure atomizer. With prefilming airblast nozzles the situation is more complex, but the key factors involved in the attainment of good atomization have been known for some time. From detailed experimental studies carried out on several different atomizer configurations, Lefebvre and Miller [53] identified the following essential features of successful nozzle design.

1. The liquid should first be spread into a thin continuous sheet of uniform thickness.
2. Finest atomization is obtained by producing a liquid sheet of minimum thickness. In practice, this means that the atomizing lip should have the largest possible diameter.
3. The annular liquid sheet formed at the atomizing lip should be exposed on both sides to air at the highest possible velocity. Thus, the atomizer should be designed to achieve minimum loss of total pressure in the airflow passages and maximum air velocity at the atomizing lip.

In most prefilming nozzles the liquid flows through a number of equispaced

Table 5.2 Number of teeth required to maintain ligament formation as a function of rotational velocity and disk diameter [45]

Diameter (cm)	Rotational velocity (rpm)	No. teeth	No. teeth per centimeter
5	1000	37.6	2.4
5	3000	94.6	6.0
5	5000	145.4	9.3
10	1000	56.6	1.8
10	3000	142.5	4.5
10	5000	219.0	7.0
20	1000	85.2	1.4
20	3000	214.5	3.4
20	5000	330.0	5.2

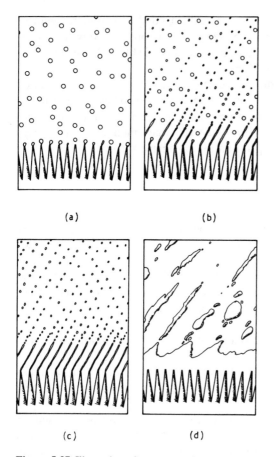

(a) (b)

(c) (d)

Figure 5.37 Illustration of rotary atomization, using a toothed rim. (a) Direct formation; (b) direct and ligament formation; (c) ligament formation; (d) sheet formation. *(Christensen and Steely [45].)*

tangential ports into a weir, from which it spreads over a prefilming surface before being discharged at the atomizing lip. This arrangement is shown schematically in Fig. 5.38. The purpose of the weir is to protect the liquid from the high-velocity air until the discrete jets of liquid flowing into the weir have lost their separate identities and have merged to form a fully annular reservoir behind the weir. As the liquid flows out of this reservoir, it is exposed to the high-velocity airstream, which causes it to spread out over the prefilming surface to form a thin continuous film at the atomizing lip. Normally, this film retains a significant proportion of its tangential velocity component, so when it flows over the prefilming lip it is flung radially outward to form a hollow conical spray. This "natural" spray cone is, of course, observed only in the absence of atomizing air. Under normal operating conditions, the liquid film produced at the prefilmer lip is rapidly disintegrated by the atomizing air, which dictates the effective cone angle of the ensuing spray. Nevertheless, a wide natural spray angle is a desirable design feature

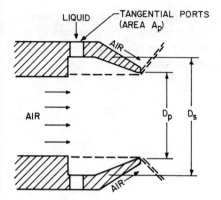

Figure 5.38 Schematic diagram of prefilmer.

because it increases the relative velocity between the liquid and the atomizing air and also increases the interaction between the liquid and the outer airstream. This is one of the reasons why it is customary, in the design of prefilming airblast atomizers, to arrange for the outer airflow rate to exceed the inner airflow rate.

Theoretical values of discharge coefficient for prefilming airblast atomizers are plotted against A_p/D_sD_p in Fig. 5.39. The values shown are the results of

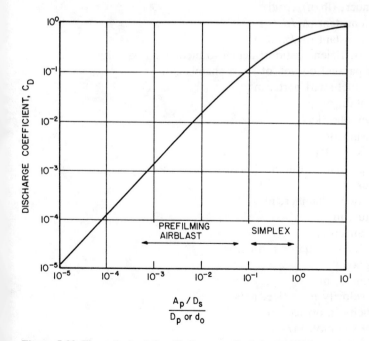

Figure 5.39 Theoretical relationship between discharge coefficient and atomizer dimensions *(Simmons [54]).*

calculations carried out by Simmons [54] using Eqs. (5.25) and (5.26), due to Giffen and Maraszew [4]. It is clear from this figure that typical values of the discharge coefficient for prefilming airblast nozzles are appreciably lower than those normally associated with pressure-swirl nozzles.

NOMENCLATURE

A_a	air core area, m^2
A_p	total inlet ports area, m^2
A_o	discharge orifice area, m^2
A_s	swirl chamber area, m^2
C	cavitation number
C_D	discharge coefficient
D	disk diameter, m
D_m	diametral distance between centerlines of inlet ports (see [28])
D_p	inlet port diameter, m
D_s	swirl chamber diameter, m
d_o	discharge orifice diameter, m
FN	flow number $[\dot{m}_L/(\Delta P_L \rho_L)^{0.5}]$, m^2
FN_{UK}	flow number, (UK gal/h)/(psid)$^{0.5}$
FN_{US}	flow number, (lb/h)/(psid)$^{0.5}$
K	atomizer constant $(A_p/d_o D_s)$
K_1	atomizer constant $(A_p/\pi r_o R_s)$
K_v	velocity coefficient, ratio of actual to theoretical velocity
L_s	length of parallel portion of swirl chamber, m
L_p	length of inlet swirl ports, m
l_o	orifice length, m
$\dot{m}_L$	liquid flow rate, kg/s
P	total pressure, Pa
p	static pressure, Pa
ΔP_L	injection pressure differential across nozzle, Pa
R	disk radius, m
R_s	radius of swirl chamber, m
r	local radius, m
r_a	radius of air core, m
r_o	radius of final discharge orifice, m
SG	specific gravity
t	film thickness, m
U	resultant velocity in orifice, m/s
u	axial velocity in orifice, m/s
v	local flow velocity, m/s
v_i	inlet velocity to swirl chamber, m/s
v_r	radial velocity of liquid, m/s

X	A_a/A_o
z	number of teeth
μ_L	liquid dynamic viscosity, kg/m s
ν_L	liquid kinematic viscosity, m^2/s
ρ_L	liquid density, kg/m^3
σ	surface tension, kg/s^2
ω	rotational speed, radian/s
θ	spray cone half angle, degrees

REFERENCES

1. Bird, A. L., Some Characteristics of Nozzles and Sprays for Oil Engines, *Trans. Second World Power Conference,* Berlin, Vol. 8, Section 29, No. 82, 1930, p. 260.
2. Gellales, A. G., Effect of Orifice Length/Diameter Ratio on Fuel Sprays for Compression Ignition Engines, NACA Report No. 402, 1931.
3. Schweitzer, P. H., Mechanism of Disintegration of Liquid Jets, *J. Appl. Phys.,* Vol. 8, 1937, pp. 513–521.
4. Giffen, E., and Muraszew, A., *Atomization of Liquid Fuels,* Chapman & Hall, London, 1953.
5. Bergwerk, W., Flow Pattern in Diesel Nozzle Spray Holes, *Proc. Inst. Mech. Eng.,* Vol. 173, No. 25, 1959, pp. 655–660.
6. Spikes, R. H., and Pennington, G. A., Discharge Coefficient of Small Submerged Orifices, *Proc. Inst. Mech. Eng.,* Vol. 173, No. 25, 1959, pp. 661–665.
7. Lichtarowicz, A., Duggins, R. K., and Markland, E., Discharge Coefficients for Incompressible Non-cavitating Flow Through Long Orifices, *J. Mech. Eng. Sci.,* Vol. 7, No. 2, 1965, pp. 210–219.
8. Arai, M., Shimizu, M., and Hiroyasu, H., Breakup Length and Spray Angle of High Speed Jet, *Proceedings of the 3rd International Conference on Liquid Atomization and Spray Systems (ICLASS),* London, 1985, pp. IB/4/1–10.
9. Zucrow, M. J., Discharge Characteristics of Submerged Jets, Bull. No. 31, Engineering Experimental Station, Purdue University, West Lafayette, Ind., 1928.
10. Nakayama, Y., Action of the Fluid in the Air Micrometer: First Report, Characteristics of Small Diameter Nozzle and Orifice, *Bull. Jpn. Soc. Mech. Eng.,* Vol. 4, 1961, pp. 516–524.
11. Asihmin, V. I., Geller, Z. I., and Skobel'cyn, Yu. A., Discharge of a Real Fluid from Cylindrical Orifices (in Russian), *Oil Ind.,* Vol. 9, Moscow, 1961.
12. Ruiz, F., and Chigier, N., The Mechanics of High Speed Cavitation, *Proceedings of the 3rd International Conference on Liquid Atomization and Spray Systems,* London, 1985, pp. VIB/3/1–15.
13. Varde, K. S., and Popa, D. M., Diesel Fuel Spray Penetration at High Injection Pressure, SAE Paper 830448, 1984.
14. Watson, E. A., unpublished report, Joseph Lucas Ltd., London, 1947.
15. Taylor, G. I., The Mechanics of Swirl Atomizers, *Seventh International Congress of Applied Mechanics,* Vol. 2, Pt. 1, 1948. pp. 280–285.
16. Taylor, G. I., The Boundary Layer in the Converging Nozzle of a Swirl Atomizer, *Q. J. Mech. Appl. Math.,* Vol. 3, Pt. 2, 1950, pp. 129–139.
17. Binnie, A. M., and Harris, D. P., The Application of Boundary Layer Theory to Swirling Liquid Flow Through a Nozzle, *Q. J. Mech. Appl. Math.,* Vol. 3, Pt. 1, 1950, pp. 80–106.
18. Hodgekinson, T. G., Porton Technical Report No. 191, 1950.
19. Dombrowski, N., and Hassan, D., The Flow Characteristics of Swirl Centrifugal Spray Pressure Nozzles with Low Viscosity Liquids, *AIChE J.,* Vol. 15, 1969, p. 604.
20. Jones, A. R., Design Optimization of a Large Pressure-Jet Atomizer for Power Plant, *Proceedings of the 2nd International Conference on Liquid Atomization and Spray Systems,* Madison, Wis., 1982, pp. 181–185.

21. Radcliffe, A., The Performance of a Type of Swirl Atomizer, *Proc. Inst. Mech. Eng.*, Vol. 169, 1955, pp. 93–106.
22. Carlisle, D. R., Communication on The Performance of a Type of Swirl Atomizer, by A. Radcliffe, *Proc. Inst. Mech. Eng.*, Vol. 169, 1955, p. 101.
23. Eisenklam, P., Atomization of Liquid Fuels for Combustion, *J. Inst. Fuel*, Vol. 34, 1961, pp. 130–143.
24. Rizk, N. K., and Lefebvre, A. H., Internal Flow Characteristics of Simplex Swirl Atomizers, *AIAA. J. Propul. Power*, Vol. 1, No. 3, 1985, pp. 193–199.
25. Tipler, W., and Wilson, A. W., Combustion in Gas Turbines, Paper B9, *Proceedings of the Congress International des Machines a Combustion (CIMAC)*, Paris, 1959, pp. 897–927.
26. Elkotb, M. M., Rafat, N. M., and Hanna, M. A., The Influence of Swirl Atomizer Geometry on the Atomization Performance, *Proceedings of the 1st International Conference on Liquid Atomization and Spray Systems*, Tokyo, 1978, pp. 109–115.
27. Joyce, J. R., Report ICT15, Shell Research Ltd., London, 1947.
28. Babu, K. R., Narasimhan, M. V., and Narayanaswamy, K., Correlations for Prediction of Discharge Rate, Cone Angle, and Air Core Diameter of Swirl Spray Atomizers, *Proceedings of the 2nd International Conference on Liquid Atomization and Spray Systems*, Madison, Wis., 1982, pp. 91–97.
29. Lefebvre, A. H., *Gas Turbine Combustion*, Hemisphere, Washington, D.C., 1983.
30. Simmons, H. C., and Harding, C. F., Some Effects of Using Water as a Test Fluid in Fuel Nozzle Spray Analysis, ASME Paper 80-GT-90, presented at ASME Gas Turbine Conference, New Orleans, 1980.
31. Sankaran Kutty, P., Narasimhan, M. V., and Narayanaswamy, K., Design and Prediction of Discharge Rate, Cone Angle, and Air Core Diameter of Swirl Chamber Atomizers, *Proceedings of the 1st International Conference on Liquid Atomization and Spray Systems*, Tokyo, 1978, pp. 93–100.
32. Narasimhan, M. V., Sankaran Kutty, P., and Narayanaswamy, K., Prediction of the Air Core Diameter in Swirl Chamber Atomizers, unpublished report, 1978.
33. Jasuja, A. K., Atomization of Crude and Residual Fuel Oils, *ASME J. Eng. Power*, Vol. 101, No. 2, 1979, pp. 250–258.
34. Rizk, N. K., and Lefebvre, A. H., Influence of Liquid Film Thickness on Airblast Atomization, *ASME J. Eng. Power*, Vol. 102, No. 3, 1980, pp. 706–710.
35. Simmons, H. C., The Prediction of Sauter Mean Diameter for Gas Turbine Fuel Nozzles of Different Types, *ASME J. Eng. Power*, Vol. 102, No. 3, 1980, pp. 646–652.
36. Rizk, N. K., and Lefebvre, A. H., Spray Characteristics of Simplex Swirl Atomizers, *Prog. Aeronaut. Astronaut.*, Vol. 95, 1985, pp. 563–580.
37. Suyari, M., and Lefebvre, A. H., Film Thickness Measurements in a Simplex Swirl Atomizer, *AIAA J. Propul. Power*, Vol. 2, No. 6, 1986, pp. 528–533.
38. Sankaran Kutty, P., Narasimhan, M. V., and Narayanaswamy, K., Prediction of the Coefficient of Discharge of Swirl Chamber Atomizers, unpublished work, 1978.
39. Rizk, N. K., and Lefebvre, A. H., Prediction of Velocity Coefficient and Spray Cone Angle for Simplex Swirl Atomizers, *Proceedings of the 3rd International Conference on Liquid Atomization and Spray Systems*, London, 1985, pp. III C/2/1–16.
40. Rizk, N. K., and Lefebvre, A. H., Influence of Liquid Properties on the Internal Flow Characteristics of Simplex Swirl Atomizers, *Atomization Spray Technol.*, Vol. 2, No. 3, 1986, pp. 219–233.
41. Hinze, J. O., and Milborn, H., Atomization of Liquids by Means of a Rotating Cup, *J. Appl. Mech.*, Vol. 17, No. 2, 1950, pp. 145–153.
42. Adler, C. R., and Marshall, W. R., Performance of Spinning Disk Atomizers, *Chem. Eng. Prog.*, Vol. 47, 1951, pp. 515–601.
43. Fraser, R. P., Dombrowski, N., and Routley, J. H., The Production of Uniform Liquid Sheets from Spinning Cups; The Filming by Spinning Cups; The Atomization of a Liquid Sheet by an Impinging Air Stream, *Chem. Eng. Sci.*, Vol. 18, 1963, pp. 315–321, 323–337, 339–353.
44. Walton, W. H., and Prewitt, W. C., The Production of Sprays and Mists of Uniform Drop Size

by Means of Spinning Disc Type Sprayers, *Proc. Phys. Soc.,* Vol. 62, Pt. 6, June 1949, pp. 341–350.

45. Christensen, L. S., and Steely, S. L., Monodisperse Atomizers for Agricultural Aviation Applications, NASA CR-159777, February 1980.

46. Tanasawa, Y., Miyasaka, Y., and Umehara, M., Effect of Shape of Rotating Disks and Cups on Liquid Atomization, *Proceedings of the 1st International Conference on Liquid Atomization and Spray Systems,* Tokyo, 1978, pp. 165–172.

47. Emslie, A. G., Benner, F. T., and Peck, L. G., *J. Appl. Phys.,* Vol. 29, 1958, p. 858.

48. Oyama, Y., and Endou, K., On the Centrifugal Disk Atomization and Studies on the Atomization of Water Droplets, *Kagaku Kogaku,* Vol. 17, 1953, pp. 256–260, 269–275 (in Japanese, English summary).

49. Matsumoto, S., Belcher, D. W., and Crosby, E. J., Rotary Atomizers: Performance Understanding and Prediction, *Proceedings of the 3rd International Conference on Liquid Atomization and Spray Systems,* London, 1985, pp. IA/1/1–21.

50. Nikolaev, V. S., Vachagin, K. D., and Baryshev, Yu. N., *Int. Chem. Eng.,* Vol. 7, 1967, p. 595.

51. Bruin, S., Velocity Distributions in a Liquid Film Flowing Over a Rotating Conical Surface, *Chem. Eng. Sci.,* Vol. 24, No. 11, 1969, p. 1647–1654.

52. Matsumoto, S., Saito, K., and Takashima, Y., *Bull Tokyo Inst. Technol.,* Vol. 116, 1973, p. 85.

53. Lefebvre, A. H., and Miller, D., The Development of an Air Blast Atomizer for Gas Turbine Application, CoA-Report-AERO-193, College of Aeronautics, Cranfield, Bedford, England, June 1966.

54. Simmons, H. C., Parker Hannifin Report, BTA 136, 1981.

55. Eisenklam, P., On Ligament Formation from Spinning Disks and Cups, *Chem. Eng. Sci.,* Vol. 19, 1964, pp. 693–694.

ATOMIZER PERFORMANCE

INTRODUCTION

The spray properties of importance include mean drop size, drop size distribution, radial and circumferential patternation, droplet number density, cone angle, and penetration. The *quality* or *fineness* of an atomization process is usually described in terms of a mean drop size. Various definitions of mean drop size are available (see Chapter 3), of which the most widely used is the Sauter mean diameter, which represents the volume/surface ratio of the spray. This definition of mean drop size has special significance for heat and mass transfer applications, such as spray drying and the combustion of liquid fuel sprays.

Unfortunately, the physical processes involved in atomization are not yet sufficiently well understood for mean diameters to be expressed in terms of equations derived from basic principles. The simplest case of the breakup of a liquid jet has been studied theoretically for more than a hundred years, but the results of these studies have failed to predict the spray characteristics to a satisfactory level of accuracy. The situation in regard to the complex sprays produced by more sophisticated types of atomizers is, understandably, even worse. As the physical structure and dynamics of a spray are the result of many interwoven complex mechanisms, none of which are fully understood, it is hardly surprising that mathematical treatments of atomization have so far defied successful development. In consequence, the majority of investigations into the drop size distributions produced in atomization have, of necessity, been empirical in nature. Nevertheless, they have yielded a considerable body of useful information from which a number of general conclusions on the effects of liquid properties, gas properties, and injector dimensions on mean drop size can be drawn.

The properties of a liquid most relevant to atomization are surface tension, viscosity, and density. For a liquid injected into a gaseous medium, the only thermodynamic property generally considered of importance is the gas density. The liquid flow variables of importance are the velocity of the liquid jet or sheet and the turbulence in the liquid stream. The gas flow variables to be considered are the absolute velocity and the relative gas-to-liquid velocity. The importance of this parameter is unfortunate, since the actual velocity field in the atomization region cannot be determined accurately. Even when liquid is sprayed into so-called quiescent air, the air velocity adjacent to the spray can attain quite high values due to the momentum transfer from the liquid to the surrounding air [1]. The turbulence characteristics of the air or gas may also influence atomization, but no systematic study of this effect has yet been undertaken.

For plain-orifice injectors, the key geometric variables are the orifice length and diameter. Final orifice diameter is also of prime importance for pressure-swirl atomizers. For prefilming-type airblast atomizers, the dimensions that have most influence on mean drop size are the prefilmer diameter and the hydraulic mean diameter of the atomizer air duct at the exit plane. For rotary atomizers, the only dimension of significance is the diameter of the rotating disk or cup. For all types of pneumatic or twin-fluid injectors, another variable affecting atomization is the mass ratio of liquid to air, or steam.

The absence of any general theoretical treatment of the atomization process has led to the evolution of empirical equations to express the relationship between the mean drop size in a spray and the variables of liquid properties, gas properties, flow conditions, and atomizer dimensions. Many of the drop size equations published in the literature are of somewhat dubious value. In some cases they are based on data obtained using experimental techniques that are no longer considered valid. With other equations, the insertion of appropriate values of gas and liquid properties yields improbable values of SMD. The equations selected for inclusion in this chapter are considered to be the best available for engineering calculations of mean drop sizes for the types of atomizers described in Chapter 4.

PLAIN–ORIFICE ATOMIZER

In plain-orifice atomizers, disintegration of the jet into drops is promoted by an increase in flow velocity, which increases both the level of turbulence in the issuing jet and the aerodynamic drag forces exerted by the surrounding medium; it is opposed by an increase in liquid viscosity, which delays the onset of atomization by resisting breakup of the ligaments. Merrington and Richardson's [2] experiments on sprays injected from a plain circular orifice into stagnant air yielded the following relationship for mean drop size.

$$SMD = \frac{500 d_o^{1.2} v_L^{0.2}}{U_L} \qquad (6.1)$$

Most of the research carried out on plain-orifice atomizers has been directed toward the types of injectors employed in compression ignition (diesel) engines. With these injectors, jet breakup is due mainly to aerodynamic interaction with a highly turbulent jet. In an early study, Panasenkov [3] examined the influence of turbulence on the breakup of a liquid jet and determined mean drop sizes for jet Reynolds numbers ranging from 1000 to 12,000. Drop sizes were correlated in terms of discharge orifice diameter and liquid Reynolds number as

$$\text{MMD} = 6d_o\text{Re}_L^{-0.15} \tag{6.2}$$

Harmon's [4] equation for SMD takes account of ambient gas properties as well as liquid properties. We have

$$\text{SMD} = 3330d_o^{0.3}\mu_L^{0.07}\rho_L^{-0.648}\sigma^{-0.15}U_L^{-0.55}\mu_G^{0.78}\rho_G^{-0.052} \tag{6.3}$$

An unusual feature of this equation is that an increase in surface tension is predicted to give finer atomization.

From their study on the effects of ambient gas density on drop size distributions, Giffen and Lamb [5] drew the following conclusions:

1. The fineness and uniformity of a spray are improved by an increase in gas density, but the rate of improvement decreases at high gas densities.
2. Gas density appears to exert little effect on the minimum drop size in the spray.
3. The maximum drop size is reduced at higher gas densities according to the relationship $D_{\max} \propto \rho_G^{-0.2}$.
4. The improvement in atomization with increase in gas density is due mainly to a reduction in the number of large drops and is probably caused by subdivision of these big drops.

In 1955 Miesse [6] published his analysis of the results of previous theoretical and experimental studies on the disintegration of liquid jets. His equation for the "best-fit" correlation of the available experimental data is

$$D_{0.999} = d_o\text{We}_L^{-0.333}(23.5 + 0.000395\text{Re}_L) \tag{6.4}$$

At about the same time, Tanasawa and Toyoda [7] proposed the following equation:

$$\text{SMD} = 47d_oU_L^{-1}\left(\frac{\sigma}{\rho_G}\right)^{0.25}\left[1 + 331\frac{\mu_L}{(\rho_L\sigma d_o)^{0.5}}\right] \tag{6.5}$$

More recent equations for the mean drop sizes produced by the diesel-type injectors are that of Hiroyasu and Katoda [8], which is of the form

$$\text{SMD} = 2330\rho_A^{0.121}Q^{0.131}\Delta P_L^{-0.135} \tag{6.6}$$

and that of Elkotb [9]

$$\text{SMD} = 3.08\nu_L^{0.385}(\sigma\rho_L)^{0.737}\rho_A^{0.06}\Delta P_L^{-0.54} \tag{6.7}$$

The above equations for plain-orifice atomizers, as listed in Table 6.1, apply strictly to the injection of liquids into quiescent air. Two other cases of practical importance are (1) injection into a coflowing or contraflowing stream of air and (2) transverse injection across a flowing stream of air.

When a plain-orifice atomizer is oriented normal to the airflow, the larger drops penetrate farther into the airstream and the spectrum of drop sizes produced in the flowing airstream is skewed radially, as illustrated in Fig. 6.1. This distortion of the spray is not necessarily disadvantageous; for example, the ignition of a turbojet combustor is facilitated by locating the igniter in a region of small drops.

PRESSURE–SWIRL ATOMIZERS

Because of its wide range of applications, the pressure-swirl atomizer has attracted the attention of many research workers and has been the subject of considerable theoretical and experimental studies. However, despite these efforts, our knowledge of pressure-swirl atomization is still unsatisfactory. The physics is not well understood, the available data and correlations are of questionable validity, and there is little agreement between the various investigators as to the exact relationships between liquid properties, nozzle dimensions, and mean drop size. This unsatisfactory situation has several causes: the great complexity of the atomization process; differences in the design, size, and operating conditions of the nozzles tested; and the inaccuracies and limitations associated with drop size measurement techniques.

Most of the reported studies have concentrated on small-scale pressure-swirl atomizers of the type used in aircraft gas turbines, the only exception of note being the comprehensive experimental study carried out by Jones [10] on pressure-swirl nozzles of the large scale used in electrical power generation. In these var-

Table 6.1 Drop size equations for plain-orifice atomizers

Investigators	Equation
Merrington and Richardson [2]	$\mathrm{SMD} = \dfrac{500 d_o^{1.2} v_L^{0.2}}{U_L}$
Panasenkov [3]	$\mathrm{MMD} = 6 d_o \mathrm{Re}_L^{-0.15}$
Harmon [4]	$\mathrm{SMD} = 3330 d_o^{0.3} \mu_L^{0.07} \rho_L^{-0.648} \sigma^{-0.15} U_L^{-0.15} \mu_G^{0.78} \rho_G^{-0.052}$
Miesse [6]	$D_{0.999} = d_o \mathrm{We}_L^{-0.333}(23.5 + 0.000395 \mathrm{Re}_L)$
Tanasawa and Toyoda [7]	$\mathrm{SMD} = 47 d_o U_L^{-1} \left(\dfrac{\sigma}{\rho G}\right)^{0.25} \left[1 + 331 \dfrac{\mu_L}{(\rho_L \sigma d_o)^{0.5}}\right]$
Hiroyasu and Katoda [8]	$\mathrm{SMD} = 2330 \rho_A^{0.121} Q^{0.131} \Delta P^{-0.135}$
Elkobt [9]	$\mathrm{SMD} = 3.08 v_L^{0.385} (\sigma \rho L)^{0.737} \rho_A^{0.06} \Delta P_L^{-0.54}$

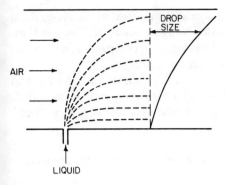

Figure 6.1 Influence of cross-stream airflow on drop size distribution.

ious studies emphasis has been placed on the variation of drop size with liquid properties and operating conditions.

Effect of Variables on Mean Drop Size

The main factors governing the atomization quality of pressure-swirl atomizers are liquid properties, the physical properties of the air or gas into which the liquid is injected, the liquid injection pressure, and the size of the atomizer as indicated by its flow number.

Liquid properties. The liquid properties of importance are surface tension and viscosity. In practice, the significance of surface tension is diminished by the fact that most commercial fuels exhibit only slight differences in this property. This is also true for density. However, viscosity varies by almost two orders of magnitude in some applications, so its effect on atomization quality can be quite large.

Surface tension. Simmons and Harding [11] studied the difference in atomizing performance between water and kerosine for six simplex fuel nozzles. These two liquids have virtually the same viscosity and a 30% difference in density, but the surface tension of water is higher by a factor of 3. Thus, according to Simmons and Harding, any significant differences in SMD must be due to differences in surface tension rather than density. For the important practical case of Weber number less than unity, it was found that SMD $\propto \sigma^a$, where a has a value of 0.19. This result is not supported by the findings of Kennedy [12], who used simplex nozzles featuring much higher flow numbers and observed a far stronger dependence of SMD on surface tension, namely SMD $\propto \sigma$. However, Jones's [10] studies on nozzles of high flow number led to a value for a of 0.25. This is fairly close to Simmons's result and is identical to the value derived by Lefebvre [13] from dimensional analysis of published data on SMD.

Wang and Lefebvre [14] used two simplex nozzles of different flow numbers to examine the influence of surface tension on SMD. They compared the results obtained with water and diesel oil and made a small adjustment to the diesel data, using Eq. (6.18) to compensate for the diesel oil's higher viscosity. Thus in Figs. 6.2 and 6.3 the higher SMD values exhibited by water, in comparison with diesel oil, are due entirely to the higher surface tension of water.

Analysis of the data contained in Figs. 6.2 and 6.3 indicates a value for a of around 0.25. Thus, if Kennedy's [12] exceptionally high value of a is neglected, the evidence acquired in the most recent experimental studies suggests a "best" value for a of around 0.25.

Viscosity. Published data on the effect of viscosity on mean drop size are usually expressed in the form

$$\text{SMD} \propto \mu_L^b \tag{6.8}$$

Values for b of 0.16, 0.20, 0.118, 0.16, 0.06, and 0.215 have been reported by Jasuja [15], Radcliffe [16], Dodge and Biaglow [17], Jones [10], Simmons [18], and Knight [19], respectively. Wang and Lefebvre [14] blended diesel oil (DF-2) with polybutene in different concentrations to achieve a wide range of viscosities. Some of their results, showing the adverse effect of increase in liquid viscosity on atomization quality, are presented in Figs. 6.4 and 6.5. Of special interest in these figures is that viscosity appears to have less effect on SMD for nozzles of lower spray cone angle.

The influence of liquid viscosity on SMD is shown more directly in Fig. 6.6, which illustrates the adverse effects of increases in viscosity and flow number on mean drop size. This figure also indicates that the dependence of SMD on viscosity diminishes with increase in flow number.

Liquid flow rate. All the reported data show that mean drop sizes increase at higher flow rates [10, 15, 16, 20]. Some of the results obtained by Wang and

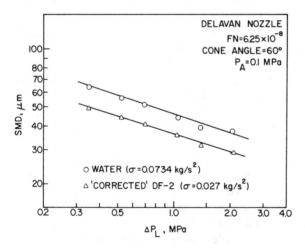

Figure 6.2 Influence of surface tension on mean drop size *(Wang and Lefebvre [14])*.

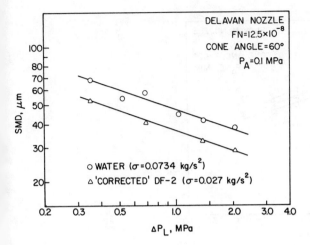

Figure 6.3 Influence of surface tension on mean drop size *(Wang and Lefebvre [14])*.

Lefebvre [14] for water are shown in Figs. 6.7 and 6.8 for 90° and 60° spray cone angles, respectively. By using three similar atomizers with different flow numbers, it was possible to study the influence of flow rate on mean drop size while maintaining either flow number or nozzle pressure differential constant. For any given flow number, both figures show that mean drop size declines with increase in flow rate. This is due to the increase in injection pressure differential that accompanies an increase in flow rate. If the pressure differential is kept constant, the flow rate can be raised only by increasing the flow number, i.e., by switching to a larger nozzle. This impairs atomization quality, especially at low injection pressures, as shown in Figs 6.7 and 6.8.

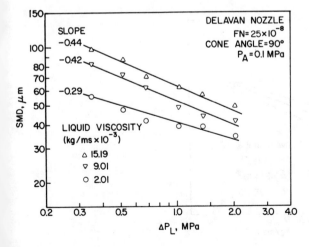

Figure 6.4 Influence of viscosity on SMD for a cone angle of 90° *(Wang and Lefebvre [14])*.

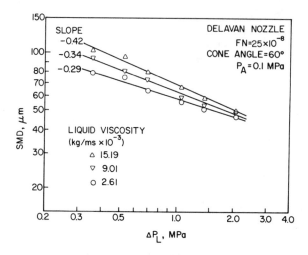

Figure 6.5 Influence of viscosity on SMD for a cone angle of 60° *(Wang and Lefebvre [14])*.

Flow number. Figures 6.7 and 6.8 also serve to demonstrate the influence of flow number on mean drop size. They show that an increase in flow number produces a coarser spray. However, the influence of flow number is seen to diminish with increase in liquid pressure differential, becoming negligibly small at pressure differentials of around 2 MPa (290 psid). Further evidence on the effect of flow number on mean drop size is provided in Figs. 6.9 and 6.10. Figure 6.9 was obtained using a diesel oil (DF-2), and Fig. 6.10 shows similar results for diesel oil after blending with polybutene to produce a large increase in viscosity. Comparison of Figs. 6.9 and 6.10 reveals that the effect of a reduction in flow

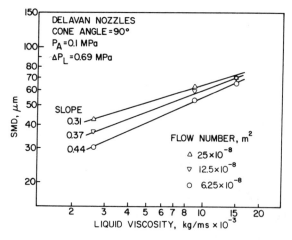

Figure 6.6 Influence of viscosity on SMD for various flow numbers *(Wang and Lefebvre [14])*.

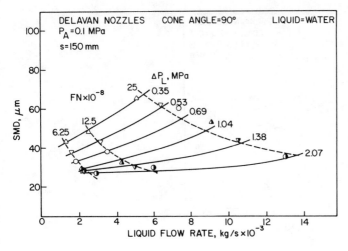

Figure 6.7 Relationship between SMD, flow number, and injection pressure for water; spray cone angle = 90° *(Wang and Lefebvre [14])*.

number, which is always conducive to better atomization, is much less for the liquid of higher viscosity.

Nozzle pressure differential. Increase in liquid pressure differential causes the liquid to be discharged from the nozzle at a higher velocity, which promotes a finer spray. The beneficial effect of raising the nozzle pressure differential on

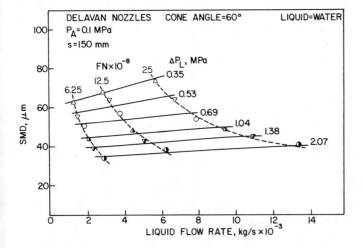

Figure 6.8 Relationship between SMD, flow number, and injection pressure for water; spray cone angle = 60° *(Wang and Lefebvre [14])*.

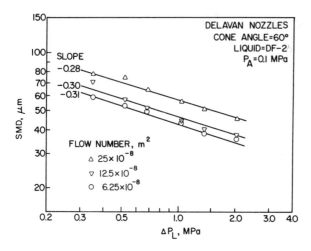

Figure 6.9 Influence of flow number on SMD for low-viscosity liquid *(Wang and Lefebvre [14])*.

atomization quality is clearly illustrated in Figs. 6.2 to 6.10. The effect may be expressed quantitatively as

$$SMD \propto \Delta P_L^d \tag{6.9}$$

Figure 6.9 illustrates the effects of liquid pressure differential on mean drop size for DF-2, while Fig. 6.10 shows similar results for a liquid of higher viscosity. These figures, are plotted in logarithmic form to facilitate data analysis. The slopes indicated in these and other figures denote the exponent d in the expression $SMD \propto \Delta P_L^d$. Figures 6.4, 6.5, 6.9, and 6.10 show values of d ranging

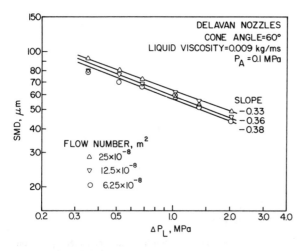

Figure 6.10 Influence of flow number on SMD for high-viscosity liquid *(Wang and Lefebvre [14])*.

from -0.28 to -0.44. Other reported values of d are -0.275, -0.35, -0.4, and -0.43, from Simmons [18], Abou-Ellail et al. [21], Radcliffe [16], and Jasuja [15], respectively.

Air properties. The two properties of interest are air pressure and air temperature. From a fundamental standpoint, they are usually regarded only as components of density. The results of various experimental studies on the influence of air density on SMD have been reviewed by Lefebvre [13] and Dodge and Biaglow [17].

De Corso [22] was among the first to investigate the influence of ambient air pressure on the spray characteristics of simplex swirl atomizers. Using a fuel of similar physical properties to diesel oil (DF-2), he measured an increase in drop size in going from 0.1 MPa (14.5 psia) to 0.79 MPa (114.5 psia). According to De Corso, "a continuing decrease in drop size with increase in ambient pressure would be expected, since the drag force on a drop increases with increasing density. Thus as ambient pressure rises, a decrease would be expected in the critical drop size, i.e., the maximum size that can withstand breakup, as indicated by Hinze [23] and Lane [24]." De Corso attributed his actual observed increase in drop size to increased coalescence of the spray drops as the ambient pressure is increased. Previous work by De Corso and Kemeny [25] had shown a reduction in spray cone angle with increase in ambient pressure, due to the action of induced gas currents, which tend to collapse the spray into a small volume and thereby provide more opportunity for coalescence.

Neya and Sato [26] also examined the influence of ambient air pressure on mean drop size and spray cone angle. Using water as the test fluid, they obtained results similar to those of De Corso and Kemeny [25] in regard to the contraction of spray cone angle with increase in air pressure. Over a pressure range of 0.1 to 0.5 MPa (1 to 5 atm) they observed a marked rise in the SMD with increase in P_A (SMD $\propto P_A^{0.27}$).

Neya and Sato also invoked droplet coalescence to explain this increase of SMD with P_A, but they asserted that under certain conditions an additional factor that should be considered is a change in the atomization process. Based on instantaneous snapshots of the spray, they concluded that the waviness of the initial liquid sheet is intensified with increasing P_A and the sheet breakup length is shortened.

Rizk and Lefebvre [27] also observed a decline in spray quality with increase in P_A. Their explanation for this result was that contraction of the spray angle reduces the volume of air that interacts with the spray. In consequence, the aerodynamic drag forces created by the spray induce a more rapid acceleration of this smaller air volume in the direction of spray motion, thereby reducing the relative velocity between the drops and the air surrounding them. As mean drop size is inversely proportional to this relative velocity, the effect of a reduction of spray angle is to increase the mean drop size.

In a recent study of the influence of air pressure on mean drop size, Wang and Lefebvre [28] flowed liquids of different viscosities through several Delavan simplex nozzles having various values of spray cone angle and flow number. Some of the results acquired during this investigation are shown in Figs. 6.11 to 6.13.

Figure 6.11 shows the influence of ambient air pressure on mean drop size for three nozzles having 90° cone angles and flow numbers of 6.25×10^{-8}, 12.5×10^{-8}, and $25 \times 10^{-8} m^2$. The liquid employed is diesel oil (DF-2), and the data correspond to a liquid injection pressure differential of 0.69 MPa (100 psi). For the nozzle of highest flow number, the SMD rises fairly steeply with increase in P_A up to a maximum value around 0.4 MPa (4 atm), beyond which any further increase in P_A causes the SMD to decline. For the nozzle of lowest flow number, the initial increase of SMD with P_A is also quite steep up to an air pressure of around 0.4 MPa. However, further increase in P_A above this pressure level appears to have little influence on SMD.

Figures 6.12 and 6.13 shows similar data for sprays of higher viscosity and lower cone angle. The general SMD levels are much higher, but in all cases the SMD again rises with P_A, reaching maximum values at around 0.3 MPa. Further increase in P_A above 0.3 MPa causes atomization quality to improve appreciably.

A characteristic feature of Figs. 6.11 to 6.13 is that SMD increases with ambient pressure up to a maximum value and then declines with any further increase in pressure. Possible explanations for the observed initial rise in SMD with increase in P_A include: (1) reduction in spray angle [25, 26, 29], which promotes drop coalescence and also increases the proportion of large drops in the spray sampling beam, and (2) reduction in sheet breakup length, so that drops are formed from a thicker sheet initially and consequently are larger.

If possible errors arising from measurement techniques are neglected, then Figs. 6.11 to 6.13 show that the forces inhibiting atomization are predominant at low pressures and the SMD rises with increase in P_A. With continuing increase

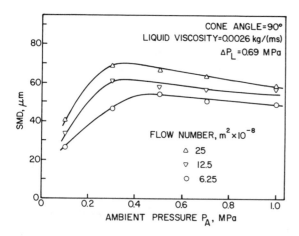

Figure 6.11 Influence of ambient air pressure and nozzle flow number on mean drop size *(Wang and Lefebvre [28])*.

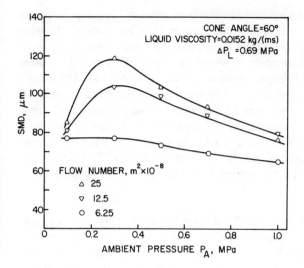

Figure 6.12 Influence of ambient air pressure and nozzle flow number on mean drop size *(Wang and Lefebvre [28])*.

in P_A, the contraction in spray angle, which is initially quite steep, starts to diminish and eventually becomes zero. This allows the disintegration forces to become dominant, so that any further increase in P_A above the critical value causes the SMD to decline. Thus, when SMD values are plotted against P_A, their characteristic shape is one that shows SMD rising up to a maximum value and then falling with further increase in P_A. During this latter stage the variation of SMD with P_A roughly corresponds to the relationship SMD $\propto P_A^{-0.25}$.

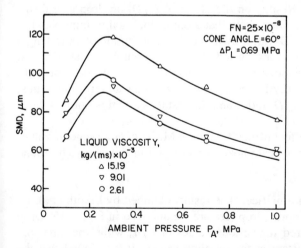

Figure 6.13 Influence of ambient air pressure and liquid viscosity on mean drop size *(Wang and Lefebvre [28])*.

Abou-Ellail et al. [21] have studied the effect of ambient air temperature on mean drop size. Their results are interesting because they are contrary to what might be expected from purely theoretical considerations. Since an increase in air temperature lowers the Weber number, it would be expected to impede disintegration and give rise to larger drops. However, these workers found that drop sizes diminished with increasing air temperature according to the relationship SMD $\propto T_A^{-0.56}$. Abou-Ellail et al. do not indicate whether the supply lines and spray nozzle were cooled to maintain a constant liquid temperature. As Dodge and Biaglow pointed out, if the liquid temperature is allowed to increase with air temperature, the drop size distribution will be shifted to smaller sizes. The results obtained by Dodge and Biaglow [17] using Jet-A and DF-2 fuels showed no effect of air temperature on SMD.

Atomizer dimensions. It is now generally accepted that the dimension of most importance for atomization is the thickness of the liquid sheet as it leaves the final orifice [13]. Theory [30–32] predicts, and experiment confirms, that mean drop size is roughly proportional to the square root of liquid sheet thickness. Thus, provided the other key parameters that are known to affect atomization are maintained constant, an increase in atomizer size will reduce atomization quality according to the relationship SMD $\propto d_o^{0.5}$. Although this relationship is generally supported by the results of dimensional analysis, the influence of atomizer size per se has not yet been subjected to systematic investigation. However, nozzle size is generally closely related to nozzle flow number, and it is well established that increase in flow number produces coarser atomization, as illustrated in Figs. 6.7 to 6.10.

Length/diameter ratio of swirl chamber. The effect of this ratio was studied by Elkotb et al. [33] as part of a wide investigation on the effects of swirl atomizer geometry on spray drop size. Results on the effect of L_s/D_s on drop size distribution are shown in Fig. 6.14. They show that atomization quality improves initially with an increase in the L_s/D_s ratio, owing to the contribution of extra length toward eliminating the flow striations caused by the finite number of swirl ports. However, once these flow striations have been damped out, the effect of additional length is to raise the energy losses and thereby impair atomization. Thus, one would expect an optimal L_s/D_s ratio, and the results show this optimal value to be around 2.75 [33].

Jones [10] examined the effect of L_s/D_s ratio on MMD over a range of values from 0.31 to 1.26. For the large-scale nozzles employed in this study, he found that MMD $\propto (L_s/D_s)^{0.07}$.

Length/diameter ratio of final orifice. The results obtained by Elkotb et al. on the effect of the l_o/d_o ratio on spray drop size are summarized in Fig. 6.15. Drop size distributions were measured for various values of l_o/d_o while the other parameters were maintained constant at the baseline values. It was found that the SMD diminished continuously over the range of l_o/d_o from 2.82 to 0.4. An op-

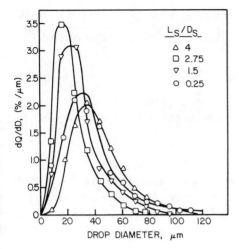

Figure 6.14 Effect of swirl chamber length/diameter ratio on drop size distribution *(Elkotb et al. [33])*.

timal value of l_o/d_o was not encountered, so presumably further improvement in atomization quality could be gained by further reduction in l_o/d_o below 0.4. This conclusion is supported by the results of Jones [10], which showed, over a range of l_o/d_o ratios from 0.1 to 0.9, that MMD $\propto (l_o/d_o)^{0.03}$.

Drop Size Relationships

Owing to the complexity of the various physical phenomena involved in pressure-swirl nozzles, the study of atomization has been pursued principally by empirical methods, yielding correlations for mean drop size of the form

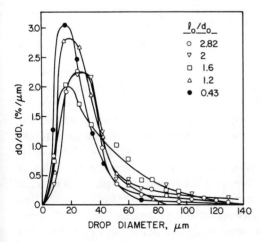

Figure 6.15 Effect of final orifice length/diameter ratio on drop size *(Elkotb et al. [33])*.

$$SMD \propto \sigma^a v^b \dot{m}_L^c \Delta P_L^d \tag{6.10}$$

One of the earliest and most widely quoted expressions is that of Radcliffe [16]:

$$SMD = 7.3\sigma^{0.6} v_L^{0.2} \dot{m}_L^{0.25} \Delta P_L^{-0.4} \tag{6.11}$$

This equation was derived from analysis of experimental data obtained by Needham [34], Joyce [35], and Turner and Moulton [36]. Subsequent work by Jasuja [15] yielded the expression

$$SMD = 4.4\sigma^{0.6} v_L^{0.16} \dot{m}_L^{0.22} \Delta P_L^{-0.43} \tag{6.12}$$

However, the variation of surface tension in these experiments was very small and was accompanied by wide variations in viscosity. Thus the exponent 0.6 has no special significance in Eqs. (6.11) and (6.12). Babu et al. [37] used regression analysis to determine the following equations for kerosine-type fuels.

For ΔP_L less than 2.8 MPa (400 psi)

$$SMD = 133 \frac{FN^{0.64291}}{\Delta P_L^{0.22565} \rho_L^{0.3215}} \tag{6.13}$$

For ΔP_L above 2.8 MPa (400 psi)

$$SMD = 607 \frac{FN^{0.75344}}{\Delta P_L^{0.19936} \rho_L^{0.3767}} \tag{6.14}$$

The high degree of precision implied in these equations should be regarded with caution in view of the well-known uncertainties in all methods of drop size measurement.

From a series of tests conducted on 25 different fuels, using six different simplex nozzles of large flow number, Kennedy [12] derived the following correlating parameter for nozzles operating at Weber numbers larger than 10:

$$SMD = 10^{-3}\sigma_L(6.11 + 0.32 \times 10^5 \, FN\sqrt{\rho_L}$$
$$- 6.973 \times 10^{-3}\sqrt{\Delta P_L} + 1.89 \times 10^{-6}\Delta P_L) \tag{6.15}$$

In estimating Weber number, Kennedy [12] used the film thickness in the final orifice, as given by Simmons's equation (5.39), as the characteristic dimension.

Equation (6.15) implies a very strong dependence of mean drop size on surface tension, while viscosity appears to have no effect at all. Kennedy attributes this, and other differences between his results and those of other workers, to the larger Weber numbers resulting from his use of nozzles of exceptionally high flow number. According to Kennedy, "for Weber numbers greater than 10, a different atomization process occurs, i.e., shear-type breakup, which results in much finer atomization than predicted by previously-reported correlations." However, Jones [10], using large industrial nozzles of much higher flow numbers than those employed by Kennedy, found the effects of surface tension and viscosity on mean drop size to be fully consistent with all previous observations on small-scale nozzles.

Jones [10] used a high-speed photographic technique to investigate the effects of changing liquid properties, operational variables, and geometric parameters on the drop sizes produced by large pressure-swirl atomizers. Variations in atomizer geometry were achieved by using a series of specially designed nozzle pieces that enabled 50 different geometric configurations to be studied. Analysis of the experimental data yielded the following equation for mean drop size:

$$\text{MMD} = 2.47 m_L^{0.315} \Delta P_L^{-0.47} \mu_L^{0.16} \mu_A^{-0.04} \sigma^{0.25} \rho_L^{-0.22}$$

$$\times \left(\frac{l_o}{d_o}\right)^{0.03} \left(\frac{L_s}{D_s}\right)^{0.07} \left(\frac{A_p}{D_s d_o}\right)^{-0.13} \left(\frac{D_s}{d_o}\right)^{0.21} \tag{6.16}$$

The ranges of variables covered in this investigation are listed in Table 5.1.

Lefebvre's [13] analysis of the flow processes in the final orifice of a simplex nozzle led to the following equation for SMD:

$$\text{SMD} = A\sigma^{0.25} \mu_L^{0.25} \rho_L^{0.125} d_o^{0.5} \rho_A^{-0.25} \Delta P_L^{-0.375} \tag{6.17}$$

The value of 0.25 for the surface tension exponent is reasonably close to Simmons's estimates (0.16 to 0.19) and is in exact agreement with the experimental findings of Jones [10] and Wang and Lefebvre [14]. The exponent of 0.25 for viscosity agrees fairly well with all reported values. No reliable experimental data exist to test the exponent for liquid density, but all the evidence suggests that the effect of this property on SMD is quite small. The d_o exponent of 0.5 confirms the theoretical value. The exponents for injection pressure differential and air density of -0.375 and -0.25, respectively, are identical to the experimental values obtained in the investigation by Abou-Ellail et al. [21]. Substituting $d_o \propto \dot{m}_L^{0.5}/(\Delta P_L/\rho_L)^{0.25}$ into Eq. (6.17) and using Jasuja's [15] data to determine the value of A gives

$$\text{SMD} = 2.25\sigma^{0.25} \mu_L^{0.25} \dot{m}_L^{0.25} \Delta P_L^{-0.5} \rho_A^{-0.25} \tag{6.18}$$

Recent developments. It is clearly unreasonable to expect any single-term parameter for SMD of the types described above, all of which are essentially derivatives or modifications of Eq. (6.10), to describe adequately the complex fluid dynamic processes involved in atomization. Further objections to this type of formula can be made on purely practical grounds. For example, it takes no account of spray cone angle, which is known to affect mean drop size. Moreover, it implies that a nonviscous liquid should yield a spray of infinitely small drops, which seems highly unlikely from both theoretical and practical considerations.

In an attempt to overcome some of the deficiencies of existing SMD formulas and to explain some of the apparent anomalies that careful measurements often reveal, Lefebvre [38] has proposed an alternative form of equation for the mean drop sizes produced by pressure-swirl atomizers. This equation is not the result of a mathematical treatment of the subject but is based on considerations of the basic mechanisms involved in pressure-swirl atomization. It starts from the notion that the disintegration of a liquid jet or sheet issuing from a nozzle is not caused solely by aerodynamic forces but must be the result, at least in part, of turbulence or other disruptive forces within the liquid itself. These disturbances within the flow have a strong influence on sheet disintegration, especially in the first stage of atomization. Subsequently, and to some extent simultaneously, the relative velocity between the liquid and the surrounding gas plays an important role in atomization through its influence on the development of waves on the initially smooth surface and the production of unstable ligaments. Any increase in this relative velocity causes a reduction in the size of the ligaments, so that when they disintegrate much smaller drops are produced.

As the process of atomization in pressure-swirl nozzles is highly complex, it is convenient to subdivide it into two main stages. The first stage represents the generation of surface instabilities due to the combined effects of hydrodynamic and aerodynamic forces. The second stage is the conversion of surface protuberances into ligaments and then drops. It is recognized that this subdivision of the total atomization process into two separate and distinct stages represents an oversimplification of the mechanisms involved. However, it allows the formulation of an equation for SMD as

$$\text{SMD} = \text{SMD}_1 + \text{SMD}_2 \tag{6.19}$$

Here, SMD_1 represents the first stage of the atomization process. Its magnitude depends partly on the Reynolds number, which provides a measure of the disruptive forces present within the liquid sheet. These forces are enhanced by increases in liquid velocity and sheet thickness and are diminished by an increase in liquid viscosity. SMD_1 is also influenced by the Weber number, which governs the development of capillary waves (ripples) on the liquid surface. The rate of growth of these perturbations into projections large enough to break off and form ligaments is dependent on the ratio of the aerodynamic forces at the liquid/air interface to the consolidating surface tension forces in the liquid, i.e., on Weber number.

It appears, therefore, that a suitable form of SMD_1 might be one that combines both Reynolds and Weber numbers into a single parameter. Analysis of the experimental data leads to the following expression for SMD_1:

$$\frac{\text{SMD}_1}{t_s} \propto (\text{Re} \times \sqrt{\text{We}})^{-x} \tag{6.20}$$

where Re is $\rho_L U_L t_s / \mu_L$, We is $\rho_A U_R^2 t_s / \sigma$, and t_s is the initial sheet thickness at the nozzle exit.

SMD_1 should not be confused with the Z or Ohnesorge number [39], which is denoted as $\sqrt{\text{We}}/\text{Re}$ and represents the *ratio* of surface tension to viscous forces operating on the liquid. In contrast, SMD_1 is intended to represent the manner in which surface tension and viscous forces *act together* in opposing the disruptive actions of the hydrodynamic and aerodynamic momentum forces [38].

For both pressure and airblast nozzles the relative velocity between the liquid and the surrounding gas has a profound effect on atomization. It generates the projections on the liquid surface that are a prerequisite to atomization and also furnishes the energy needed to convert these projections into ligaments and then drops. However, another important factor in atomization, as discussed above, is the contribution made to sheet or jet disintegration by the instabilities created within the liquid itself, which are very dependent on liquid velocity. In airblast atomization, where high-velocity air impacts on a slow-moving liquid, the only factor promoting atomization is the relative velocity between the air and the liquid. This is also vitally important in pressure atomization but, by achieving this relative velocity through liquid motion instead of air motion, an important advantage is gained in that the liquid now makes an additional and independent contribution to its own disintegration, an effect that is either absent or negligibly small in airblast and air-assist atomization.

These arguments highlight the special importance of velocity in pressure atomization. The velocity at which the liquid is discharged from the nozzle has two separate effects on atomization. One important effect, which is dependent on the *absolute* velocity U_L, is in generating the turbulence and instabilities within the bulk liquid that contribute to the first stage of the atomization process. The other effect, which depends on the *relative* velocity U_R, is in promoting the atomization mechanisms that occur on the liquid surface and in the adjacent ambient gas. This means that Reynolds number, which is related to the bulk liquid, should be based on U_L, while Weber number, which is associated with events occurring through the action of the surrounding gas on the liquid surface, should be based on U_R.

Dividing $\sqrt{We}$ by Re to produce the Z number has the effect of eliminating velocity from both We and Re. This severely curtails its useful application to pressure atomization, where velocity is of paramount importance. In contrast, *multiplying* $\sqrt{We}$ by Re not only enhances the role of velocity but also provides a more accurate description of how SMD is affected by changes in liquid properties as well as nozzle size and geometry.

From simple geometric considerations, the initial thickness of the liquid sheet, t_s, after it is discharged from the nozzle can be related to the film thickness within the final orifice, t, by the equation

$$t_s = t \cos \theta \tag{6.21}$$

where θ is half the spray cone angle.

Substituting for t_s from Eq. (6.21) into Eq. (6.20), along with the appropriate terms for Re and We, gives

$$SMD_1 \propto \left(\frac{\sigma^{0.5} \mu_L}{\rho_A^{0.5} \rho_L U_R U_L} \right)^x (t \cos \theta)^{1-1.5x} \tag{6.22}$$

For the normal case of a nozzle spraying fluid into stagnant or slow-moving air, $U_R \simeq U_L$. Also, $0.5 \rho_L U_L^2 = \Delta P_L$. Hence Eq. (6.22) can be simplified to

$$SMD_1 \propto \left(\frac{\sigma^{0.5} \mu_L}{\rho_A^{0.5} \Delta P_L} \right)^x (t \cos \theta)^{1-1.5x} \tag{6.23}$$

The term SMD_2 represents the final stage of the atomization process, in which the high relative velocity induced at the liquid/air interface by the rapidly evolving conical sheet causes the surface protuberances generated in the first stage to become detached and break down into ligaments and then drops. This final disintegration is opposed by surface tension forces, but Reynolds number is no longer relevant. Thus, we have

$$\frac{SMD_2}{t_s} \propto We^{-y} \propto \left(\frac{\sigma}{\rho_A U_R^2 t_s}\right)^y \qquad (6.24)$$

Substituting for t_s from Eq. (6.21) converts Eq. (6.24) into

$$SMD_2 \propto \left(\frac{\sigma}{\rho_A U_R^2}\right)^y (t \cos \theta)^{1-y} \qquad (6.25)$$

or, since $U_R \simeq U_L$ and $\Delta P_L = 0.5 \rho_L U_L^2$,

$$SMD_2 \propto \left(\frac{\sigma \rho_L}{\rho_A \Delta P_L}\right)^y (t \cos \theta)^{1-y} \qquad (6.26)$$

Equation (6.19) thus becomes

$$SMD = A \left(\frac{\sigma^{0.5} \mu_L}{\rho_A^{0.5} \Delta P_L}\right)^x (t \cos \theta)^{1-1.5x} + B \left(\frac{\sigma \rho_L}{\rho_A \Delta P_L}\right)^y (t \cos \theta)^{1-y} \qquad (6.27)$$

where A and B are constants whose values depend on nozzle design.

Wang and Lefebvre [14] carried out a detailed experimental study of the factors governing the mean drop sizes produced by pressure-swirl atomizers. Extensive measurements of SMD were made using six simplex nozzles of different sizes and spray cone angles. Several different liquids were employed to provide a range of viscosity from 10^{-6} to 18×10^{-6} kg/m s (1 to 18 cS) and a range of surface tension from 0.027 to 0.0734 kg/s^2 (27 to 73.4 dyn/cm). The large amount of SMD data acquired in this investigation indicated values for x and y of 0.5 and 0.25, respectively. Moreover, the best fit to the experimental data was obtained using values of A and B of 4.52 and 0.39, respectively.

Substituting these values of x, y, A, and B into Eq. (6.27) gives

$$SMD = 4.52 \left(\frac{\sigma \mu_L^2}{\rho_A \Delta P_L^2}\right)^{0.25} (t \cos \theta)^{0.25} + 0.39 \left(\frac{\sigma \rho_L}{\rho_A \Delta P_L}\right)^{0.25} (t \cos \theta)^{0.75} \qquad (6.28)$$

Inspection of this equation reveals several interesting features of pressure-swirl atomization:

1. Increase in spray cone angle reduces mean drop size. This result is fully consistent with the experimental data shown in Figs. 6.7 and 6.8 for water and in Figs. 6.4 and 6.5 for special liquids of high viscosity. In both sets of figures, the comparison is made between spray angles of 90° and 60°.

Equation (6.28) also suggests that an increase in spray angle should cause a stronger dependence of SMD on both viscosity and pressure differential ΔP_L. It does this by causing SMD_2 to decline more than SMD_1, thereby increasing the relative importance of the contribution made by SMD_1 to the overall mean drop size. Both these effects are evident in the data shown in Figs. 6.4 and 6.5. Comparison of these two figures shows that widening the spray angle not only improves atomization quality but also increases the dependence of SMD on both viscosity and liquid injection pressure. Also worthy of note in these figures is that the exponent of ΔP_L increases with increase in viscosity. This again stems directly from the added importance imparted to SMD_1 by increases in liquid viscosity.

2. Equation (6.28) indicates a surface tension exponent of 0.25. This agrees exactly with the findings of Jones [10] and is fairly close to Simmons and Harding's [11] value of 0.19. The data shown in Figs. 6.2 and 6.3 for water ($\sigma = 0.0734$ kg/s^2) and "corrected" DF-2 ($\sigma = 0.027$ kg/s^2) also suggest a surface tension exponent of 0.25.

3. As liquid density appears only in SMD_2, its influence on mean drop size should generally be quite small, depending on the contribution of SMD_2 to SMD. The density of commercial hydrocarbon fuels is normally between 760 and 900 kg/m^3, which means that changes in fuel type are not usually accompanied by large changes in density. This fact, combined with the small dependence of SMD on ρ_L shown in Eq. (6.28), implies that for most practical purposes the influence of density can usually be neglected when considering the effect of a change in fuel type on SMD.

4. As t is directly proportional to flow number [see Eq. (5.39)], Eq. (6.28) shows that the impact of a change in flow number on SMD should be less for liquids of high viscosity. This prediction is borne out by the experimental data plotted in Figs. 6.9 and 6.10 for liquid viscosities of 0.0026 and 0.009 kg/m s, respectively.

The same figures also show a stronger dependence of SMD on injection pressure for the higher-viscosity liquid. This is consistent with Eq. (6.28), which indicates exponents for ΔP_L varying from -0.25 for nonviscous liquids to -0.5 for liquids of high viscosity.

5. The influence of viscosity on mean drop size is demonstrated directly in Fig. 6.6. This figure illustrates the diminishing effect of flow number on SMD with increase in viscosity, as discussed above, and also shows that SMD becomes less dependent on viscosity with increase in flow number, as predicted by Eq. (6.28).

6. A further conclusion to be drawn from Eq. (6.28) is that the impact on SMD of changes in nozzle dimensions and operating conditions will vary depending on the level of liquid viscosity. From inspection of the equation it is evident that for liquids of high viscosity SMD will be more dependent on injection pressure differential ΔP_L and less dependent on nozzle flow number (via t) and spray cone angle than liquids of low viscosity.

Values of film thickness for insertion into Eq. (6.28) may be obtained from Eqs. (5.37) or (5.40) or from Eq. (5.44), with the value of the constant reduced from 3.66 to 2.7 following the recommendation of Suyari and Lefebvre [40], to give

$$t = 2.7 \left[\frac{d_o F N \mu_L}{(\Delta P_L \rho_L)^{0.5}} \right]^{0.25} \tag{6.29}$$

Drop size equations for pressure-swirl atomizers are given in Table 6.2.

ROTARY ATOMIZERS

The process of centrifugal atomization has been studied by several workers, including Hinze and Milborn [41], Dombrowski and Fraser [42], and Tanasawa et al. [43]. It is described in reviews of atomization methods by Dombrowski and Munday [44], Christensen and Steely [45], Eisenklam [46], Matsumoto et al. [47], and Fraser et al. [48].

From the mechanisms for rotary atomization as outlined in Chapter 5, one would expect the mean drop size to diminish with increasing rotational speed and increase with flow rate and liquid viscosity. This expectation is confirmed by various theoretical and experimental investigations of the factors governing the

Table 6.2 Drop size relationships for pressure-swirl atomizers

Investigators	Equation	Remarks
Radcliffe [16]	$\text{SMD} = 7.3\sigma^{0.6} v_L^{0.2} \dot{m}_L^{0.25} \Delta P_L^{-0.4}$	No effect of nozzle dimensions or air properties
Jasuja [15]	$\text{SMD} = 4.4\sigma^{0.6} v_L^{0.16} \dot{m}_L^{0.22} \Delta P_L^{-0.43}$	No effect of nozzle dimensions or air properties
Babu et al. [37]	$\text{SMD} = 133 \dfrac{FN^{0.64291}}{\Delta P_L^{0.22565} \rho_L^{0.3215}}$	For $\Delta P_L < 2.8$ MPa
Babu et al. [37]	$\text{SMD} = 607 \dfrac{FN^{0.75344}}{\Delta P_L^{0.19936} \rho_L^{0.3767}}$	For $\Delta P_L > 2.8$ MPa
Jones [10]	$\text{MMD} = 2.47\dot{m}_L^{0.315} \Delta P_L^{-0.47} \mu_L^{0.16} \mu_A^{-0.04} \sigma^{0.25} \rho L^{-0.22} S$ $\times \left(\dfrac{l_o}{d_o} \right)^{0.03} \left(\dfrac{L_s}{D_s} \right)^{0.07} \left(\dfrac{A_p}{D_s d_o} \right)^{-0.13} \left(\dfrac{D_s}{d_o} \right)^{0.21}$	Suitable for large-capacity nozzles
Lefebvre [13]	$\text{SMD} = 2.25 \, \sigma^{2.25} \mu_L^{0.25} \dot{m}_L^{0.25} \Delta P_L^{-0.5} \rho_A^{-0.25}$	
Wang and Lefebvre [14]	$\text{SMD} = 4.52 \left(\dfrac{\sigma \mu_L^2}{\rho_A \Delta P_L^2} \right)^{0.25} (t \cos \theta)^{0.25}$	Includes effect of spray cone angle
	$+ 0.39 \left(\dfrac{\sigma \rho_L}{\rho_A \Delta P_L} \right)^{0.25} (t \cos \theta)^{0.75}$	Values of film thickness obtained from Eq. (5.37), (5.40), or (6.29)

atomization properties of rotary atomizers, which have resulted in a number of drop size relationships.

Karim and Kumar [49] used a photographic technique to measure the drop sizes produced by rotating cup atomizers. Their experiments were carried out with gasoline, using cups having exit diameters of 16, 19, and 20 mm and semicone angles of 10°, 20°, and 30°. Figure 6.16 is typical of the results obtained. It shows that, for a given rotational speed, the mean diameter of the drops formed increases almost linearly as the feed rate of the liquid is increased.

Figure 6.17 shows the influence of rotational speed on mean drop size. The data plotted in this figure were obtained with a 16-mm-diameter cup and semicone angle of 10° for various liquid temperatures. It can be seen that the mean diameter decreases by over an order of magnitude, down to about 10 μm, as the rotational speed is increased. Similar trends are usually observed for other liquids and other cup geometries.

The adverse effect on atomization quality of lowering the liquid temperature is attributed by Karim and Kumar to the associated increase in viscosity, which decreases both the turbulence level and the rate of disintegration of the liquid into drops. Moreover, the surface tension of the liquid increases as the temperature goes down, which opposes the formation of any distortion or irregularity on the liquid surface, thereby delaying the formation of ligaments and drops.

In many practical applications of rotary atomizers, the drops produced are exposed to appreciable air movement. Figure 6.18 shows how the mean drop diameter varies as the air velocity parallel to the axis of the atomizer increases

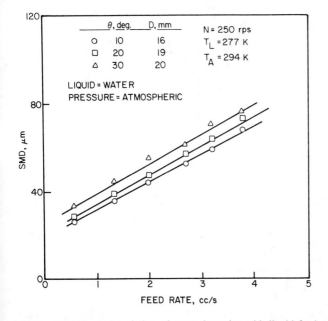

Figure 6.16 Typical variation of mean drop size with liquid feed rate *(Karim and Kumar [49])*.

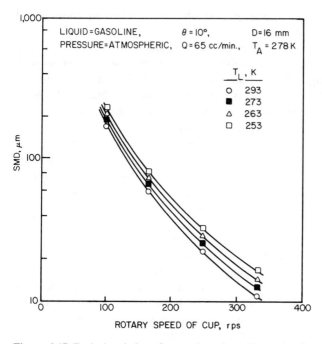

Figure 6.17 Typical variation of mean drop size with rotational speed *(Karim and Kumar [49])*.

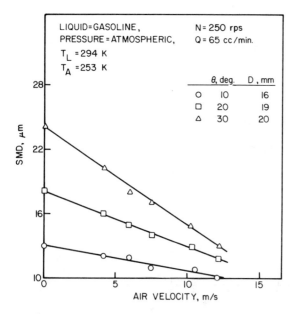

Figure 6.18 Typical variation of mean drop size with cross-flowing air velocity *(Karim and Kumar [49])*.

for a constant value of liquid flow rate. It can be seen that the mean diameter decreases with increase of air velocity. The extent of this decrease appears to be very dependent on the geometry of the atomizing cup. Lowering the airstream temperature, as shown in Fig. 6.19, tends to decrease the mean drop diameter almost linearly. This decrease in drop diameter is attributed to the increased density of the air, which increases the frictional forces on the liquid surface, thereby increasing the tendency for splitting up of the liquid into finer drops.

Drop Size Equations for Smooth Flat Vaneless Disks

The various modes whereby drops are produced by rotary atomization are discussed in Chapter 5. Several equations have been proposed for predicting mean drop sizes in each mode.

Direct drop formation. One of the earliest relationships for correlating the drop sizes produced by a flat disk rotary atomizer is the following due to Bar [50]:

$$D_{0.999} = \frac{1.07}{N}\left(\frac{\sigma}{d\rho_L}\right)^{0.5}$$

(6.30)

According to this equation the maximum drop size produced by direct drop formation increases with increase in surface tension and/or reduction in liquid

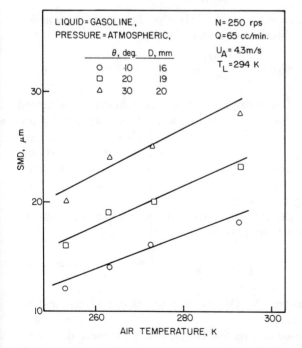

Figure 6.19 Typical variation of mean drop size with air temperature *(Karim and Kumar [49])*.

density and is inversely proportional to the speed of rotation. A notable feature of Eq. (6.30) is the absence of any term to account for the influence of liquid viscosity on drop size. This omission was remedied some 40 years later by Tanasawa et al. [51]. Their modified equation for Sauter mean diameter is

$$SMD = \frac{27}{N}\left(\frac{\sigma}{d\rho_L}\right)^{0.5}\left(1 + 0.003\frac{Q}{dv_L}\right) \qquad (6.31)$$

According to Matsumoto et al. [52], the dimensionless mean drop size produced in this mode of drop formation is solely dependent on Weber number. We have

$$\frac{SMD}{d} = 1.6\,We^{-0.523} \qquad (6.32)$$

where $We = \rho_L d^3\omega^2/8\sigma$.

Atomization by ligament formation. Walton and Prewett [53] were among the first to investigate the production of drops by ligament formation at the periphery of a rotating disk. Their equation for maximum drop size is almost identical to Bar's Eq. (6.30) for direct drop formation. We have

$$D_{0.999} = \frac{0.87}{N}\left(\frac{\sigma}{d\rho_L}\right)^{0.5} \qquad (6.33)$$

Another well-known expression for the drop sizes produced via ligament formation is the following due to Oyama and Endou [54]:

$$SMD = \frac{0.177Q^{0.2}}{Nd^{0.3}} \qquad (6.34)$$

This equation was derived from analysis of drop size measurements carried out with water as the working fluid. It contains no terms to denote liquid properties, so its range of useful application is limited to water. However, it does have the merit of indicating that drop sizes should increase with increase in liquid flow rate, which is consistent with experimental observations. This effect of higher flow rate may be attributed to thickening of the liquid sheet, which on disintegration forms ligaments of larger diameter.

The relationship between ligament diameter and mean drop size has been expressed by Matsumoto and Takashima [55] as

$$\frac{SMD}{d_1} = (1.5\pi)^{1/3}\left(1 + 3\frac{We}{Re^2}\right)^{1/6} \qquad (6.35)$$

where d_1 is the ligament diameter, We (Weber number) is $\rho_L d^3\omega^2/8\sigma$, and Re (Reynolds number) is $\rho_L d^2\omega/4\mu$.

Tanasawa et al. [51] proposed an equation for predicting the mean drop sizes produced via ligament formation that takes full account of variations in disk diameter, rotational speed, liquid flow rate, and liquid properties. We have

$$\text{SMD} = 0.119 \frac{Q^{0.1}\sigma^{0.5}}{Nd^{0.5}\rho_L^{0.4}\mu_L^{0.1}} \tag{6.36}$$

An interesting feature of this equation is that drop sizes are predicted to increase slightly with reduction in liquid viscosity. Usually, of course, reduction in liquid viscosity leads to finer atomization. An equation for mean drop size that is more consistent with this general rule is the following, proposed by Kayano and Kamiya [56]:

$$\text{SMD} = 0.26N^{-0.79}Q^{0.32}d^{-0.69}\rho_L^{-0.29}\sigma^{0.26}(1 + 1.027\mu_L^{0.65}) \tag{6.37}$$

Atomization by sheet disintegration. According to Fraser and Eisenklam [57], an appropriate drop size relationship for this mode of atomization is the following:

$$D_{10} = \frac{0.76}{N}\left(\frac{\sigma}{d\rho_L}\right)^{0.5} \tag{6.38}$$

It is of interest to note the close similarity between this equation for sheet disintegration and Eqs. (6.30) and (6.33), which relate to direct drop formation and atomization via ligament formation, respectively.

A similar relationship, with an additional term to include the influence of liquid flow rate on mean drop size, is the following, due to Tanasawa et al. [51]:

$$\text{SMD} = 15.6 \frac{Q^{0.5}}{N}\left(\frac{\sigma}{d^2\rho_L}\right)^{0.4} \tag{6.39}$$

Drop Size Equations for Vaneless Disks

For vaneless disks that do not fall in the category of smooth and flat, the drop size equations corresponding to the three modes of atomization are the following:

Direct drop formation [90]:

$$D_{10} = \left[\frac{3Q\mu_L}{2\pi\rho_L(\pi dN)^2 \sin\theta}\right]^{1/3} \tag{6.40}$$

Atomization by ligament formation [41]:

$$D_{10} = \left[0.77\frac{Q}{Nd}\left(\frac{\rho_L N^2 d^3}{\sigma}\right)^{-5/12}\left(\frac{\rho_L\sigma d}{\mu_L^2}\right)^{-1/16}\right]^{0.5} \tag{6.41}$$

Atomization by sheet disintegration [108]:

$$D_{10} = 4.42\left(\frac{\sigma}{\omega^2\rho_L d}\right)^{0.5} \tag{6.42}$$

Drop Size Equations for Atomizer Wheels

Some commercial rotary atomizers consist of disks or wheels fitted with vanes. For these devices, additional geometric variables to be considered are the number of vanes n and the vane height h. According to Masters [58], suitable drop size equations for atomizer wheels are the following.

Friedman et al. [59]:

$$\text{SMD} = 0.44d\left(\frac{\dot{m}_p}{\rho_L Nd^2}\right)^{0.6}\left(\frac{\mu_L}{\dot{m}_p}\right)^{0.2}\left(\frac{\sigma\rho_L nh}{\dot{m}_p^2}\right)^{0.1} \tag{6.43}$$

Herring and Marshall [60]:

$$\text{SMD} = \frac{3.3 \times 10^{-9}K\dot{m}_L^{0.24}}{(Nd)^{0.83}(nh)^{0.12}} \tag{6.44}$$

where $K = 8.5 \times 10^5$ to 9.5×10^5. See Masters' Table 6.11 [58].

Fraser et al. [61]:

$$\text{SMD} = 0.483N^{-0.6}\rho_L^{-0.5}\left(\frac{\mu_L\dot{m}_L}{d}\right)^{0.2}\left(\frac{\sigma}{nh}\right)^{0.1} \tag{6.45}$$

Scott et al. [62]:

$$\text{SMD} = 6.3 \times 10^{-4}\dot{m}_p^{0.171}(ndN)^{-0.537}\mu_L^{-0.017} \tag{6.46}$$

Drop size equations for smooth flat vaneless disks, other types of vaneless disks, and atomizer wheels are listed in Tables 6.3, 6.4, and 6.5, respectively. Other equations for the mean drop sizes produced by rotary atomizers may be found in the literature, but not all of them can be recommended for prediction purposes, because they were derived from data sources that include more than one mechanism of drop formation.

AIR–ASSIST ATOMIZERS

In attempting to review published work on air-assist atomizer performance, a major difficulty involved is that of differentiating between air-assist and airblast atomizers, since both types of nozzle employ high-velocity air to achieve atomization and their mean geometric features are always roughly the same. For gas turbine applications, the most common method of differentiating between the two systems is on the basis of air velocity. Air-assist nozzles employ very high air velocities that usually necessitate an external supply of high-pressure air, while the lower air velocity requirements of airblast systems can usually be met by utilizing the pressure differential across the combustor liner. However, as most of the reported studies on air atomization have covered fairly wide ranges of air

Table 6.3 Drop size equations for smooth flat vaneless disks

Investigators	Equation	Remarks
Bar [50]	Direct drop formation $$D_{0.999} = \frac{1.07}{N}\left(\frac{\sigma}{d\rho_L}\right)^{0.5}$$	No effect of liquid viscosity
Tanasawa et al. [51]	$$SMD = \frac{27}{N}\left(\frac{\sigma}{d\rho_L}\right)^{0.5}\left(1 + 0.003\frac{Q}{d\nu_L}\right)$$	Includes all liquid properties
Matsumoto et al. [52]	$$SMD = 1.6d\left(\frac{8\sigma}{\rho_L d^3 \omega^2}\right)^{0.523}$$	Mean drop size dependent on Weber number
Walton and Prewett [53]	Atomization by ligament formation $$D_{0.999} = \frac{0.87}{N}\left(\frac{\sigma}{d\rho_L}\right)^{0.5}$$	No effect of liquid viscosity
Oyama and Endou [54]	$$SMD = \frac{0.177Q^{0.2}}{Nd^{0.3}}$$	Limited to water. Includes liquid flow rate
Matsumoto and Takashima [55]	$$\frac{SMD}{d_1} = (1.5\pi)^{1/3}\left(1 + \frac{3We}{Re^2}\right)^{1/6}$$	d_1 = ligament diameter We = Weber number = $\rho_L d^3 \omega^2 / 8\sigma$ Re = Reynolds number = $\rho_L d^2 \omega / 4\mu$
Tanasawa et al. [51]	$$SMD = 0.119\frac{Q^{0.1}\sigma^{0.5}}{Nd^{0.5}\rho_L^{0.4}\mu_L^{0.1}}$$	Includes all relevant parameters
Kayano and Kamiya [56]	$$SMD = 0.26N^{-0.79}Q^{0.32}d^{-0.69}\rho_L^{-0.29}\sigma^{0.26}$$ $$(1 + 1.027\,\mu_L^{0.65})$$	Includes all relevant parameters
Fraser and Eisenklam [57]	Atomization by sheet disintegration $$D_{10} = \frac{0.76}{N}\left(\frac{\sigma}{d\rho_L}\right)^{0.5}$$	
Tanasawa et al. [51]	$$SMD = 15.6\frac{Q^{0.5}}{N}\left(\frac{\sigma}{d^2\rho_L}\right)$$	Includes effect of liquid flow rate

velocity, the decision to describe any given atomizer as "air-assist" or "air-blast" is sometimes almost arbitrary.

In general, air-assist nozzles require less air than airblast nozzles, which makes them especially attractive for use in combustion systems that are required to operate with minimum excess air over a wide turndown ratio. They are usually classified into *internal-mixing* and *external-mixing* types. Internal-mixing nozzles have the advantage of wide flexibility and are particularly useful for atomizing highly viscous liquids and liquid slurries. However, their aerodynamic and fluid dynamic flow patterns are highly complex, due to the intense mixing of gas and liquid within the mixing chamber.

Table 6.4 Drop size equations for vaneless disks

Investigators	Equation
Fraser et al. [90]	Direct drop formation $$D_{10} = \left[\frac{3Q\mu_L}{2\pi\rho_L(\pi dN)^2 \sin\theta}\right]^{1/3}$$
Hinze and Milborn [41]	Atomization by ligament formation $$D_{10} = \left[0.77\frac{Q}{Nd}\left(\frac{\rho_L N^2 d^3}{\sigma}\right)^{-5/12}\left(\frac{\rho_L \sigma d}{\mu_L^2}\right)^{-1/16}\right]^{0.5}$$
Hege [108]	Atomization by sheet disintegration $$D_{10} = 4.42\left(\frac{\sigma}{\omega^2\rho_L d}\right)^{0.5}$$

External-mixing atomizers usually employ a high-velocity stream of air or steam that flows from an annular nozzle and is arranged to impinge at some angle onto a jet or conical sheet of liquid that is created at the center of the atomizer. External-mixing air-assist nozzles have many features in common with airblast atomizers, and the drop size equations that have been established for one type of nozzle are generally relevant to the other.

Internal-Mixing Nozzles

Wigg's [63] analysis of the mechanism of airblast atomization highlighted the importance of the kinetic energy of the atomizing air and indicated that the energy difference between the inlet air and the emerging spray was the dominant factor affecting mean drop size. In a later work [64], Wigg used the wax spray data of Clare and Radcliffe [65] and Wood [66], obtained on the NGTE atomizer shown in Fig. 4.42, to derive the following dimensionless expression for MMD:

Table 6.5 Drop size equations for atomizer wheels

Investigators	Equation
Friedman et al. [59]	$$SMD = 0.44d\left(\frac{\dot{m}_L}{\rho_L Nd^2}\right)^{0.6}\left(\frac{\mu_L}{\dot{m}_L}\right)^{0.2}\left(\frac{\sigma\rho_L nh}{\dot{m}_L^2}\right)^{0.1}$$
Herring and Marshall [60]	$$SMD = \frac{3.3 \times 10^{-9} K\dot{m}_L^{0.24}}{(Nd)^{0.83}(nh)^{0.12}}$$
Fraser et al. [61]	$$SMD = 0.483N^{-0.6}\rho_L^{-0.5}\left(\frac{\mu_L \dot{m}_L}{d}\right)^{0.2}\left(\frac{\sigma}{nh}\right)^{0.1}$$
Scott et al. [62]	$$SMD = 6.3 \times 10^{-4}\dot{m}_L^{0.171}(ndN)^{-0.537}\mu_L^{-0.017}$$

$$\text{MMD} = 20v_L^{0.5}\dot{m}_L^{0.1}\left(1 + \frac{\dot{m}_L}{\dot{m}_A}\right)^{0.5} h^{0.1}\sigma^{0.2}\rho_A^{-0.3}U_R^{-1.0} \tag{6.47}$$

where h is the height of the air annulus in meters.

Mullinger and Chigier [67] studied the performance of an "internal-mixing" twin-fluid atomizer of the type illustrated in Fig. 4.43, in which liquid fuel is injected into a mixing chamber along with compressed air or steam. Some atomization occurs within the mixing chamber, but most of the liquid emerges from the atomizer in the form of a sheet that is then shattered into drops by the atomizing fluid. They found good agreement between their experimental data and predictions based on Eq. (6.47), even at air/liquid mass ratios as low as 0.005, which is one-tenth the lowest value tested by Wigg [63].

Application of Eq. (6.47) to water spray data [68, 69] gave a poor correlation; this was attributed by Wigg to recombination or coalescence of drops in the spray. To account for this effect, he suggested that whenever the conditions are conducive to drop coalescence, the value of MMD calculated from Eq. (6.47) should be multiplied by the empirical expression

$$1 + 5.0\left(\frac{\dot{m}_L}{\dot{m}_A}\right)^{0.6}\dot{m}_L^{0.1} \tag{6.48}$$

From studies carried out on the atomization of water, with air as the atomizing fluid, Sakai et al. [70] derived the following empirical formula for mean drop size.

$$\text{SMD} = 14 \times 10^{-6}d_o^{0.75}\left(\frac{\dot{m}_L}{\dot{m}_A}\right)^{0.75} \tag{6.49}$$

This equation applies over a range of water flow rates from 30 to 100 kg/h and values of $\dot{m}_L/\dot{m}_A$ from 5 to 100.

Inamura and Nagai [71] flowed air at uniform velocity through a vertically mounted cylindrical nozzle and used a thin annular slot, 1 mm in width, to inject liquid along the inside wall of the nozzle. Water, ethanol, and glycerin solutions were used to investigate the effects of liquid properties on atomization. Drop sizes were measured from spray samples collected on oil-coated glass slides.

For low liquid flow rates and low air velocities, Inamura and Nagai observed disturbances of large wavelength in the liquid flowing along the wall, causing the liquid film to break down into drops through the formation of unstable ligaments at the end of the wall. They called this process "atomization by ligament formation." Increase in air velocity and liquid flow rate caused the wavelength of the disturbances to diminish, so that the liquid emerged at the nozzle edge in the form of a thin continuous film. They termed the resulting atomization of this film by the high-velocity air "atomization by film formation." Some of the results obtained for the effects on SMD of variations in air velocity and air/liquid mass ratio are shown in Fig. 6.20. The curves drawn in this figure exhibit a change in slope, which is attributed by Inamura and Nagai to the transition from one mode

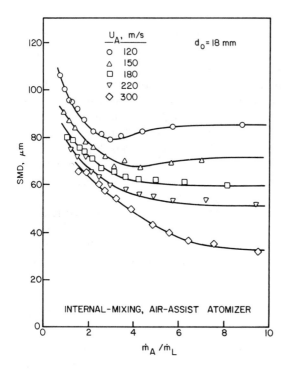

Figure 6.20 Variation of mean drop size with air/liquid mass ratio *(Inamura and Nagai [71])*.

of atomization to the other. It is of interest to note that the value of $\dot{m}_A/\dot{m}_L$ at which this transition occurs increases with increase in air velocity.

The influence of viscosity on SMD is illustrated in Fig. 6.21. The data plotted in this figure confirm that this type of atomizer is no different from all other types of twin-fluid atomizer in demonstrating that increase in liquid viscosity leads to coarser atomization, regardless of air velocity.

External-Mixing Nozzles

Inamura and Nagai [71] also investigated the performance of an external-mixing atomizer. The nozzle they used is designed to produce a flat circular sheet of liquid whose thickness can be varied from 0 to 0.7 mm by rotation of a screw. This liquid sheet is then deflected downward and atomized by an annular air jet, whose angle of impingement, relative to the nozzle axis, can also be varied. By using various ethanol and glycerin solutions, Inamura and Nagai were able to examine the effects of surface tension and viscosity on mean drop size. Analysis of the experimental data yielded the following dimensionally correct empirical equation for mean drop size:

$$\frac{\text{SMD}}{t} = \left[1 + \frac{16850\,\text{Oh}^{0.5}}{\text{We}(\rho_L/\rho_A)}\right]\left[1 + \frac{0.065}{(\dot{m}_A/\dot{m}_L)^2}\right] \tag{6.50}$$

where t = initial film thickness = $D_o\text{h}/D_{\text{an}}$

$\quad D_o$ = outer diameter of pressure nozzle

$\quad D_{\text{an}}$ = diameter of annular gas nozzle

$\quad h$ = slot width of pressure nozzle

$\quad \text{Oh}$ = stability number = $(\mu_L^2/\rho_L t\sigma)^{0.5}$

$\quad \text{We}$ = Weber number = $\rho_A U_A^2 t/\sigma$

Elkotb et al. [72] used 40 different nozzle configurations to examine the effects of nozzle geometry and operating variables on the characteristics of kerosine sprays. The kerosine employed had the following specifications: ρ_L = 800 kg/ m³, σ = 0.0304 kg/s², μ_L = 0.00335 kg/m s. Mean drop sizes were measured by the slide-sampling method. Up to 4000 drops were used to determine the SMD of any given spray.

Of special interest in this study are the data obtained on the influence of ambient air pressure on mean drop size. These results are shown in Fig. 6.22. Correlation of the experimental data led to the following dimensionally correct equation for mean drop size.

$$\text{SMD} = 51 d_o \text{Re}^{-0.39} \text{We}^{-0.18} \left(\frac{\dot{m}_L}{\dot{m}_A}\right)^{0.29} \tag{6.51}$$

where $\text{Re} = \rho_L U_R d_o/\mu_L$ and $\text{We} = \rho_L d_o U_R^2/\sigma$.

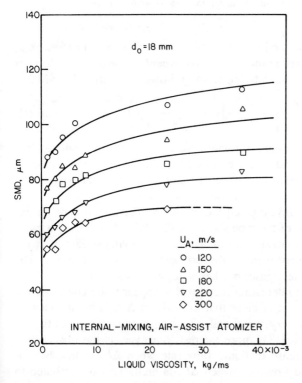

Figure 6.21 Variation of mean drop size with liquid viscosity (*Inamura and Nagai [71]*).

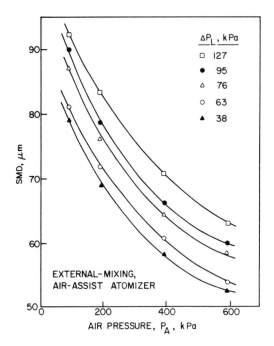

Figure 6.22 Influence of ambient air pressure on mean drop size *(Elkotb et al. [72])*.

In an attempt to bridge the gap between pressure atomizers and air atomizers, Simmons [18] carried out a large number of measurements of SMD on both types of atomizer. His analysis of these data led to the following equation for SMD.

$$\text{SMD} = C\left(\frac{\rho_L^{0.25}\mu_L^{0.06}\sigma^{0.375}}{\rho_A^{0.375}}\right)\left(\frac{\dot{m}_L}{\dot{m}_L U_L + \dot{m}_A U_A}\right)^{0.55} \tag{6.52}$$

In the above equation C is a constant whose value depends on nozzle design. The liquid properties ρ_L, μ_L, and σ are expressed relative to those of standard calibrating fluid [MIL-C-7024II] at room temperature, while air density is expressed as the value relative to sea level, room temperature conditions.

Suyari and Lefebvre [73] investigated the atomizing performance of an external-mixing, air-assist nozzle of the type shown in Fig. 6.23. Essentially it comprises a simplex pressure-swirl nozzle surrounded by a coflowing stream of swirling air. Drop sizes were measured using a Malvern particle sizing instrument. The liquids employed included water, gasoline, kerosine, and diesel oil.

Figure 6.24 exhibits some interesting features that appear to be characteristic of this type of atomizer. First, it is clear that increase in ΔP_A, i.e., increase in atomizing air velocity, has a very beneficial effect on atomization quality, especially at low liquid flow rates. Of more interest, however, is the observation that SMD decreases with increase in liquid flow rate when ΔP_A is low but increases with increase in liquid flow rate when ΔP_A is high. This is attributed to the fact that when atomizing air velocities are low ($\Delta P_A/P_A < 3\%$) the system

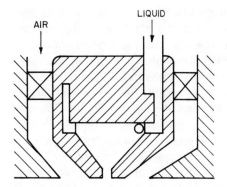

Figure 6.23 Schematic diagram of external-mixing air-assist nozzle *(Suyari and Lefebvre [73]).*

operates primarily as a simplex pressure-swirl nozzle. Thus increase in liquid flow rate reduces SMD by increasing the injection pressure. However, when the same nozzle is operating in an airblast mode, i.e., when $\Delta P_A/P_A$ is high (>4%), increase in liquid flow rate impairs atomization quality by lowering both the atomizing air/liquid ratio and the relative velocity between the air and the liquid.

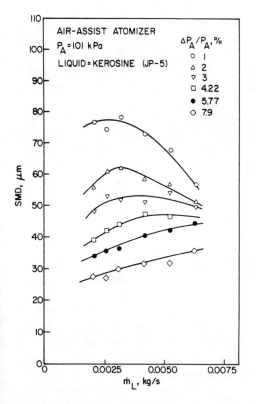

Figure 6.24 Influence of liquid flow rate and atomizing air velocity on mean drop size *(Suyari and Lefebvre [73]).*

These considerations help to explain the shape of the curve drawn in Fig. 6.24 for a value of $\Delta P_A/P_A$ of 2%. This curve shows SMD increasing with increase in liquid flow rate until a maximum value is reached, beyond which any further increase in $\dot{m}_L$ causes SMD to decline. The unusual shape of this curve can be explained in terms of the relative velocity between the conical sheet of liquid produced by the simplex atomizer and the surrounding coflowing air. At low liquid flow rates the air velocity is higher than the liquid velocity. Thus, in this regime any increase in liquid flow rate, i.e., any increase in liquid velocity, will impair atomization quality by reducing the relative velocity between the air and the liquid. At the peak of the curve, where atomization quality is poorest, the relative velocity is zero and the interaction between the liquid sheet and the surrounding air is minimal. Beyond this point the liquid velocity exceeds the air velocity, and any further increase in liquid flow rate serves to improve atomization by increasing the relative velocity between the liquid and the air.

These arguments suggest that for this type of nozzle there will always be one value of liquid flow rate for which SMD is a maximum. The particular value of $\dot{m}_L$ at which this occurs will be that for which the liquid velocity just equals the air velocity. Now in most practical forms of air-assist atomizer, the air and the liquid issuing from the nozzle both have radial and tangential as well as axial components of velocity. In this complex two-phase flow field the term "relative velocity" is difficult to define. Nevertheless, for any given air velocity there will always be one particular value of liquid velocity at which the level of interaction between the fuel and the air is a minimum. With increases in atomizing air velocity, the condition of zero relative velocity will obviously be reached at higher levels of liquid velocity. Thus one would expect the value of $\dot{m}_L$ at which the SMD attains its maximum value to increase with increase in $\Delta P_A/P_A$, and this is evident in the changing shapes of the curves drawn in Fig. 6.24.

Figure 6.24 demonstrates that atomization quality is improved by an increase in liquid injection pressure when the system is operating in the pressure-swirl mode and by an increases in atomizing air velocity and/or air/liquid ratio when it is operating as an air atomizer. This is fully consistent with all previous experience on pressure-swirl and airblast nozzles. However, it also shows that the relative velocity between the air and the liquid has an additional independent effect over and above the effects of air velocity and liquid velocity.

The results obtained when operating with JP 4 fuel at atmospheric pressure are shown in Fig. 6.25, in which measured values of SMD are plotted against liquid injection pressure differential, ΔP_L. One interesting feature of this figure is that except at the lowest liquid flows, where liquid pressure alone is too low to achieve good atomization, the atomization quality is actually higher without atomizing air than when air is supplied at a pressure differential of 2%. In fact, over a wide range of liquid pressures, an atomizing air pressure differential of at least 3% is needed to provide the same atomization quality that is achieved with no air assistance at all. Nevertheless, the very substantial improvements in atomization quality that are derived from air assistance at low liquid injection pressures are also apparent in Fig. 6.25.

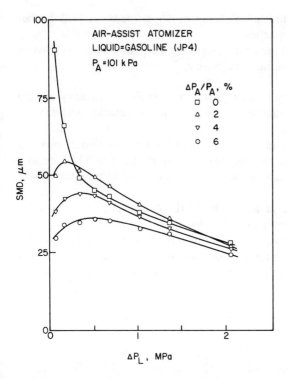

Figure 6.25 Influence of liquid pressure differential and atomizing air velocity on mean drop size *(Suyari and Lefebvre [73]).*

From analysis of their experimental data, Suyari and Lefebvre [73] concluded that the two key factors governing the atomization quality achieved with external-mixing air-assist atomizers, when operating with low-viscosity liquids, are air dynamic pressure, $0.5\rho_A U_A^2$ (as indicated by ΔP_A), and the relative velocity between the liquid sheet produced by the pressure-swirl nozzle and the surrounding air. Thus relative velocity appears to play an important and independent role, presumably through its influence on the development of capillary waves (ripples) on the initially smooth liquid surface and the production of unstable ligaments. As the relative velocity increases, the size of the ligaments decreases, their life becomes shorter, and, on their collapse when exposed to the high-velocity airstream, much smaller drops are formed, in accordance with Rayleigh theory [74].

These findings raise doubts concerning the validity of some existing equations for SMD. For example, Eq. (6.52) suggests that increase in either U_L or U_A should be beneficial to atomization quality. However, the results show that whereas increase in U_A is usually conducive to better atomization, especially at low liquid injection pressures, increase in U_L can either raise or lower SMD, depending on the initial values of U_L and U_A. If the initial value of U_L is less than U_A, then an increase in U_L will impair atomization by reducing the relative velocity between the liquid and the air. Only if the initial value of U_L exceeds U_A will further increase in U_L lead to better atomization.

From the results of their study on external-mixing air-assist atomizers, Suyari and Lefebvre [73] drew the following conclusions.

1. For any given liquid, the key factors governing atomization quality are the dynamic pressure of the atomizing air and the relative velocity between the liquid and the surrounding air.
2. For any given value of air velocity, continual increase in liquid injection pressure from an initial value of zero produces an increase in SMD up to a maximum value, beyond which further increase in injection pressure causes SMD to decline.
3. The liquid injection pressure at which the SMD attains its maximum value increases with increase in atomizing air velocity.
4. Whereas increase in air velocity is usually beneficial to atomization quality, increase in liquid velocity may help or hinder atomization, depending on whether it increases or decreases the relative velocity between the liquid and the surrounding air.

The expressions for mean drop size presented above are summarized in Table 6.6.

AIRBLAST ATOMIZERS

Although the principle of airblast atomization has been known for a long time, strong interest in the design and performance of airblast nozzles was not evoked until the mid-1960s, when their potential for achieving significant reductions in soot formation and exhaust smoke in gas turbine engines of high pressure ratio started to become recognized.

Most of the systems now in service are of the *prefilming* type, in which the fuel is first spread out into a thin continuous sheet and then subjected to the atomizing action of high-velocity air. In other designs, the fuel is injected into the high-velocity airstream in the form of one or more discrete jets. The atomization performance of prefilming nozzles is generally superior to that of plain-jet nozzles [75, 76], but they are fully effective only when both sides of the liquid sheet are exposed to the air. This requirement introduces a complication in design, since it usually means arranging for two separate airflows through the atomizer. For this reason, the plain-jet type of airblast atomizer, in which the fuel is not transformed into a thin sheet but instead is injected into the high-velocity airstream in the form of discrete jets, is sometimes preferred.

Plain Jet

The first major study of airblast atomization was conducted over 40 years ago by Nukiyama and Tanasawa [68] on a plain-jet airblast atomizer, as illustrated in

Table 6.6 Drop size equations for air-assist atomizers

Investigators	Atomizer type	Equations	Remarks
Wigg [64]	Internal mixing	$\mathrm{MMD} = 20 v_L^{0.5} \dot{m}_L^{0.1}\left(1 + \dfrac{\dot{m}_L}{\dot{m}_A}\right)^{0.5} h^{0.1} \sigma^{0.2} \rho_A^{-0.3} U_R^{-1.0}$	h = height of air annulus, m
Sakai et al. [70]	Internal mixing	$\mathrm{SMD} = 14 \times 10^{-6} d_o^{0.75}\left(\dfrac{\dot{m}_L}{\dot{m}_A}\right)^{0.75}$	d_o = nozzle diameter
Inamura and Nagai [71]	External mixing	$\dfrac{\mathrm{SMD}}{t} = \left[1 + \dfrac{16850\,\mathrm{Oh}^{0.5}}{\mathrm{We}(\rho_L/\rho_A)}\right]\left[1 + \dfrac{0.065}{(\dot{m}_A/\dot{m}_L)^2}\right]$	$\mathrm{Oh} = \left(\dfrac{\mu_L^2}{\rho_L t \sigma}\right)^{0.5}$
Elkotb et al. [72]	External mixing	$\mathrm{SMD} = 51 d_o\,\mathrm{Re}^{-0.39}\mathrm{We}^{-0.18}\left(\dfrac{\dot{m}_L}{\dot{m}_A}\right)^{0.29}$	$\mathrm{Re} = \dfrac{\rho_L U_R d_o}{\mu_L}$ $\mathrm{We} = \dfrac{\rho_L d_o U_R^2}{\sigma}$
Simmons [18]	General	$\mathrm{SMD} = C\left(\dfrac{\rho_L^{0.25} \mu_L^{0.06} \sigma^{0.375}}{\rho_A^{0.375}}\right)\left(\dfrac{\dot{m}_L}{\dot{m}_L U_L + \dot{m}_A U_A}\right)^{0.55}$	

Fig. 6.26. The drop sizes were measured by collecting samples of the spray on oil-coated glass slides. Drop size data were correlated by the following empirical equation for the SMD:

$$\mathrm{SMD} = \frac{0.585}{U_R}\left(\frac{\sigma}{\rho_L}\right)^{0.5} + 53\left(\frac{\mu_L^2}{\sigma \rho_L}\right)^{0.225}\left(\frac{Q_L}{Q_A}\right)^{1.5} \qquad (6.53)$$

It is of interest to note that the right-hand side of Eq. (6.53) is expressed as the sum of two separate terms, one of which is dominated by relative velocity and surface tension and the other by viscosity.

Although Eq. (6.53) is not dimensionally correct, it nevertheless allows some useful conclusions to be drawn. For example, for liquids of low viscosity it shows that mean drop size is inversely proportional to the relative velocity between the air and the liquid, while for large values of air/liquid ratio the influence of viscosity on SMD becomes negligibly small.

Equation (6.53) can be made dimensionally correct by introducing a term to denote length, raised to the 0.5 power. One obvious choice for this length is the diameter of the liquid orifice or air nozzle. However, from tests carried out with different sizes and shapes of nozzles and orifices, Nukiyama and Tanasawa concluded that these factors have virtually no effect on mean drop size. Thus, the absence of atomizer dimensions is a notable feature of Eq. (6.53).

Another significant omission is air density, which was kept constant (at the normal atmospheric value) in all experiments. This represents a serious limitation, since it prohibits the application of Eq. (6.53) to the many types of atomizers that are required to operate over wide ranges of air pressure and temperature.

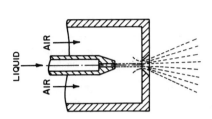

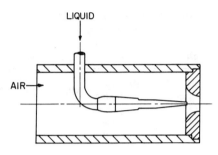

Figure 6.26 Nukiyama-Tanasawa plain-jet airblast atomizer [68].

Figure 6.27 Lorenzetto-Lefebvre plain-jet airblast atomizer [75].

The performance of plain-jet atomizers was investigated in detail by Lorenzetto and Lefebvre [75], using a specially designed system in which liquid physical properties, air/liquid ratio, air velocity, and atomizer dimensions could be varied independently over wide ranges and examined for their separate effects on spray quality. A cross-sectional drawing of the atomizer is shown in Fig. 6.27.

Some of the results of this study are shown in Figs. 6.28 to 6.32. They show that atomization quality deteriorates with increases in viscosity and surface tension, in the same manner as with pressure-swirl and rotary atomizers. For liquids of low viscosity, it was found that mean drop size is inversely proportional to the relative velocity between the air and the liquid at the nozzle exit. It was also observed, in agreement with Nukiyama and Tanasawa [68], that nozzle dimensions appear to have little influence on mean drop size for liquids of low viscosity. For liquids of high viscosity, SMD varies roughly in proportion to $d_o^{0.5}$.

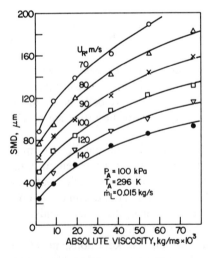

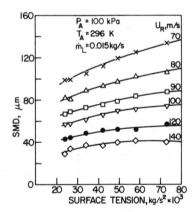

Figure 6.28 Variation of mean drop size with liquid viscosity for a plain-jet airblast atomizer *(Lorenzetto and Lefebvre [75]).*

Figure 6.29 Variation of mean drop size with surface tension for a plain-jet airblast atomizer *(Lorenzetto and Lefebvre [75]).*

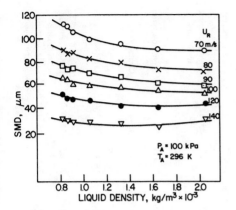

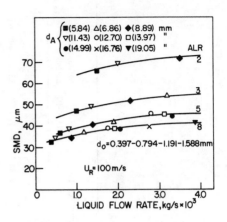

Figure 6.30 Variation of mean drop size with liquid density for a plain-jet airblast atomizer *(Lorenzetto and Lefebvre [75]).*

Figure 6.31 Variation of mean drop size with liquid flow rate for a plain-jet airblast atomizer *(Lorenzetto and Lefebvre [75]).*

The atomizing performance of prefilming and plain-jet airblast nozzles was compared by plotting some of the measured values of SMD along with the experimental data obtained by Rizkalla and Lefebvre [77] under comparable operating conditions. The results of such a comparison are presented in Fig. 6.32. It is clear that the plain-jet atomizer performs less satisfactorily than the prefilming type, especially under the adverse condition of low air/liquid ratio (ALR) and/ or low air velocity.

The following dimensionally correct expression for the SMD was derived from an analysis of the experimental data:

$$\text{SMD} = 0.95 \frac{(\sigma \dot{m}_L)^{0.33}}{U_R \rho_L^{0.37} \rho_A^{0.30}} \left(1 + \frac{1}{\text{ALR}}\right)^{1.70} + 0.13 \left(\frac{\mu_L^2 d_o}{\sigma \rho_L}\right)^{0.5} \left(1 + \frac{1}{\text{ALR}}\right)^{1.70} \quad (6.54)$$

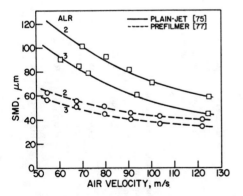

Figure 6.32 Comparison of atomizing performance of plain-jet and prefilming types of airblast atomizer *(Lorenzetto and Lefebvre [75]).*

Table 6.7 Test conditions employed in experimental studies

Type of atomizer	Investigators	Liquids employed	Liquid properties			Air properties			Air/liquid mass ratio	SMD × 10⁶	Method of measurement
			$\sigma \times 10^3$	ρ_L	$\mu_L \times 10^3$	$P_A \times 10^{-2}$	T_A	U_A			
Prefilmer	Lefebvre and Miller [80]	Water, kerosine	27.7–73.5	784–1000	1.0–1.29	1.0	295	167–122	3–9	42–90	Collected on coated slide
	Rizkalla and Lefebvre [77]	Water, kerosine, special solutions	24–73	780–1500	1.0–44	1.0–8.5	296–424	70–125	2–11	30–120	Light-scattering technique
	Jasuja [15]	Residual fuel oil	27–74	784–1000	1.0–8.6	1.0	295	55–135	1–8	30–140	Light-scattering technique
	El-Shanawany and Lefebvre [82]	Water, kerosine, special solutions	26–74	784–1000	1.0–44	1.0–8.5	295	60–190	0.5–5.0	25–125	Light-scattering technique
Spinning cup, prefilm	Fraser et al. [90]	Various oils	29–35	810–830	5–165		295	29–198	0.17–4.0	20–320	Light absorption
Plain jet	Nukiyama and Tanasawa [68]	Water, gasoline, alcohol, heavy fuel oil	19–73	700–1200	1.0–5.0	1.0	295	60–340	1–14	15–19	Coated slide
	Lorenzetto and Lefebvre [75]	Water, kerosine, special solutions	26–76	794–2180	1.0–76	1.0	295	60–180	1–16	20–130	Light-scattering technique
	Jasuja [78]	Residual fuel oils	21–74	784–1000	1.0–53	1.0	295	70–135	2–18	35–120	Light-scattering technique
	Rizk and Lefebvre [79]	Kerosine, gas oil, blended oils	27–29	780–840	1.3–3.0	1–7.7	298	10–120	2–8	15–110	Light-scattering technique

Category	Investigator	Liquid									Technique
Air-assist	Wigg [64]	Water	73.5	1000	1.0	1.0	295	300–340	1–3	50–115	Immersion technique
	Sakai et al. [70]	Water	73.5	1000	1.0	1.0	298		0.001–0.1	90–500	Collected on oil-coated slide
	Inamura and Nagai [71]	Water	73.5	1000	1.0	1.0	298	0–300	1–10	35–140	Collection on coated slide
	Elkotb et al. [72]	Kerosine	30	800	3.35	1.0–6.0	295		0.5–2	20–65	
	Simmons [18]	Calibrating fluid MIL-F-70411	25	763	1.2	1.0	295		0.5–6	20–300	Parker Hannifin spray analyzer
Miscellaneous	Gretzinger and Marshall [88]	Aqueous solutions	50		1.0	1.6–4	295	Up to sonic	1–16	5–30	Microscope counting
	Ingebo and Foster [84]	Water, iso-octane, benzene, JP 5	16–71		0.45–1.93	0.4–1.64	298	30–210	5–100	25–80	High-speed photography
	Kim and Marshall [89]	Wax melts	30–50	800–960	1.0–5.0	0.76–2.0	298	75–393	0.06–40	10–160	Microscope counting
	Weiss and Worsham [87]	Wax melts	18–22	806–828	3.2–11.3	1.0–5.0	298	60–300	4–25	17–100	Micromerograph and sifting
	Ingebo [86]	Water	73.5	1000	1.0	1.0–21	295	45–220		50–200	Radiometer

This expression was shown to be accurate to within 8% over the broad range of air and liquid properties indicated in Table 6.7.

Further studies at Cranfield on plain-jet atomization were made by Jasuja [78] using a single-nozzle configuration, as illustrated in Fig. 4.49. In this nozzle the liquid flows through a number of radially drilled plain circular holes, from which it emerges in the form of discrete jets that enter a swirling airstream. These jets then undergo in-flight disintegration without any further preparation such as prefilming. Jasuja derived the following equation for drop size correlations:

$$\text{SMD} = 0.022 \left(\frac{\sigma}{\rho_A U_A^2} \right)^{0.45} \left(1 + \frac{1}{\text{ALR}} \right)^{0.5}$$
$$+ 0.00143 \left(\frac{\mu_L^2}{\sigma \rho_L} \right)^{0.4} \left(1 + \frac{1}{\text{ALR}} \right)^{0.8} \tag{6.55}$$

It should be noted that this equation contains no dimensions and is not dimensionally correct.

Rizk and Lefebvre [79] have also examined the effects of air and liquid properties and atomizer dimensions on mean drop size for the type of nozzle shown in Fig. 6.27. Tests were conducted using two geometrically similar atomizers having liquid orifice diameters of 0.55 and 0.75 mm. The liquids employed and the ranges of test conditions covered are shown in Table 6.7.

Figure 6.33, showing the effects of air velocity, air pressure, air/liquid ratio, and nozzle dimensions on SMD, is typical of the results obtained, all of which generally confirm the results of previous studies on airblast atomizers. They show that increases in air pressure, air velocity, and air/liquid ratio all tend to lower the mean drop size. An empirical, dimensionally correct equation for mean drop size was derived as follows:

$$\frac{\text{SMD}}{d_o} = 0.48 \left(\frac{\sigma}{\rho_A U_R^2 d_o} \right)^{0.4} \left(1 + \frac{1}{\text{ALR}} \right)^{0.4} + 0.15 \left(\frac{\mu_L^2}{\sigma \rho_L d_o} \right)^{0.5} \left(1 + \frac{1}{\text{ALR}} \right) \tag{6.56}$$

This equation was shown to provide an excellent data correlation, especially for low-viscosity liquids. Comparison with the results obtained by other workers is hindered by the different methods adopted for presenting the data. However, a comparison is provided for liquids of low viscosity, as illustrated in Table 6.8. The values for the exponents of air pressure, air velocity, and surface tension listed in this table are reasonably consistent with each other, but there are appreciable differences in regard to the effect of nozzle dimensions on SMD.

Prefilming

Plain-jet airblast atomizers are not very suitable for many combustion applications because they tend to concentrate the fuel drops within a small region just downstream of the nozzle. Much more preferable is the prefilming type of airblast atomizer, which can be designed to disperse the fuel drops throughout the entire combustion zone.

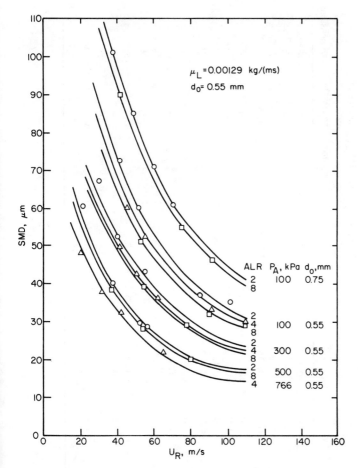

Figure 6.33 Variation of mean drop size with air pressure, air/liquid ratio, and fuel nozzle diameter for a plain-jet airblast atomizer *(Rizk and Lefebvre [79])*.

The prefilming concept for airblast atomization evolved from the experimental studies of Lefebvre and Miller in the early 1960s [80]. These workers carried out a large number of tests, using both water and kerosine, on several different atomizer configurations. Drop sizes were measured by collecting the drops on magnesium oxide-coated slides. Their experimental data generally confirmed the results obtained with other types of twin-fluid atomizer in regard to the important influence of air velocity on mean drop size. They also showed that variation in air/liquid ratio over the range from 3 to 9 had little effect on atomization quality. However, the main conclusion from this study was that "Minimum drop sizes are obtained by using atomizers designed to provide maximum physical contact between the air and the liquid. In particular it is important to arrange for the liquid sheet formed at the atomizing lip to be subjected to high velocity air on *both* sides. This also ensures that the drops remain airborne and avoids deposition of liquid on solid surfaces."

Table 6.8 Summary of data on effect of variables on mean drop size for liquids of low viscosity

Type of twin-fluid atomizer	Investigators	Power dependence on mean drop size of variable indicated						
		Velocity U_A	Air density ρ_A	Liquid density ρ_L	Surface tension σ	Air/liquid ratio $(1 + \dot{m}_L/\dot{m}_A)$	Liquid/air ratio $(\dot{m}_L/\dot{m}_A)$	Dimensions
Air assist	Wigg [64]	−1.0	−0.3	0	0.2	0.5	—	$h^{0.1}$
	Sakai et al. [70]	—	—	—	—	—	0.75	$d_o^{0.75}$
	Elkotb et al. [72]	−0.75	−0.39	−0.18	0.18	—	0.29	$d^{0.43}$
	Simmons [18]	−0.55	−0.375	.25	0.375	—	0.55	—
Plain jet	Nukiyama and Tanasawa [68]	−1.0	0	−0.5	0.5		0	0
	Lorenzetto and Lefebvre [75]	−1.0	−0.3	−0.37	0.33	1.70	—	—
	Jasuja [78]	−0.9	−0.45	0	0.45	0.5	—	$d_o^{0.55}$
	Rizk and Lefebvre [79]	−0.8	−0.4	0	0.4	0.4	—	$d_o^{0.6}$
Prefilmer	Rizkalla and Lefebvre [77]	−1.0	−1.0	0.5	0.5	1.0	—	$D_p^{0.5}$
	Jasuja [15]	−1.0	−1.0	0.5	0.5	0.5	—	0
	Lefebvre [91]	−1.0	−0.5	0	0.5	1.0	—	$D_p^{0.5}$
	El-Shanawany and Lefebvre [82]	−1.2	−0.7	0.1	0.6	1.0	—	$D_p^{0.4}$
Miscellaneous	Ingebo and Foster [84]	−0.75	−0.25	−0.25	0.25	0	0	$D_p^{0.5}$
	Weiss and Worsham [87]	−1.33	−0.30	—	—	—	—	$d_o^{0.16}$
	Gretzinger and Marshall [88]	−0.15	−0.15	0	0	—	0.6	$L^{-0.15}$
	Kim and Marshall [89]	−1.44	0.72	−0.16	0.41	0	0	0
	Fraser et al. [90]	—	−0.5	0	0.5	—	—	—

246

Lefebvre and Miller also stressed the importance of spreading the liquid into the thinnest possible sheet before subjecting it to airblast action, on the grounds that "Any increase in the thickness of the liquid sheet flowing over the atomizing lip will tend to increase the thickness of the ligaments which, upon disintegration, will then yield drops of larger size."

The main value of this study is not so much in the experimental data obtained, which are necessarily suspect due to the fairly primitive measuring techniques employed, but in its identification of the essential features of successful prefilming atomizer design.

Rizkalla and Lefebvre [77] used the light-scattering technique in a detailed investigation of prefilming airblast atomization. A cross-sectional drawing of the atomizer employed is shown in Fig. 6.34. In this design the liquid flows through six equispaced tangential ports and into a weir, from which it spills over the prefilming surface before being discharged at the atomizing lip. To subject both sides of the liquid to high-velocity air, two separate airflow paths are provided. One airstream flows through a central circular passage and is deflected radially outward by a pintle before striking the inner surface of the liquid sheet. The other airstream flows through an annular passage surrounding the main body of the atomizer. This passage has its minimum flow area in the plane of the atomizing lip, to impart a high velocity to the air where it meets the outer surface of the liquid sheet.

From inspection of all the data obtained on the effects of air and liquid properties on atomization quality [77], some of which are shown in Figs. 6.35 to 6.40, Rizkalla and Lefebvre drew certain general conclusions concerning the main factors governing mean drop size. For liquids of low viscosity, the key factors are surface tension, air velocity, and air density. From results obtained over a wide test range, it was concluded that liquid viscosity has an effect that is quite separate and independent from that of surface tension. This suggested a form of equation in which size is expressed as the sum of two terms—the first dominated by surface tension and air momentum and the second by liquid viscosity. Dimensional analysis was used to derive the following equation, the various constants and indices being deduced from the experimental data.

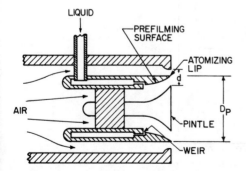

Figure 6.34 Prefilming type of airblast atomizer [77, 82]. Note that $D_h = 2d$.

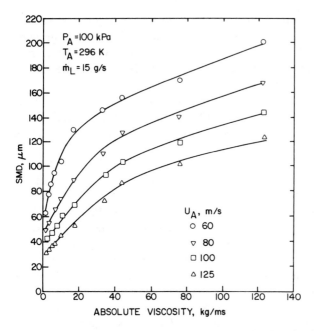

Figure 6.35 Variation in mean drop size with liquid viscosity for a prefilming airblast atomizer *(Rizkalla and Lefebvre [77])*.

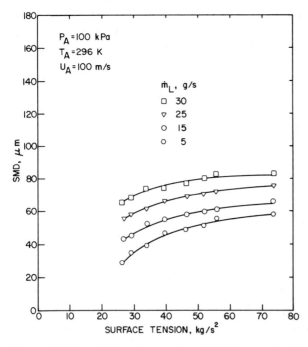

Figure 6.36 Variation in mean drop size with surface tension for a prefilming airblast atomizer *(Rizkalla and Lefebvre [77])*.

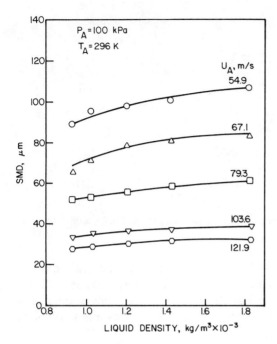

Figure 6.37 Variation in mean drop size with liquid density for a prefilming airblast atomizer *(Rizkalla and Lefebvre [77]).*

$$\text{SMD} = 3.33 \times 10^{-3} \frac{(\sigma \rho_L D_p)^{0.5}}{\rho_A U_A} \left(1 + \frac{1}{\text{ALR}}\right)$$

$$+ 13.0 \times 10^{-3} \left(\frac{\mu_L^2}{\sigma \rho_L}\right)^{0.425} D_p^{0.575} \left(1 + \frac{1}{\text{ALR}}\right)^2 \quad (6.57)$$

where L_c, the characteristic dimension of the atomizer, is set equal to D_p.

For liquids of low viscosity, such as water and kerosine, the first term pre-

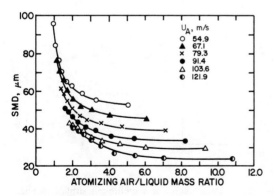

Figure 6.38 Variation of mean drop size with air/liquid ratio for a prefilming airblast atomizer *(Rizkalla and Lefebvre [77]).*

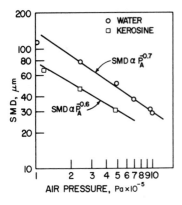

Figure 6.39 Variation of mean drop size with ambient air pressure for a prefilming airblast atomizer *(Rizkalla and Lefebvre [77]).*

dominates; the SMD thus increases with increases in liquid surface tension, liquid density, and atomizer dimensions and decreases with increases in air velocity, air/liquid ratio, and air density. For liquids of high viscosity the second term acquires greater significance; in consequence, the SMD becomes less sensitive to variations in air velocity and density.

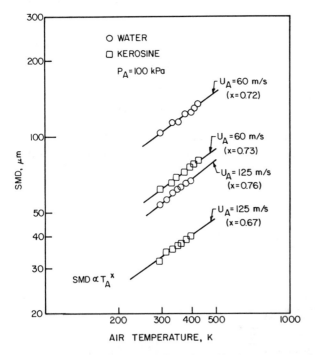

Figure 6.40 Variation of mean drop size with air temperature for a prefilming airblast atomizer *(Rizkalla and Lefebvre [77]).*

In view of the growing interest in "alternative" fuels for gas turbines, Jasuja [15] carried out an experimental study of the airblast atomization characteristics of kerosine, gas oil, and various blends of gas oil with residual fuel oil. He used the airblast nozzle developed by Bryan et al. [81], as illustrated in Fig. 6.41, and measured mean drop sizes by the light-scattering technique. The experimental data correlated well with the equation

$$SMD = 10^{-3}\left(1 + \frac{1}{ALR}\right)^{0.5}\left[\frac{(\sigma\rho_L)^{0.5}}{\rho_A U_A} + 0.06\left(\frac{\mu_L^2}{\sigma\rho_A}\right)^{0.425}\right] \quad (6.58)$$

This equation is very similar to Eq. (6.57), except for the absence of a term representing the atomizer dimensions and a somewhat lower dependence of the SMD on air/liquid ratio.

The effect of atomizer scale on mean drop size was examined by El-Shanawany and Lefebvre [82]. They used three geometrically similar nozzles having cross-sectional areas in the ratio of 1 : 4 : 16. The basic atomizer design was almost identical to that used by Rizkalla and Lefebvre [77], shown in Fig. 6.34, except for some modifications to the inner airflow passage to ensure that its cross-sectional area diminished gradually and continuously toward the atomizing lip. These changes eliminated any possibility of airflow separation within the passage and encouraged the liquid film to adhere to the prefilming surface until it reached the atomizing lip.

El-Shanawany and Lefebvre's experiments were mainly confined to water and kerosine, but some specially prepared liquids of high viscosity were also used. Mean drop sizes were measured by the light-scattering technique in its improved form, as developed by Lorenzetto [75]. Their results for water, obtained over a

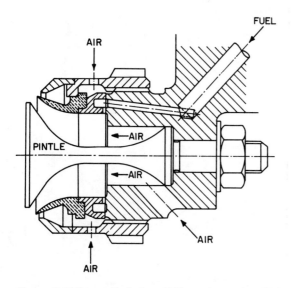

Figure 6.41 Bryan, Godbole, and Norster's atomizer [81].

wide range of air velocities with all three nozzles, are shown in Fig. 6.42. They demonstrate that SMD $\propto L_c^{0.43}$. From an analysis of all the experimental data, El-Shanawany and Lefebvre [82] concluded that the mean drop sizes produced by prefilming airblast atomizers could be correlated by the following dimensionally correct equation:

$$\frac{SMD}{D_h} = \left(1 + \frac{1}{ALR}\right)\left[0.33\left(\frac{\sigma}{\rho_A U_A^2 D_p}\right)^{0.6}\left(\frac{\rho_L}{\rho_A}\right)^{0.1} + 0.068\left(\frac{\mu_L^2}{\rho_L \sigma D_p}\right)^{0.5}\right] \quad (6.59)$$

Wittig and colleagues [83] have also examined the spray characteristics of a prefilming type of airblast atomizer in which the liquid sheet is injected into the outer airstream. Their results are generally consistent with those obtained for other types of prefilming airblast atomizers.

Miscellaneous Types

Several important studies of airblast atomization have been carried out on nozzle designs or nozzle configurations that do not fit neatly into either of the two main categories of plain-jet or prefilming systems. For example, Ingebo and Foster [84] used a plain-jet type of airblast atomizer, featuring cross-current air injection, to examine the breakup of isooctane, JP 5, benzene, carbon tetrachloride, and water. These workers derived an empirical relationship to correlate their experimental data, which they expressed in the following form:

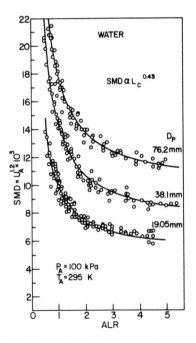

Figure 6.42 Effect of linear scale on mean drop size for water *(El-Shanawany and Lefebvre [82])*.

$$\frac{\text{SMD}}{d_o} = 5.0(\text{We Re})^{-0.25} \tag{6.60}$$

Substituting for We and Re gives

$$\text{SMD} = 5.0\left(\frac{\sigma\mu_L d_o^2}{\rho_A U_R^3 \rho_L}\right)^{0.25} \tag{6.61}$$

According to Ingebo [85], the above equations are valid for We Re $< 10^6$. For We Re $> 10^6$ Ingebo recommends the following expression for mean drop size:

$$\frac{\text{SMD}}{d_o} = 37(\text{We Re})^{-0.4} \tag{6.62}$$

or

$$\text{SMD} = 37\left(\frac{\sigma\mu_L d_o^{0.5}}{\rho_A U_R^3 \rho_L}\right)^{0.4} \tag{6.63}$$

Only water jets were used in deriving these dimensionally correct empirical expressions.

Equations (6.60)–(6.63) are highly relevant to the design of various air-breathing propulsion engines, such as ramjets and turbojet afterburners. These engines commonly employ radial fuel injection from plain-orifice atomizers into high-velocity, cross-flowing streams of air or gas.

In situations where transverse penetration of fuel into the gas stream is unnecessary or undesirable, a splash-plate type injector is generally preferred. With this device a round fuel jet is arranged to impinge at the center of a small plate. As the fuel flows over the edge of the plate, it is atomized by the high-velocity gas stream in which it is immersed. Essentially, the device functions as a simple prefilming airblast atomizer.

Ingebo [86] has studied the atomizing performance of this type of injector for air pressures ranging from 0.10 to 2.1 MPa (1 to 20 atm). His tests were confined to water, for which the effects of liquid velocity and air properties on mean drop size conformed to the relationship

$$\text{SMD} = (2.67 \times 10^4 U_L P_A^{-0.33} + 4.11 \times 10^6 \rho_A U_A P_A^{-0.75})^{-1} \tag{6.64}$$

Weiss and Worsham [87] investigated both cross-stream and costream injection of liquid jets into high-velocity air. Their experiments covered ranges of air velocity from 60 to 300 m/s, orifice diameters from 1.2 to 4.8 mm, liquid viscosities from 0.0032 to 0.0113 kg/m s, and air densities from 0.74 to 4.2 kg/m^3. The most important factor controlling the mean drop size was found to be the relative velocity between the air and the liquid. They argued that the drop size distribution in a spray depends on the range of excitable wavelengths on the surface of a liquid sheet, the shorter-wavelength limit being due to viscous damping, while the longer wavelengths are limited by inertia effects. Based on these assumptions, they concluded that

$$\text{SMD} \propto \left(\frac{U_L^{0.08} d_o^{0.16} \mu_L^{0.34}}{\rho_A^{0.30} U_R^{1.33}} \right) \tag{6.65}$$

Their experimental data confirmed the dependence predicted for air velocity but gave slightly different indices for the effects of liquid properties.

Gretzinger and Marshall [88] studied the characteristics of mean drop size and drop size distribution for two different designs of airblast atomizer. One was a "converging" nozzle of the type employed by Nukiyama and Tanasawa [68], in which the liquid is first brought into contact with the atomizing airstream at the throat of an air nozzle. The other, described as an "impingement" nozzle, consisted of a central circular air tube surrounded by an annular liquid duct. An impinger, mounted on a rod centered in the air tube, permitted adjustment of the airflow pattern and produced corresponding changes in the spray cone angle. Three orifices of 2.4, 2.8, and 3.2 mm diameter were employed.

Sprays of aqueous solutions of black dye were first dried and then collected in mineral oil and counted using a light microscope. The experiments covered a liquid flow range from 2 to 17 liters/h. Over the range of drop diameters from 5 to 30 μm, their experimental data were correlated by the equation

$$\text{MMD} = 2.6 \times 10^{-3} \left(\frac{\dot{m}_L}{\dot{m}_A} \right)^{0.4} \left(\frac{\mu_A}{\rho_A U_A L} \right)^{0.4} \tag{6.66}$$

for the convergent nozzle and

$$\text{MMD} = 1.22 \times 10^{-4} \left(\frac{\dot{m}_L}{\dot{m}_A} \right)^{0.6} \left(\frac{\mu_A}{\rho_A U_A L} \right)^{0.15} \tag{6.67}$$

for the impingement nozzle, where L is the diameter of the wetted periphery between the air and liquid streams.

The results of this investigation generally confirmed previous findings in regard to the beneficial effects of increases in air/liquid mass ratio, air velocity, and air density on atomization quality. However, Gretzinger and Marshall [88] emphasized that discretion should be exercised in applying their correlations to other nozzle types, especially for fluids whose physical properties are dissimilar to those used in their tests.

An interesting feature of Eqs. (6.66) and (6.67) is that they contain an air viscosity term, although the effect of air viscosity on mean drop size was not studied directly. More surprising, perhaps, is the result that increase in nozzle size reduces the mean drop size. This was interpreted to mean that a longer wetted perimeter creates a thinner liquid film and hence a finer spray.

Kim and Marshall [89] used an airblast nozzle of versatile design that provided two different configurations for the atomizing airflow. In one form, the atomizing air converged and expanded through an annulus around a liquid nozzle. In the alternative form, a secondary air nozzle was inserted axially in the liquid nozzle as shown in Fig. 6.43. Thus, an annular liquid sheet was produced between

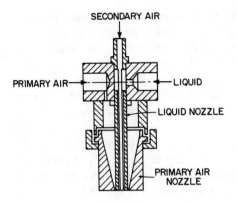

SECONDARY AIR

PRIMARY AIR —

— LIQUID

— LIQUID NOZZLE

PRIMARY AIR
NOZZLE

Figure 6.43 Kim and Marshall's atomizer [89].

two airstreams, an arrangement that previous work by Lefebvre and Miller [80] had shown to be very successful in producing fine sprays. Drop size measurements were carried out on melts of wax mixtures over a range of liquid viscosity of 0.001 to 0.050 kg/m s, relative air velocity of 75 to 393 m/s, air/liquid mass flow ratio of 0.06 to 40, liquid density of 800 to 960 kg/m^3, and air density of 0.93 to 2.4 kg/m^3.

For the convergent single airblast nozzle, it was found that

$$\text{MMD} = 5.36 \times 10^{-3} \frac{\sigma_L^{0.41} \mu_L^{0.32}}{(\rho_A U_R^2)^{0.57} A^{0.36} \rho_L^{0.16}}$$
$$+ 3.44 \times 10^{-3} \left(\frac{\mu_L^2}{\rho_L \sigma}\right)^{0.17} \left(\frac{\dot{m}_A}{\dot{m}_L}\right)^m \frac{1}{U_R^{0.54}} \quad (6.68)$$

where A is the flow area of the atomizing airstream in square meters and

$$m = \begin{cases} -1 \text{ for } \dot{m}_A/\dot{m}_L < 3 \\ -0.5 \text{ for } \dot{m}_A/\dot{m}_L > 3 \end{cases}$$

For the double concentric airblast atomizer, it was found that

$$\text{MMD} = 2.62 \frac{\sigma^{0.41} \mu_L^{0.32}}{(\rho_A U_R^2)^{0.72} \rho_L^{0.16}} + 1.06 \times 10^{-3} \left(\frac{\mu_L}{\rho_L \sigma}\right)^{0.17} \left(\frac{\dot{m}_A}{\dot{m}_L}\right)^m \frac{1}{U_R^{0.54}} \quad (6.69)$$

Kim and Marshall [89] concluded that the important operating variables in airblast atomization are the ALR and the dynamic force $\rho_A U_R^2$. Increasing either (or both) reduces the mean drop size. The recommended operating range for the ALR is from 0.1 to 10. Below the lower limit the atomization deteriorates, while above the upper limit air energy is wasted. It should be noted that these operating limits of ALR are far wider than those indicated by the various Cranfield studies [77, 80, 82], all of which demonstrated a marked deterioration in atomization quality at air/liquid ratios below 2.0.

It is of interest to note that the atomizer with concentric double air nozzles, i.e., the arrangement in which the liquid sheet is entrained between two air-streams, yields only a slight improvement in atomization quality over the single air nozzle. Thus, the superior performance of the former type, as observed by Lefebvre and Miller [80], was not confirmed in these experiments.

Fraser et al. [90] studied the spray characteristics of an airblast system in which a spinning cup produced flat circular liquid sheets of sensibly uniform thickness. Disintegration of the liquid sheet was accomplished by a high-velocity airstream that flowed through an annular gap located axially symmetrically to the cup. They compared their experimental data with those of other workers and reported that prefilming atomizers are superior to plain-jet atomizers in producing finer sprays and that the controlled production of thin liquid sheets is a prerequisite to fine atomization. Their correlation for drop size data was derived as

$$SMD = 6 \times 10^{-6} + 0.019 \left[\frac{\sigma^{0.5} v_r^{0.21}}{\rho_A^{0.5}(aD_L + a^2)^{0.25}} \right]$$

$$\times \left[1 + 0.065 \left(\frac{\dot{m}_L}{\dot{m}_A} \right)^{1.5} \right] \left[\frac{Q_L}{U_p^2(0.5U_r^2 - U_r + 1)} \right]^{0.5} \qquad (6.70)$$

where v_r = kinematic viscosity ratio relative to water
$\quad a$ = radial distance from cup lip, m
$\quad D_L$ = diameter of cup at lip, m
$\quad U_p$ = cup peripheral velocity, m/s
$\quad U_r$ = air/liquid velocity ratio = U_A/U_p

Fraser et al. also concluded that the air/liquid mass ratio has little effect on SMD for values exceeding 1.5. This is appreciably lower than the values of between 4 and 5 reported in subsequent investigations [77, 80, 82].

Many expressions for correlating and predicting the mean drop sizes of air-blast atomizers have been proposed. Those believed to be the most significant are listed in Tables 6.9, 6.10, and 6.11 for plain-jet, prefilming, and other types of nozzles.

Effect of Variables on Mean Drop Size

The main factors influencing the mean drop size of the spray are liquid properties, air properties, and atomizer geometry.

Liquid properties. The liquid properties of importance in airblast atomization are viscosity, surface tension, and density. The adverse effect on spray quality of an increase in viscosity is demonstrated in Fig. 6.35, in which the SMD is plotted against viscosity for various levels of air velocity at a constant liquid flow rate of 15 g/s. The influence of surface tension on the SMD is illustrated in Fig. 6.36, which also shows the adverse effect of an increase in liquid flow rate on atomization quality.

Table 6.9 Drop size equations for plain-jet airblast atomizers

Investigators	Equations	Remarks
Nukiyama and Tanasawa [68]	$$SMD = 0.585 \left(\frac{\sigma}{\rho_L U_R^2}\right)^{0.5} + 53 \left(\frac{\mu_L^2}{\sigma\rho_L}\right)^{0.225}\left(\frac{Q_L}{Q_A}\right)^{1.5}$$	No effect of nozzle dimensions or air density
Lorenzetto and Lefebvre [75]	$$SMD = 0.95 \left[\frac{(\sigma\dot{m}_L)^{0.33}}{\rho_L^{0.37}\rho_A^{0.30} U_R}\right]\left(1 + \frac{\dot{m}_L}{\dot{m}_A}\right)^{1.70}$$ $$+ 0.13 \left(\frac{\mu_L^2 d_o}{\sigma\rho_L}\right)^{0.5}\left(1 + \frac{\dot{m}_L}{\dot{m}_A}\right)^{1.70}$$	For low-viscosity liquids SMD is independent of initial jet diameter d_o
Jasuja [78]	$$SMD = 0.022\left(\frac{\sigma}{\rho_A U_A^2}\right)^{0.45}\left(1 + \frac{1}{ALR}\right)^{0.5}$$ $$+ 0.00143\left(\frac{\mu_L^2}{\sigma\rho_L}\right)^{0.4}\left(1 + \frac{1}{ALR}\right)^{0.8}$$	For cross-current breakup
Rizk and Lefebvre [79]	$$SMD = 0.48d_o\left(\frac{\sigma}{\rho_A U_R^2 d_o}\right)^{0.4}\left(1 + \frac{1}{ALR}\right)^{0.4}$$ $$+ 0.15d_o\left(\frac{\mu_L^2}{\sigma\rho_L d_o}\right)^{0.5}\left(1 + \frac{1}{ALR}\right)$$	Coflowing air and liquids

Liquid density affects drop size in a fairly complex manner. For example, with prefilming atomizers, the distance to which the coherent liquid sheet extends downstream of the atomizing lip increases with density, so that ligament formation occurs under conditions of lower relative velocity between the air and the liquid. Moreover, for any given flow rate, an increase in liquid density produces a more compact spray that is less exposed to the atomizing action of the high-velocity air. These two effects combine to increase the mean drop size. However, an increase in liquid density can also improve atomization by reducing the thickness of the sheet produced at the atomizing lip of prefilming systems ($t \propto \rho_L^{-0.4}$) and by increasing the relative velocity U_R for plain-jet nozzles. The net effect of these conflicting factors is that the influence of liquid density on SMD is fairly small, as illustrated in Fig. 6.37.

Air properties. Of all the various factors influencing mean drop size, air velocity is undoubtedly the most important. This is very clear from inspection of Figs. 6.35, 6.37, and 6.38. For low-viscosity liquids, the SMD is roughly inversely proportional to the air velocity. This underlines the importance, in airblast atomizer design, of arranging for the liquid to be exposed to the highest possible air velocity consistent with the available pressure drop.

The effects of air pressure and temperature on mean drop size are illustrated

Table 6.10 Drop size equations for prefilming airblast atomizers

Investigators	Equation	Remarks
Rizkalla and Lefebvre [77]	$$\text{SMD} = 3.33 \times 10^{-3} \frac{(\sigma \rho_L t)^{0.5}}{\rho_A U_A}\left(1 + \frac{\dot{m}_L}{\dot{m}_A}\right)$$ $$+ 13.0 \times 10^{-3}\left(\frac{\mu_L^2}{\sigma \rho_L}\right)^{0.425} t^{0.575}\left(1 + \frac{\dot{m}_L}{\dot{m}_A}\right)^2$$	L_c equated to t
Jasuja [15]	$$\text{SMD} = 10^{-3} \frac{(\sigma \rho_L)^{0.5}}{\rho_A U_A}\left(1 + \frac{\dot{m}_L}{\dot{m}_A}\right)^{0.5}$$ $$+ 0.6 \times 10^{-4}\left(\frac{\mu_L^2}{\sigma \rho_A}\right)^{0.425}\left(1 + \frac{\dot{m}_L}{\dot{m}_A}\right)^{0.5}$$	No effect of nozzle dimensions
Lefebvre [91]	$$\frac{\text{SMD}}{L_c} = A\left(\frac{\sigma}{\rho_A U_A^2 D_p}\right)^{0.5}\left(1 + \frac{\dot{m}_L}{\dot{m}_A}\right)$$ $$+ B\left(\frac{\mu_L^2}{\sigma \rho_L D_p}\right)^{0.5}\left(1 + \frac{\dot{m}_L}{\dot{m}_A}\right)$$	"Basic" equation for prefilming airblast atomizers L_c = characteristic dimensions D_p = prefilmer diameter A, B = constants whose values depend on atomizer design
El-Shanawany and Lefebvre [82]	$$\frac{\text{SMD}}{D_h} = \left(1 + \frac{\dot{m}_L}{\dot{m}_A}\right)\left[0.33\left(\frac{\sigma}{\rho_A U_A^2 D_p}\right)^{0.6}\left(\frac{\rho_L}{\rho_A}\right)^{0.1}\right.$$ $$\left. + 0.068\left(\frac{\mu_L^2}{\sigma \rho_L D_p}\right)^{0.5}\right]$$	L_c equated to D_h
Fraser et al. [90]	$$\text{SMD} = 6 \times 10^{-6} + 0.019\left[\frac{\sigma^{0.5} v_r^{0.21}}{\rho_A^{0.5}(a D_L + a^2)^{0.25}}\right]$$ $$\left[1 + 0.065\left(\frac{\dot{m}_L}{\dot{m}_A}\right)^{1.5}\right]\left[\frac{Q_L}{U_p^2(0.5 U_r^2 - U_r + 1)}\right]^{0.5}$$	a = radial distance from cup lip U_p = cup peripheral velocity U_r = air/liquid velocity ratio

in Figs. 6.39 and 6.40, respectively, for kerosine and water. These results, taken together, strongly suggest that for prefilming atomizers the SMD is proportional to air density to the -0.6 power.

For prefilming atomizers the SMD increases with an increase in atomizer scale (size) according to the relationship SMD $\propto L_c^{0.4}$.

Analysis of Drop Size Relationships

From inspection of all the experimental data on prefilming airblast atomizers, some general conclusions concerning the effects of air and liquid properties on mean drop size can be drawn. For liquids of low viscosity, such as water and kerosine, the main factors governing the SMD are surface tension, air density,

Table 6.11 Drop size equations for miscellaneous airblast atomizers

Investigators	Equation	Remarks
Gretzinger and Marshall [88]	$\text{MMD} = 2.6 \times 10^{-3} \left(\dfrac{\mu_A}{\rho_A U_A L}\right)^{0.4} \left(\dfrac{\dot{m}_L}{\dot{m}_A}\right)^{0.4}$	L = diameter of wetted periphery Drop size independent of liquid viscosity but dependent on air viscosity
Ingebo and Foster [84]	$\text{SMD} = 5.0 \left(\dfrac{\sigma \mu_L d_0^2}{\rho_A U_R^3 \rho_L}\right)^{0.25}$	For cross-current breakup
Kim and Marshall [89]	$\text{MMD} = 5.36 \times 10^{-3} \left[\dfrac{\sigma^{0.41} \mu_L^{0.32}}{(\rho_A U_R^2)^{0.57} A^{0.36} \rho_L^{0.16}}\right]$ $+ 3.44 \times 10^{-3} \left(\dfrac{\mu_L^2}{\sigma \rho_L}\right)^{0.17} \left(\dfrac{\dot{m}_A}{\dot{m}_L}\right)^m \dfrac{I}{U_R^{0.54}}$	A = flow area of atomizing airstream $m = -1$ at $\dot{m}_A/\dot{m}_L < 3$ $m = -0.5$ at $m_A/m_L > 3$
Weiss and Worsham [87]	$\text{SMD} \propto \left(\dfrac{U_L^{0.08} d_o^{0.16} \mu_L^{0.34}}{\rho_A^{0.30} U_R^{1.33}}\right)$	Cross-stream and costream injection
Ingebo [86]	$\text{SMD} = (2.67 \times 10^4 \, U_L P_A^{-0.33}$ $+ 4.11 \times 10^6 \, \rho_A U_A P_A^{-0.75})^{-1}$	Splash plate injector

and air velocity; for liquids of high viscosity the effects of air properties are less significant, and the SMD becomes more dependent on the liquid properties, especially viscosity.

The fact that viscosity has an effect quite separate from that of air velocity, as observed experimentally by Nukiyama and Tanasawa [68], Kim and Marshall [89], Lefebvre et al. [77, 80, 82], and Jasuja [15], suggests a form of equation in which the SMD is expressed as the sum of two terms, one dominated by air velocity and air density and the other by liquid viscosity. For prefilming types of atomizer, it can be shown [91] that

$$\frac{\text{SMD}}{L_c} = \left[A'\left(\frac{\sigma}{\rho_A U_A^2 D_p}\right)^{0.5} + B'\left(\frac{\mu_L^2}{\sigma \rho_L D_p}\right)^{0.5}\right]\left(1 + \frac{1}{\text{ALR}}\right) \quad (6.71)$$

where A' and B' are constants whose value depends on the atomizer design and must be determined experimentally.

Equation (6.71) is the "basic" equation for the mean drop size of the prefilming airblast atomizer. It shows that the SMD is proportional to the characteristic dimension L_c, which represents the linear scale of the atomizer. Equation (6.71) also shows that mean drop size is inversely proportional to the square root of the prefilmer lip diameter D_p. This is not surprising; if other parameters are kept constant, an increase in D_p will reduce the liquid film thickness at the atomizing lip and hence the SMD. The implication of this result for atomizer design is clear—for any given atomizer size (i.e., for any given value of L_c), the ratio D_p/L_c should be made as large as possible.

In practice, some secondary factors, such as liquid stream Reynolds number and airstream Mach number, affect the atomization process in a manner that is not yet fully understood. Thus, the ability of Eq. (6.71) to predict mean drop diameters can be improved by raising the exponent of the term $\sigma_L/\rho_A U_A^2 D_p$ from 0.5 to 0.6. Moreover, high-speed photographs of the atomization process have shown that for prefilming atomizers the effect of an increase in ρ_L/ρ_A is to extend the distance downstream of the atomizing lip at which sheet disintegration occurs, so that it takes place in a region of lower relative velocity [92]. Analysis of the experimental data shows that this effect may be accommodated by introducing into Eq. (6.71) the dimensionless term $(\rho_L/\rho_A)^{0.1}$. With these modifications Eq. (6.71) becomes

$$\frac{\text{SMD}}{L_c} = A\left(\frac{\sigma}{\rho_A U_A^2 D_p}\right)^{0.6}\left(\frac{\rho_L}{\rho_A}\right)^{0.1}\left(1 + \frac{1}{\text{ALR}}\right) + B\left(\frac{\mu_L^2}{\sigma \rho_L D_p}\right)^{0.5}\left(1 + \frac{1}{\text{ALR}}\right) \quad (6.72)$$

A logical choice for the characteristic dimension L_c is the hydraulic mean diameter of the atomizer air duct at its exit plane (see Fig. 6.34). Equating L_c to D_h and substituting into Eq. (6.72) the values of A and B derived from the experimental data of reference [82] yields Eq. (6.59).

Summary of Main Points

From the experimental data obtained by many workers on many different types of airblast atomizer, the following conclusions are drawn:

1. The mean drop size of the spray increases with increasing liquid viscosity and surface tension and with decreasing air/liquid ratio. Ideally, the air/liquid mass ratio should exceed 3, but little improvement in atomization quality is gained by raising this ratio above a value of about 5.

2. Liquid density appears to have little effect on the mean drop size. For prefilming nozzles, the mean drop size increases slightly with an increase in liquid density; for plain-jet nozzles, the opposite effect is observed.

3. The air properties of importance in airblast atomization are density and velocity. In general, the mean drop size is roughly inversely proportional to air velocity. The effect of air density may be expressed as SMD $\propto \rho_A^{-n}$, where n is about 0.3 for plain-jet atomizers and between 0.6 and 0.7 for prefilming types.

4. For plain-jet nozzles, the initial liquid jet diameter has little effect on mean drop size for liquids of low viscosity, but for high-viscosity liquids the atomization quality deteriorates with increasing jet size.

5. For prefilming atomizers, the mean drop size increases with increasing atomizer scale (size) according to the relationship SMD $\propto L_c^{0.4}$.

6. For any given size of prefilming atomizer (i.e., for any fixed value of L_c), the finest atomization is obtained by making the prefilmer lip diameter D_p as large as possible. This is because an increase in D_p reduces the liquid film thickness t, which in turn reduces the mean drop size (SMD $\propto t^{0.4}$).

7. Minimum drop sizes are obtained by using atomizers designed to provide maximum physical contact between the air and the liquid. With prefilming systems the best atomization is obtained by producing the thinnest possible liquid sheet of uniform thickness. It is also important to ensure that the liquid sheet formed at the atomizing lip is subjected to high-velocity air on both sides. This not only provides the finest atomization but also eliminates droplet deposition on adjacent solid surfaces.

8. The performance of prefilming atomizers is superior to that of plain-jet types, especially under adverse conditions of low air/liquid ratio and/or low air velocity.

9. At least two different mechanisms are involved in airblast atomization, and the relative importance of each depends mainly on the liquid viscosity. When a low-viscosity liquid is injected into a low-velocity airstream, waves are produced on the liquid surface; this surface becomes unstable and disintegrates into fragments. The fragments then contract into ligaments, which in turn break down into drops. With increased air velocity, the liquid sheet disintegrates earlier, so ligaments are formed nearer the lip. These ligaments tend to be thinner and shorter and disintegrate into smaller drops. With liquids of high viscosity, the wavy-surface mechanism is no longer present. Instead, the liquid is drawn out from the atomizing lip in the form of long ligaments. When atomization occurs, it does so well downstream of the atomizing lip in regions of relatively low velocity. In consequence, drop sizes tend to be larger.

The effects of the major variables on mean drop size are summarized for low-viscosity liquids in Table 6.8.

EFFERVESCENT ATOMIZER

Some of the results of drop size measurements made by Lefebvre et al. [93] for the type of atomizer shown in Fig. 4.49 are illustrated in Fig. 6.44. This figure shows mean drop size plotted against gas/liquid ratio by mass (GLR) for four values of liquid injection pressure ΔP_L, namely 34.5, 138, 345, and 690 kPa. The corresponding pressures in English units are 5, 20, 50, and 100 psi. The atomization quality demonstrated in this figure is clearly quite high. Even at a water pressure of only 138 kPa (20 psi), mean drop sizes of less than 50 μm are obtained at a GLR of 0.04, i.e., when using only 1 part by mass of nitrogen for every 25 parts of water. Figure 6.44 also shows, as would be expected, that increases in gas flow rate and/or liquid injection pressure lead to significant improvements in atomization quality.

If we now consider the flow of gas through the injector orifice, for continuity we have

$$\dot{m}_G = \rho_G A_G U_G \tag{6.73}$$

where A_G is the average cross-sectional area occupied by the gas flow and U_G is the mean gas velocity through the injection orifice.

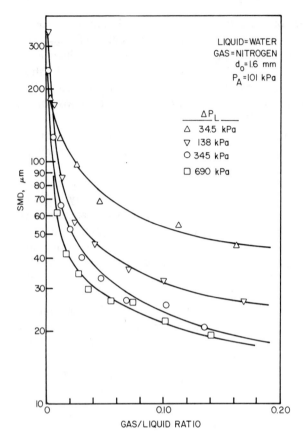

Figure 6.44 Influence of gas/liquid ratio and injection pressure on SMD for $d_o = 1.6$ mm [93].

From the above equation it follows that, for a constant gas flow rate, a reduction in ρ_G must lead to an increase in A_G or U_G, or both. Increase in gas flow area is beneficial to atomization because it reduces the area available for the liquid flow; i.e., it squeezes the liquid into thinner films and ligaments as it flows through the injector orifice. Increase in gas velocity is also beneficial to atomization because it accelerates the flow of liquid through the injector orifice, causing it to be discharged at a higher velocity. Thus, when operating at high injection pressures, atomization quality is high for two reasons. One is the high pressure drop across the exit orifice. This represents the atomization that would have been achieved in the absence of any injected gas. The second reason stems from the beneficial effects of the injected gas (1) in squeezing the liquid into fine ligaments as it flows through the injector orifice and (2) in "exploding" downstream of the nozzle exit to shatter these ligaments into small drops. With decrease in injection pressure, the "natural" atomization, i.e., the atomization that occurs solely as the result of the pressure drop across the nozzle, is impaired, but this is compensated in a large measure by the relatively bigger role played by the gas, which expands in volume with reduction in pressure, leading to increases in the number of gas bubbles in the flow. Both these effects are conducive to better atomization.

If all the gas bubbles flowing through the discharge orifice are the same size, then from simple geometric considerations it can be shown that the average minimum thickness between adjacent gas bubbles is given by

$$t_{min} = d_G \left\{ \left[\frac{\pi}{6} \left(1 + \frac{\rho_G}{\rho_L \, GLR} \right) \right]^{1/3} - 1 \right\} \qquad (6.74)$$

where d_G is the average diameter of the gas bubbles, GLR is the gas/liquid ratio by mass, and ρ_G and ρ_L are the gas and liquid densities, respectively. Equation (6.74) indicates that t_{min} is directly proportional to the average diameter of the gas bubbles, which suggests that better atomization could be achieved by generating smaller bubbles. Equation (6.74) also suggests that increase in GLR should improve atomization quality, a result that is very evident in Fig. 6.44. Another interesting feature of Eq. (6.74) is that it does not include a term to represent the injector orifice diameter. This is fully supported by the results shown in Fig. 6.45, which demonstrate that SMD is virtually independent of injector orifice diameter. From a practical viewpoint, this is a most useful characteristic, because hitherto the general tendency has been to associate small drop sizes with small nozzle dimensions. From a combustion standpoint it is clearly most advantageous to have

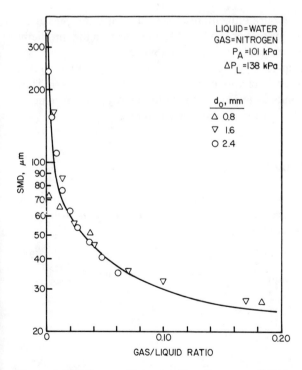

Figure 6.45 Influence of injector orifice diameter on SMD for an injection pressure of 138 kPa [93].

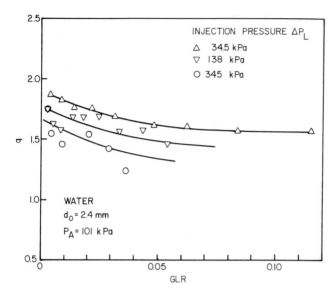

Figure 6.46 Influence of gas/liquid ratio and injection pressure on drop size distribution parameter [93].

a fuel nozzle that can provide good atomization at low fuel injection pressures, while still employing flow passages that are so large in cross section that problems of plugging by contaminants in the fuel are greatly alleviated.

Measured values of the Rosin-Rammler drop size distribution parameter q (see Chapter 3) for a 2.4-mm-diameter discharge orifice are shown in Fig. 6.46. They indicate that any change that tends to impair atomization quality, i.e., produce a larger value of SMD, also yields a more monodisperse spray.

ELECTROSTATIC ATOMIZERS

Very few measurements of the drop sizes produced in electrostatic atomization appear to have been made. A search of the literature revealed only the following expression for SMD, due to Mori et al. [94]:

$$\text{SMD} = 5.39 d_o \bar{E}^{-0.255} \bar{Q}^{0.277} \text{Re}^{-0.124} \tag{6.75}$$

where d_o = outer diameter of liquid supply tube
$\bar{E} = \epsilon E^2 d_o / \sigma$
ϵ = dielectric constant
E = intensity of electric field
$\bar{Q} = \rho_L Q^2 d_o^3 / \sigma$
Q = liquid flow rate
$\text{Re} = Q / \mu_L d_o$

ULTRASONIC ATOMIZERS

The drop formation mechanism for this type of atomizer was originally attributed to the creation and subsequent collapse of cavities caused by the intense sound wave. However, more recent opinion favors a capillary wave theory based on the observation that the mean drop size produced from thin layers of liquid on an ultrasonically excited plate is proportional to the capillary wavelength on the liquid surface. This implies that the mean drop size should be related to the ripple wavelength, which in turn is controlled by the vibration frequency. The results of experiments tend to support this hypothesis. For example, Lang [95] found that

$$D = 0.34\lambda = 0.34\left(\frac{8\pi}{\rho_L F}\right) \tag{6.76}$$

which is in close agreement with Lobdell's theoretical value of 0.36λ obtained from considerations of drop formation from high-amplitude capillary waves [96].

Another capillary wave theory was developed by Peskin and Raco [97] in terms of Taylor instability. They found that the atomization process is governed by several nondimensional parameters. Figure 6.47 gives the drop size in terms of transducer amplitude a, frequency ω_0, liquid film thickness t, surface tension, and density. The drop size is seen to be a function of frequency and film thickness. For films of large thickness, their analytical result reduces to

$$D = \left(\frac{4\pi^3\sigma}{\rho_L\omega_0^2}\right)^{1/3} \tag{6.77}$$

The ranges of drop sizes obtained by various workers are summarized in Table 6.12 from Lee et al. [98]. The operating frequency used in these studies varied from 10 to 2000 kHz.

Mochida [99] has studied a horn type of ultrasonic atomizer, operating at a frequency of 26 kHz, at flow rates up to 50 liters/h. Using distilled water and solutions of water with methanol and glycerin to obtain suitable variations in the

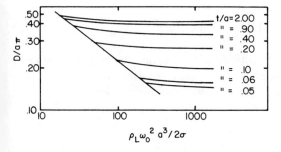

Figure 6.47 Relationship between dimensionless groups in ultrasonic atomization *(Peskin and Raco [97])*.

Table 6.12 Range of drop sizes covered by various studies employing an ultrasonic atomizer

Investigators	Drop size (μm)			
	1	10	100	1000
Wilcox and Tate [100]			———	
Crawford [101			——	
Antonevich [102]		————————		
Lang [95]		———————————		
Bisa et al. [103]		—————————————————		
McCubbin [104]		———		
Topp [105]		——		
Muromstev and Nenashev [106]	——			

Source: Lee et al. [98].

Table 6.13 Ability of ultrasonic nozzles to atomize various liquid classes

Easy	Medium difficulty	Very difficult or impossible
Water	Oil-based paints and	Heavy hydrocarbons (unless heated)
Soluble water	coatings	Certain long-chain polymer solutions,
solutions	Medium-weight oils	e.g., tablet coating
Light hydrocarbons	Coal slurries	High solids content latex paints and
Light oils	Most emulsions	coatings
Organic solvents		Latex adhesives
Short-chain polymer		Ceramic slurries
solutions		

Source: Berger [107].

liquid properties that influence atomization, he obtained the following empirical equation for SMD:

$$\text{SMD} = 0.158\left(\frac{\sigma}{\rho_L}\right)^{0.354} \mu_L^{0.303} Q_L^{0.139} \tag{6.78}$$

Further information on the performance of ultrasonic atomizers is contained in references [99] to [107]. The degree of difficulty involved in the atomization of various classes of liquids is indicated in Table 6.13.

NOMENCLATURE

A_o discharge orifice area, m^2

A_p total inlet ports area, m^2

A_s	swirl chamber area, m^2
ALR	air/liquid mass ratio
D	drop diameter, m
D_h	hydraulic mean diameter of air exit duct, m
D_p	prefilmer diameter, m
D_s	swirl chamber diameter, m
d	disk diameter, m
d_o	discharge orifice diameter, m
F	charging factor
FN	flow number $[=\dot{m}_L/(\Delta P_L \rho_L)^{0.5}]$, m^2
L_c	characteristic dimension of airblast atomizer, m
L_s	length of swirl chamber, m
MMD	mass median diameter, m
$\dot{m}$	flow rate, kg/s
N	rotational speed, rps
P	pressure, Pa
ΔP_L	injection pressure differential across nozzle, Pa
Q	volumetric flow rate, m^3/s
R	disk radius, m
Re	Reynolds number
Re_L	Reynolds number based on liquid properties $(=U_L \rho_L d_o/\mu_L)$
SMD	Sauter mean diameter, m
s	downstream distance of measuring plane from nozzle
T	temperature, K
t	film thickness in final orifice, m
t_s	initial sheet thickness at nozzle exit, m
U	velocity, m/s
We	Weber number
We_L	Weber number based on liquid properties $(=\rho_L U_L^2 d_o/\sigma)$
w	rotational speed, radians/s
λ	wavelength, m
μ	dynamic viscosity, kg/m s
ν	kinematic viscosity, m^2/s
ρ	density, kg/m^3
σ	surface tension, kg/s^2

Subscripts

A	air
L	liquid
R	air relative to liquid

REFERENCES

1. Rizk, N. K., and Lefebvre, A. H., Drop-Size Distribution Characteristics of Spill-Return Atomizers, *AIAA J. Propul. Power*, Vol. 1, No. 1, 1985, pp. 16–22.

2. Merrington, A. C., and Richardson, E. G., The Break-Up of Liquid Jets, *Proc. Phys. Soc. London,* Vol. 59, No. 33, 1947, pp. 1–13.
3. Panasenkov, N. J., *Zh. Tekh. Fiz.,* Vol. 21, 1951, p. 160.
4. Harmon, D. B., *J. Franklin Inst.,* Vol. 259, 1955, p. 519.
5. Giffen, E., and Lamb, T. A. J., The Effect of Air Density on Spray Atomization, Motor Industry Research Association Report 1953/5, 1953.
6. Miesse, C. C., Correlation of Experimental Data on the Disintegration of Liquid Jets, *Ind. Eng. Chem,* Vol. 47, No. 9, 1955, pp. 1690–1701.
7. Tanasawa, Y., and Toyoda, S., On the Atomization of a Liquid Jet Issuing from a Cylindrical Nozzle, Tech. Report of Tohoku University, Japan, No. 19-2, 1955, p. 135.
8. Hiroyasu, H., and Katoda, T., Fuel Droplet Size Distribution in a Diesel Combustion Chamber, *SAE Trans.,* Paper 74017, 1974.
9. Elkotb, M. M., Fuel Atomization for Spray Modeling, *Prog. Energy Combust. Sci.,* Vol. 8, No. 1, 1982, pp. 61–91.
10. Jones, A. R., Design Optimization of a Large Pressure-Jet Atomizer for Power Plant, *Proceedings of the 2nd International Conference on Liquid Atomization and Sprays,* Madison, Wis., 1982, pp. 181–185.
11. Simmons, H. C., and Harding, C. F., Some Effects on Using Water as a Test Fluid in Fuel Nozzle Spray Analysis, ASME Paper 80-GT-90, 1980.
12. Kennedy, J. B., High Weber Number SMD Correlations for Pressure Atomizers, ASME Paper 85-GT-37, 1985.
13. Lefebvre, A. H., *Gas Turbine Combustion,* Hemisphere, Washington, D.C., 1983.
14. Wang, X. F., and Lefebvre, A. H., Mean Drop Sizes from Pressure-Swirl Nozzles, *AIAA J. Propul. Power,* Vol. 3, No. 1, 1987, pp. 11–18.
15. Jasuja, A. K., Atomization of Crude and Residual Fuel Oils, *ASME J. Eng. Power,* Vol. 101, No. 2, 1979, pp. 250–258.
16. Radcliffe, A., Fuel Injection, *High Speed Aerodynamics and Jet Propulsion,* Vol. XI, Sect. D, Princeton University Press, Princeton, N.J., 1960.
17. Dodge, L. G., and Biaglow, J. A., Effect of Elevated Temperature and Pressure on Sprays from Simplex Swirl Atomizers, ASME Paper 85-GT-58, 1985.
18. Simmons, H. C., The Prediction of Sauter Mean Diameter for Gas Turbine Fuel Nozzles of Different Types, ASME Paper 79-WA/GT-5, 1979.
19. Knight, B. E., Communication on The Performance of a Type of Swirl Atomizer, by A. Radcliffe, *Proc. Inst. Mech. Eng.,* Vol. 169, 1955, p. 104.
20. Rizk, N. K., and Lefebvre, A. H., Spray Characteristics of Simplex Swirl Atomizers, *Dynamics of Flames and Reactive Systems,* J. R. Bowen, N. Manson, A. K. Oppenheim, and R. I. Soloukhin, eds., *Prog. Astronaut. Aeronaut.,* Vol. 95, 1985, pp. 563–580.
21. Abou-Ellail, M. M. M., Elkotb, M. M., and Rafat, N. M., Effect of Fuel Pressure, Air Pressure and Air Temperature on Droplet Size Distribution in Hollow-Cone Kerosine Sprays, *Proceedings of the 1st International Conference on Liquid Atomization and Spray Systems,* Tokyo, 1978, pp. 85–92.
22. De Corso, S. M., Effect of Ambient and Fuel Pressure on Spray Drop Size, *ASME J. Eng. Power,* Vol. 82, 1960, p. 10.
23. Hinze, J. O., Critical Speeds and Sizes of Liquid Globules, *Appl. Sci. Res.,* A-1, 1948, p. 273.
24. Lane, W. R., Shatter of Drops in Streams of Air, *Ind. Eng. Chem.,* Vol. 43, 1951, pp. 1312–1317.
25. De Corso, S. M., and Kemeny, G. A., Effect of Ambient and Fuel Pressure on Nozzle Spray Angle, *Trans. ASME,* Vol. 79, No. 3, 1957, pp. 607–615.
26. Neya, K., and Sato, S., Effect of Ambient Air Pressure on the Spray Characteristics of Swirl Atomizers, Ship Res. Inst., Tokyo, Paper 27, 1968.
27. Rizk, N. K., and Lefebvre, A. H., Spray Characteristics of Spill-Return Atomizer, *AIAA J. Propul. Power,* Vol. 1, No. 3, 1985, pp. 200–204.
28. Wang, X. F., and Lefebvre, A. H., Influence of Ambient Air Pressure on Pressure-Swirl Atom-

ization, paper presented at the 32nd ASME International Gas Turbine Conference, Anaheim, Calif., June 1987.

29. Ortman, J., and Lefebvre, A. H., Fuel Distributions from Pressure-Swirl Atomizers, *AIAA J. Propul. Power,* Vol. 1, No. 1, 1985, pp. 11–15.

30. York, J. L., Stubbs, H. E., and Tek, M. R., The Mechanism of Disintegration of Liquid Sheets, *Trans. ASME,* Vol. 75, No. 7, 1953, pp. 1279–1286.

31. Hagerty, W. W., and Shea, J. F., A Study of the Stability of Plane Fluid Sheets, *J. Appl. Mech.,* Vol. 22, 1955, pp. 509–514.

32. Dombrowski, N., and Johns, W. R., The Aerodynamic Instability and Disintegration of Viscous Liquid Sheets, *Chem. Eng. Sci.,* Vol. 18, 1963, pp. 203–214.

33. Elkotb, M. M., Rafat, N. M., and Hanna, M. A., The Influence of Swirl Atomizer Geometry on the Atomization Performance, *Proceedings of the 1st International Conference on Liquid Atomization and Spray Systems,* Tokyo, 1978, pp. 109–115.

34. Needham, H. C., Power Jets R & D Report, No. R1209, 1946.

35. Joyce, J. R., Report ICT 15, Shell Research Ltd., London, 1949.

36. Turner, G. M., and Moulton, R. W., *Chem. Eng. Prog.,* Vol. 49, 1943, p. 185.

37. Babu, K. R., Narasimhan, M. V., and Narayanaswamy, K., Prediction of Mean Drop Size of Fuel Sprays from Swirl Spray Atomizers, *Proceedings of the 2nd International Conference on Liquid Atomization and Sprays,* Madison, Wis., 1982, pp. 99–106.

38. Lefebvre, A. H., The Prediction of Sauter Mean Diameter for Simplex Pressure-Swirl Atomizers, *Atomization Spray Technol.,* Vol. 3, No. 1, 1987, pp. 37–51.

39. Ohnesorge, W., Formation of Drops by Nozzles and the Breakup of Liquid Jets, *Z. Angew. Math. Mech.,* Vol. 16, 1936.

40. Suyari, M., and Lefebvre, A. H., Film Thickness Measurements in a Simplex Swirl Atomizer, *AIAA J. Propul. Power,* Vol. 2, No. 6, 1986, pp. 528–533.

41. Hinze, J. O., and Milborn, H., Atomization of Liquids by Means of a Rotating Cup, *ASME J. Appl. Mech.,* Vol. 17, No. 2, 1950, pp. 145–153.

42. Dombrowski, N., and Fraser, R. P., A Photographic Investigation into the Disintegration of Liquid Sheets, *Philos. Trans. R. Soc. London Ser. A,* Vol. 247, No. 924, September 1954, pp. 101–130.

43. Tanasawa, Y., Miyasaka, Y., and Umehara, M., On the Filamentation of Liquid by Means of Rotating Discs, *Trans. JSME,* Vol. 25, No. 156, 1963, pp. 888–896.

44. Dombrowski, N., and Munday, G., Spray Drying, *Biochemical and Biological Engineering Science,* Vol. 2, Chapter 16, Academic Press, New York, 1968, pp. 209–320.

45. Christensen, L. S., and Steely, S. L., Monodisperse Atomizers for Agricultural Aviation Applications, NASA CR-159777, February 1980.

46. Eisenklam, P., Recent Research and Development Work on Liquid Atomization in Europe and the USA, invited paper to the 5th Conference on Liquid Atomization, Tokyo, 1976.

47. Matsumoto, S., Belcher, D. W., and Crosby, E. J., Rotary Atomizers: Performance Understanding and Prediction, *Proceedings of the 3rd International Conference on Liquid Atomization and Sprays,* London, 1985, pp. 1A/1/1–21.

48. Fraser, R. P., Eisenklam, P., Dombrowski, N., and Hasson, D., Drop Formation from Rapidly Moving Liquid Sheets, *AIChE J.,* Vol. 8, 1962, pp. 672–680.

49. Karim, G. A., and Kumar, R., The Atomization of Liquids at Low Ambient Pressure Conditions, *Proceedings of the 1st International Conference on Liquid Atomization and Spray Systems,* Tokyo, August 1978, pp. 151–155.

50. Bar, P., Dr. Eng. dissertation, Technical College, Karlsruhe, Germany, 1935.

51. Tanasawa, Y., Miyasaka, Y., and Umehara, M., Effect of Shape of Rotating Disks and Cups on Liquid Atomization, *Proceedings of the 1st International Conference on Liquid Atomization and Spray Systems,* Tokyo, 1978, pp. 165–172.

52. Matsumoto, S., Saito, K., and Takashima, Y., *J. Chem. Eng. Jpn.,* Vol. 7, 1974, p. 13.

53. Walton, W. H., and Prewett, W. G., *Proc. Phys. Soc. London Sect. B,* Vol. 62, 1949, p. 341.

54. Oyama, Y., and Endou, K., On the Centrifugal Disk Atomization and Studies on the Atom-

ization of Water Droplets, *Kagaku Kogaku,* Vol. 17, 1953, pp. 256–260, 269–275 (in Japanese, English summary).

55. Matsumoto, S., and Takashima, Y., *Kagaku Kogaku,* Vol. 33, 1969, p. 357.
56. Kayano, A., and Kamiya, T., Calculation of the Mean Size of the Droplets Purged from the Rotating Disk, *Proceedings of the 1st International Conference on Liquid Atomization and Sprays,* Tokyo, 1978, pp. 133–143.
57. Fraser, R. P., and Eisenklam, P., Liquid Atomization and the Drop Size of Sprays, *Trans. Inst. Chem. Eng.,* Vol. 34, 1956, pp. 294–319.
58. Masters, K., Spray Drying, 2nd ed., Wiley, New York, 1976.
59. Friedman, S. J., Gluckert, F. A., and Marshall, W. R., *Chem. Eng. Prog.,* Vol. 48, No. 4, 1952, p. 181.
60. Herring, W. H., and Marshall, W. R., *J. Am. Inst. Chem. Eng.,* Vol. 1, No. 2, 1955, p. 200.
61. Fraser, R. P., Eisenklam, P., and Dombrowski, N., *Br. Chem. Eng.,* Vol. 2, No. 9, 1957, p. 196.
62. Scott, M. N., Robinson, M. J., Pauls, J. F., and Lantz, R. J., *J. Pharm. Sci.,* Vol. 53, No. 6, 1964, p. 670.
63. Wigg, L. D., The Effect of Scale on Fine Sprays Produced by Large Airblast Atomizers, Report No. 236, National Gas Turbine Establishment, Pyestock, England, 1959.
64. Wigg, L. D., Drop-Size Predictions for Twin Fluid Atomizers, *J. Inst. Fuel,* Vol. 27, 1964, pp. 500–505.
65. Clare, H., and Radcliffe, A., An Airblast Atomizer for Use with Viscous Fuels, *J. Inst. Fuel,* Vol. 27, No. 165, 1954, pp. 510–515.
66. Wood, R., unpublished work at Thornton Shell Research Center, 1954.
67. Mullinger, P. J., and Chigier, N. A., The Design and Performance of Internal Mixing Multi-Jet Twin-Fluid Atomizers, *J. Inst. Fuel,* Vol. 47, 1974, pp. 251–261.
68. Nukiyama, S., and Tanasawa, Y., Experiments on the Atomization of Liquids in an Airstream, *Trans. Soc. Mech. Eng. Jpn.* Vol. 5, 1939, pp. 68–75.
69. Mayer, E., Theory of Liquid Atomization in High Velocity Gas Streams, *Am. Rocket Soc. J.,* Vol. 31, 1961, pp. 1783–1785.
70. Sakai, T., Kito, M., Saito, M., and Kanbe, T., Characteristics of Internal Mixing Twin-Fluid Atomizer, *Proceedings of the 1st International Conference on Liquid Atomization and Sprays,* Tokyo, 1978, pp. 235–241.
71. Inamura, T., and Nagai, N., The Relative Performance of Externally and Internally-Mixed Twin-Fluid Atomizers, *Proceedings of the 3rd International Conference on Liquid Atomization and Sprays,* London, July 1985, pp. IIC/2/1–11.
72. Elkotb, M. M., Mahdy, M. A., and Montaser, M. E., Investigation of External-Mixing Airblast Atomizers, *Proceedings of the 2nd International Conference on Liquid Atomization and Sprays,* Madison, Wis., 1982, pp. 107–115.
73. Suyari, M., and Lefebvre, A. H., Drop-Size Measurements in Air-Assist Swirl Atomizer Sprays, paper presented at Central States Combustion Institute Spring Meeting, NASA-Lewis Research Center, Cleveland, Ohio, May 1986.
74. Rayleigh, Lord, On the Stability of Jets, *Proc. London Math. Soc.,* Vol. 10, 1879, pp. 4–13.
75. Lorenzetto, G. E., and Lefebvre, A. H., Measurements of Drop Size on a Plain Jet Airblast Atomizer, *AIAA J.,* Vol. 15, No. 7, 1977, pp. 1006–1010.
76. Rizk, N. K., and Lefebvre, A. H., Influence of Airblast Atomizer Design Features on Mean Drop Size, *AIAA J.,* Vol. 21, No. 8, August 1983, pp. 1139–1142.
77. Rizkalla, A., and Lefebvre, A. H., The Influence of Air and Liquid Properties on Air Blast Atomization, *ASME J. Fluids Eng.,* Vol. 97, No. 3, 1975, pp. 316–320.
78. Jasuja, A. K., Plain-Jet Airblast Atomization of Alternative Liquid Petroleum Fuels under High Ambient Air Pressure Conditions, ASME Paper 82-GT-32, 1982.
79. Rizk, N. K., and Lefebvre, A. H., Spray Characteristics of Plain-Jet Airblast Atomizers, *Trans. ASME J. Eng. Gas Turbines Power,* Vol. 106, July 1984, pp. 639–644.
80. Lefebvre, A. H., and Miller, D., The Development of an Air Blast Atomizer for Gas Turbine Application, CoA-Report-AERO-193, College of Aeronautics, Cranfield, England, 1966.

81. Bryan, R., Godbole, P. S., and Norster, E. R., Characteristics of Airblast Atomizers, *Combustion and Heat Transfer in Gas Turbine Systems,* Cranfield International Symp. Ser., Vol. 11, E. R. Norster, ed., Pergamon, New York, 1971, pp. 343–359.

82. El-Shanawany, M. S. M. R., and Lefebvre, A. H., Airblast Atomization: The Effect of Linear Scale on Mean Drop Size, *J. Energy,* Vol. 4, No. 4, 1980, pp. 184–189.

83. Wittig, S., Aigner, M. Sakbani, Kh., and Sattelmayer, Th., Optical Measurements of Droplet Size Distributions: Special Considerations in the Parameter Definition for Fuel Atomizers, paper presented at AGARD meeting on Combustion Problems in Turbine Engines, Cesme, Turkey, October 1983. Also Aigner, M. and Wittig, S., Performance and Optimization of an Airblast Nozzle, Drop-Size Distribution and Volumetric Air Flow, *Proceedings of the 3rd International Conference on Liquid Atomization and Sprays,* London, July 1985, pp. IIC/3/1–8.

84. Ingebo, R. D., and Foster, H. H., Drop-Size Distribution for Cross-Current Break-Up of Liquid Jets in Air Streams, NACA TN 4087, 1957.

85. Ingebo, R. D., Capillary and Acceleration Wave Breakup of Liquid Jets in Axial Flow Airstreams, NASA Technical Paper 1791, 1981.

86. Ingebo, R. D., Atomization of Liquid Sheets in High Pressure Airflow, ASME Paper HT-WA/HT-27, 1984.

87. Weiss, M. A., and Worsham, C. H., Atomization in High Velocity Air-Streams, *J. Am. Rocket Soc.,* Vol. 29, No. 4, 1959, pp. 252–259.

88. Gretzinger, J., and Marshall, W. R. Jr., Characteristics of Pneumatic Atomization, *AIChE J.,* Vol. 7, No. 2, 1961, pp. 312–318.

89. Kim, K. Y., and Marshall, W. R., Drop-Size Distributions from Pneumatic Atomizers, *AIChE J.,* Vol. 17, No. 3, 1971, pp. 575–584.

90. Fraser, R. P., Dombrowski, N., and Routley, J. H., The Production of Uniform Liquid Sheets from Spinning Cups; The Filming by Spinning Cups; The Atomization of a Liquid Sheet by an Impinging Air Stream, *Chem. Eng. Sci.,* Vol. 18, 1963, pp. 315–321, 323–337, 339–353.

91. Lefebvre, A. H., Airblast Atomization, *Prog. Energy Combust. Sci.,* Vol. 6, 1980, pp. 233–261.

92. Rizk, N. K., and Lefebvre, A. H., Influence of Liquid Film Thickness on Airblast Atomization, *ASME J. Eng. Power,* Vol. 102, July 1980, pp. 706–710.

93. Lefebvre, A. H., Wang, X. F., and Martin, C. A., Spray Characteristics of Aerated Liquid Pressure Atomizers, *AIAA J. Propul. Power,* Vol. 4, No. 4, 1988, pp. 293–298.

94. Mori, Y., Hijikata, K., and Nagasaki, T., Electrostatic Atomization for Small Droplets of Uniform Diameter, *Trans. Jpn. Soc. Mech. Eng. Ser. B,* 1981, pp. 1881–1890.

95. Lang, R. S. J., Ultrasonic Atomization of Liquids, *J. Acoust. Soc. Am.,* Vol. 34, No. 1, 1962, pp. 6–8.

96. Lobdell, D. D., Particle Size-Amplitude Relation for the Ultrasonic Atomizer, *J. Acoust. Soc. Am.,* Vol. 43, No. 2, 1967, pp. 229–231.

97. Peskin, R. L., and Raco, R. J., Ultrasonic Atomization of Liquids, *J. Acoust. Soc. Am.,* Vol. 35, No. 9, Sept. 1963, pp. 1378–1381.

98. Lee, K. W., Putnam, A. A., Gieseke, J. A., Golovin, M. N., and Hale, J. A., Spray Nozzle Designs for Agricultural Aviation Applications, NASA CR 159702, 1979.

99. Mochida, T., Ultrasonic Atomization of Liquids, *Proceedings of the 1st International Conference on Liquid Atomization and Sprays,* Tokyo, 1978, pp. 193–200.

100. Wilcox, R. L., and Tate, R. W., Liquid Atomization in a High Intensity Sound Field, *AIChE J.,* Vol. 11, No. 1, 1965, pp. 69–72.

101. Crawford, A. E., Production of Spray by High Power Magnetostriction Transducers, *J. Acoust. Soc. Am.,* Vol. 27, 1955, p. 176.

102. Antonevich, J., Ultrasonic Atomization of Liquids, *IRE Trans. PGUE,* Vol. 7, 1959, pp. 6–15.

103. Bisa, K., Dirnagl, K., and Esche, R., Zerstaubung von Flussigkeiten mit Ultraschall, *Siemens Z.,* Vol. 28, No. 8, 1954, pp. 341–347.

104. McCubbin, T., Jr., The Particle Size Distribution in Fog Produced by Ultrasonic Radiation, *J. Acoust. Soc. Am.,* 1953, pp. 1013–1014.

105. Topp, M. N., Ultrasonic Atomization—A Photographic Study of the Mechanism of Disintegration, *Aerosol. Soc.*, Vol. 4, 1973, pp. 17–25.
106. Muromtsev, S. N., and Nenashev, V. P., The Study of Aerosols—III. An Ultrasonic Aerosol Atomizer, *J. Microbiol. Epidemiol. Immunobiol.*, Vol. 31, No. 10, 1960, pp. 1840–1846.
107. Berger, H. L., Characterization of a Class of Widely Applicable Ultrasonic Nozzles, *Proceedings of the 3rd International Conference on Liquid Atomization and Sprays,* London, July 1985, pp. 1A/2/1–13.
108. Hege, H., *Aufbereit. Tech.*, No. 3, 1969, p. 142.

EXTERNAL SPRAY CHARACTERISTICS

INTRODUCTION

In most applications the function of the atomizer is not merely to break the liquid down into small drops but also to discharge these drops into the surrounding gaseous medium in the form of a symmetrical, uniform spray. With plain-orifice atomizers the cone angle is narrow and the drops are fairly evenly dispersed throughout the entire spray volume. Sprays of this type are often described as "solid." It is possible to produce solid sprays with swirl atomizers, but for most applications the spray is usually in the form of a hollow cone of wide angle, with most of the drops concentrated at the periphery.

In both types of atomizer the liquid jet or sheet rapidly disintegrates into drops, which tend to maintain the general direction of motion of the original jet or cone. However, because of air resistance, the leading drops and the drops formed on the outside of the spray rapidly lose their momentum and form a cloud of finely atomized drops suspended around the main body of the spray. Their subsequent dispersion is governed mainly by the prevailing airflow pattern in the spray region.

Because plain-orifice atomizers produce a narrow, compact spray in which only a small proportion of drops are subjected to the effects of air resistance, the distribution of the spray as a whole is dictated mainly by the magnitude and direction of the velocity imparted to it at exit from the atomizer orifice. In contrast, with twin-fluid atomizers of the air-assist and airblast types, the drops are airborne at their inception, and their subsequent trajectories are dictated by the air movements created by air swirlers and other aerodynamic devices, which form an integral part of the nozzle configuration. Thus, a feature that these atomizers have in common with plain-orifice nozzles is a lack of sensitivity of spray geometry characteristics to the physical properties of both the liquid and the surrounding gaseous medium.

The situation in regard to pressure-swirl atomizers is entirely different. With these nozzles the initial angle of the conical sheet formed at the nozzle exit is very dependent on nozzle design features, nozzle operating conditions, and liquid properties. Furthermore, even when liquid is sprayed into quiescent air, the air currents generated by the ejector action of the spray itself have a profound influence on its physical structure. This is because the initial conical sheet incurs appreciable exposure to the influence of the surrounding air. Normally any increase in spray cone angle will increase the extent of this exposure, leading to improved atomization. This is one reason why spray cone angle is such an important characteristic of a swirl atomizer.

This chapter is devoted mainly to the external spray characteristics of plain-orifice, pressure-swirl, and airblast atomizers. As mean drop size and drop size distribution are considered elsewhere, the discussion is confined to the other important spray characteristics, namely penetration, cone angle, radial liquid distribution, and circumferential liquid distribution.

SPRAY PROPERTIES

Before describing the manner and extent to which spray characteristics are influenced by liquid properties and atomizing conditions, the main properties of the spray are described.

Dispersion

The dispersion of a spray may be expressed quantitatively if, at any given instant, the volume of liquid within the spray is known. According to one definition, the degree of dispersion may be stated as the ratio of the volume of the spray to the volume of the liquid contained within it.

The advantage of good dispersion is that the liquid mixes rapidly with the surrounding gas, and the subsequent rates of evaporation are high. With plain-orifice atomizers of narrow spray angle, the dispersion is small. With swirl atomizers, dispersion is governed mainly by other spray characteristics, such as cone angle, mean drop size, and drop size distribution, and to a lesser extent by the physical properties of the liquid and the surrounding medium. In general, the factors that increase the spray cone angle also tend to increase the spray dispersion.

Penetration

The penetration of a spray may be defined as the maximum distance it reaches when injected into stagnant air. It is governed by the relative magnitudes of two opposing forces: (1) the kinetic energy of the initial liquid jet and (2) the aerodynamic resistance of the surrounding gas. The initial jet velocity is usually high, but as atomization proceeds and the surface area of the spray increases, the kinetic energy of the liquid is gradually dissipated by frictional losses to the gas. When the drops have finally exhausted their kinetic energy, their subsequent trajectory is dictated mainly by gravity and the movement of the surrounding gas.

In general, a compact, narrow spray will have high penetration, while a well-atomized spray of wide cone angle, incurring more air resistance, will tend to have low penetration. In all cases, the penetration of a spray is much greater than that of a single drop. The first drops to be formed impart their energy to the surrounding gas, which begins to move with the spray; the gas therefore offers less resistance to the following drops, which consequently penetrate farther.

Cone Angle

A major difficulty in the definition and measurement of cone angle is that the spray cone has curved boundaries, owing to the effects of air interaction with the spray. To overcome this problem, the cone angle is often given as the angle formed by two straight lines drawn from the discharge orifice to cut the spray contours at some specified distance from the atomizer face. A satisfactory method of measuring spray cone angle is to project a silhouette of the spray onto a ground-glass screen at two or three magnifications. Alternatively, measurements of spray width can be made at several axial locations to define the spray profile; one method is to use two probes equally spaced about the nozzle centerline that are moved until they contact the edges of the spray. The latest version for swirl nozzles employs linear variable displacement transducers to determine the probe positions, from which the spray angle and skewness of the spray about the nozzle axis are calculated [1].

Patternation

The symmetry of the spray pattern produced by atomizers is an important variable in most practical applications. In spray drying, for example, an unsymmetrical pattern may cause poor liquid-gas mixing, which impairs process efficiency and product quality [2]. In painting or coating of surfaces, a uniform spray pattern is essential. Patternation is also important in combustion apparatus such as domestic and industrial oil burners. In gas turbine combustors, the fuel must be distributed uniformly to achieve high combustion efficiency, low pollutant emissions, and long turbine blade life. Unless spray dissymmetry is severe it may not be detected by visual inspection, so quantitative determination of patternation is desirable, not only in nozzle design and development but also for quality control in specific applications [2].

Radial Liquid Distribution

Patternation is measured both radially and circumferentially to determine the distribution of liquid within a spray. A typical radial patternator consists of a number of small collection tubes oriented equidistant radially from the origin of the spray. The patternator shown in Fig. 7.1 consists of 29 sampling tubes of square cross-section spaced 4.5° apart on an arc of 10 cm. It is mounted below the test nozzle with the nozzle axis located at the center of curvature, as illustrated in Fig. 7.2.

Before a spray sample is taken, the patternator is inverted to drain away any liquid in the sampling tubes. The liquid flow rate is then adjusted until the nozzle is operating at the desired conditions. The patternator is rotated to the upright position, and the sampling tubes begin to fill. When one of the tubes is about three-quarters full, the liquid supply is turned off and the patternator is rotated approximately 30° until a thin metal plate, which is attached to the patternator, blocks the spray. This is necessary due to the risk of liquid dribbling after the pump is turned off.

The volume of liquid in each tube is measured by visually locating the meniscus between lines scribed into the clear plastic of the patternator. Radial distribution curves are made by plotting liquid volume as the ordinate and the corresponding angular location of the sampling tubes as the abscissa, as illustrated in Fig. 7.2. This type of plot is useful for determining how changes in the operating parameters affect the volume flow rate of liquid at individual locations in the spray. However, comparisons cannot be made between volume flow rates at different angular positions on the same curve. This is due to the fact that each sampling tube is the same size, so specific volumes are being measured. As the distance of the sampling tubes from the center of the spray increases, the proportion of the spray measured must diminish.

To overcome this problem, each liquid volume is corrected using an "area weighting" factor. This factor represents the total number of sampling tubes that would be needed to measure all of the liquid falling at a specified radial distance from the nozzle axis. The total amount of liquid in the spray is calculated by

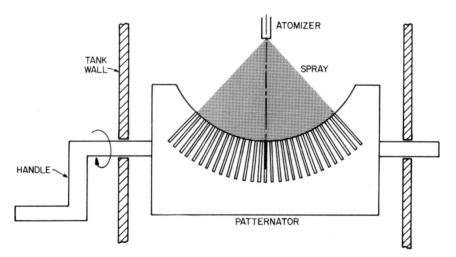

Figure 7.1 Schematic diagram of patternator for measuring radial liquid distribution.

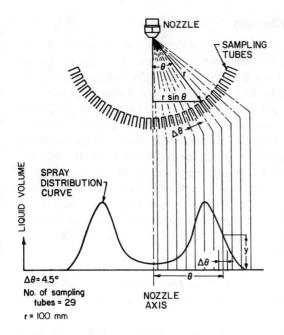

Figure 7.2 Measurement of radial liquid distribution.

adding the corrected liquid volumes. The percentage of the total spray volume measured at each angular location is found by dividing the corrected volume by the total volume. A new liquid distribution curve can then be plotted that shows the relationship between volume flow rates at different angular locations.

Equivalent Spray Angle

To describe more succinctly the effect of changes in operating parameters on liquid distribution, a radial distribution curve may be reduced to a single numerical value called the *equivalent spray angle* [3, 4]. The equivalent spray angle is the sum of two angles, $\phi = \phi_L + \phi_R$, which are calculated using the following equation:

$$\phi_L \ (\text{or} \ \phi_R) = \frac{\Sigma y\theta \ \Delta\theta \ \sin \theta}{\Sigma y \ \Delta\theta \ \sin \theta} = \frac{\Sigma y\theta \ \sin \theta}{\Sigma y \ \sin \theta} \tag{7.1}$$

Here L and R represent the left- and right-hand lobes of the liquid distribution curve, respectively, θ is the angular location of the sampling tubes, $\Delta\theta$ is the angle between the sampling tubes, and y is the liquid volume measured at the corresponding tubes. The physical meaning of the equivalent spray angle is that ϕ_L (or ϕ_R) is the value of θ that corresponds to the position of the center of mass of a material system for the left (or right) lobe of the distribution curve.

Circumferential Liquid Distribution

Quantitative measurement of circumferential liquid distribution, commonly known as *circumferential patternation,* originated many years ago in the gas turbine industry [5], when circular-sectored vessels were developed to measure symmetry about the spray axis of hollow-cone fuel nozzles. In this method the nozzle is centered above the vessel and sprays down into a cylindrical collection tray that is partitioned into a number of pie-shaped sectors, usually 12. Each sector drains into a separate sampling tube. The height of the nozzle above the patternator is adjusted for changes in cone angle to ensure that the entire spray is collected. The duration of each test is determined as the time required for one of the sampling tubes to become nearly full. After the level of liquid in each tube is measured and recorded, the values are averaged to get a mean height. The levels of the tubes are normalized against the mean, and the standard deviation of the normalized values is calculated. The normalized standard deviation is indicative of the circumferential irregularity of the nozzle spray.

According to Tate [2], the minimum/maximum sector ratio varies appreciably with the total number of sectors employed. With fewer sectors the patternation appears better. This variable should, therefore, be specified when reporting results.

Recent developments include the use of phototransistors to measure automatically the liquid levels in each sector sampling tube [1].

A less convenient technique for measuring circumferential liquid distribution is to "clock" the radial patternator around its axis to different angular locations, usually 15° or 22.5° apart. This lengthy procedure provides detailed information on both radial and circumferential liquid distributions.

Factors Influencing Spray Patternation

Tate [2] has discussed in some detail the causes of poor spray patternation, some of which may be traced to the sampling technique or apparatus. Clearly, the nozzle should not be skewed relative to the axis of the collecting vessel, which should be accurately centered and located at the appropriate distance from the nozzle. Under conditions of high liquid flow rates or, for airblast atomizers, high air velocities, results may be affected by splashing and entraiment in the collecting vessel. This can be minimized by designing radial and circumferential patternators having deep collection tubes and sectors, respectively. Pattern symmetry is also likely to be poor at low liquid flow rates, where the spray is not fully developed.

Circumferential patternation is sensitive to the dimensional relationships characteristic of a particular nozzle design. For example, for hollow-cone simplex atomizers, patternation has been correlated with the degree of eccentricity between the swirl chamber and the final discharge orifice [2]. Nozzle quality is also important, and spray patternation may be impaired by poor surface finish, orifice imperfections, plugged or contaminated flow passages, eccentric alignment of key nozzle components, and other conditions. For airblast atomizers, lack of symmetry in the various air passages and swirlers could also have an adverse effect on spray patternation.

Almost 30 years ago, Tate [2] observed that patternation criteria are quite arbitrary and empirical. This fairly describes the situation today. Although spray symmetry can be measured and expressed quantitatively, little effort has been made to relate patternation to the various performance parameters associated with the nozzle application.

PENETRATION

Spray penetration is of prime importance in diesel engines. Overpenetration of the spray leads to impingement of fuel on the combustion chamber walls. This is acceptable if the walls are hot and considerable air swirl is present, but in large engines the chambers are generally of the quiescent type, with little air swirl, so overpenetration would result in impingement on a cold wall with consequent fuel wastage. On the other hand, if spray penetration is inadequate, fuel-air mixing is unsatisfactory. Optimum engine performance is obtained when the spray penetration is matched to the size and geometry of the combustion chamber. Methods for calculating penetration are therefore essential to sound engine design.

Many researchers have studied the spray penetration in diesel engines. The results of these studies have been reviewed by Hiroyasu [6].

Sitkei [7] used dimensional analysis to derive the following expression for the penetration of a diesel jet in still air.

$$S = 0.2d_o \left(\frac{U_L t}{d_o}\right)^{0.48} \left(\frac{U_L d_o}{\nu_L}\right)^{0.3} \left(\frac{\rho_L}{\rho_A}\right)^{0.35} \tag{7.2}$$

Taylor and Walsham [8] employed both conventional and schlieren photography for tracking the penetration of single injections of gas oil into quiescent nitrogen at high pressure. Their results are correlated by the expression

$$S = 0.5d_o \left[\left(\frac{\Delta P_L}{\rho_A}\right)^{0.5} \frac{t}{d_o}\right]^{0.64} \left(\frac{l_o}{d_o}\right)^{0.18} \tag{7.3}$$

Dent [9] used jet mixing theory to derive the following equation for the correlation penetration data:

$$S = 3.01 \left[\left(\frac{\Delta P_L}{\rho_A}\right)^{0.5} d_o t\right]^{0.5} \left(\frac{295}{T_A}\right)^{0.25} \tag{7.4}$$

Hay and Jones [10] have critically reviewed the literature on jet penetration relevant to diesel injectors published before 1972. They assert that the Sitkei correlation [Eq. (7.2)] is satisfactory at low load conditions but at high load conditions gives values that are too large. The Taylor and Walsham correlation [Eq. (7.3)] is said to overestimate the effect of nozzle l_o/d_o ratio. Its use for $d_o > 0.5$

mm is not recommended, as the correlation predicts very low penetrations under these conditions.

According to Hay and Jones, the Dent correlation [Eq. (7.4)] shows satisfactory agreement with experimental data from many sources. They recommend its use under all conditions except for exceptionally large chamber pressure (P_A > 10 MPa), where it tends to predict overlarge penetrations.

More recent studies of diesel spray penetration have been made by Hiroyasu et al. [11,12]. Figure 7.3 shows a logarithmic plot, based on their experimental data, of spray tip penetration versus time for several values of injection pressure. It is of interest to note that in the early stage of injection the slope of the graph is unity, but after a short period of time it decreases to 0.5. The corresponding equations for spray tip penetration are

$$S_1 = K_1 \Delta P_L^{0.6} \rho_A^{-0.33} t \tag{7.5}$$

and

$$S_2 = K_2 \Delta P_L^{0.1} \rho_A^{-0.42} t^{0.5} \tag{7.6}$$

Hiroyasu and Arai [12] combined their experimental results with the jet disintegration theory of Levich [13] to derive the following equations for spray tip penetration.

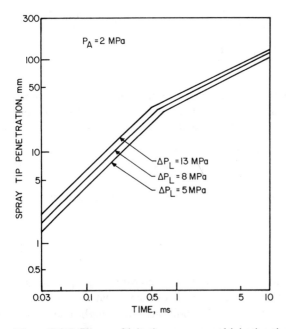

Figure 7.3 Influence of injection pressure and injection time on spray penetration (*Hiroyasu et al.* [11, 12]).

For injection times shorter than the jet breakup time t_b:

$$S = 0.39 \left(\frac{2\Delta P_L}{\rho_A}\right)^{0.5} t \qquad (7.7)$$

For $t_b < t$:

$$S = 2.95 \left(\frac{\Delta P_L}{\rho_A}\right)^{0.25} (d_o t)^{0.5} \qquad (7.8)$$

The jet breakup time is given by the dimensionally correct equation

$$t_b = 28.65 \rho_L d_o (\rho_A \Delta P_L)^{-0.5} \qquad (7.9)$$

Very little information is available on the penetration of sprays from pressure-swirl nozzles. For simplex and dual-orifice nozzles operating in gas turbine combustors, Lefebvre [14] has shown that spray penetration is inversely proportional to the cube root of the ambient gas pressure. This contraction of the fuel spray is a primary cause of increased soot formation and exhaust smoke at high combustion pressures.

SPRAY CONE ANGLE

The characteristics of the flow in pressure atomizers have been studied by several workers. Their results show that spray angle is influenced by nozzle dimensions, liquid properties, and the density of the medium into which the liquid is sprayed.

Pressure-Swirl Atomizers

The spray angles produced by pressure-swirl nozzles are of special importance in their application to combustion systems. In gas turbine combustors, the spray angle exercises a strong influence on ignition performance, flame blowout limits, and the pollutant emissions of unburned hydrocarbons and smoke. This is reflected in the numerous theoretical and experimental studies that have been carried on the factors governing spray cone angle.

Theory. According to Taylor's theory [15], spray cone angle is determined solely by the swirl chamber geometry and is a unique function of the ratio of the inlet ports area to the product of swirl chamber diameter and orifice diameter, $A_p/D_s d_o$, as shown in Fig. 7.4. This relationship is unique only for nonviscous fluids. It is modified in practice by viscous effects, which depend on the form and area of the wetted surface as expressed in the ratios D_s/d_o, L_s/D_s, and l_o/d_o. The predictions of Taylor's inviscid theory are represented by the dashed curve in Fig. 7.4, while the solid curve corresponds to experimental data obtained by Watson [16], Giffen and Massey [17], and Carlisle [18]. Agreement between theory and

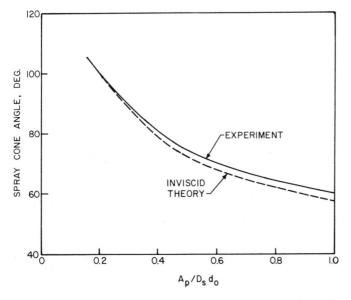

Figure 7.4 Practical relationship between spray cone angle and atomizer geometry.

experiment is clearly satisfactory for large cone angles, but the theoretical predictions are about 3° too low at a cone angle of 60°.

Giffen and Muraszew's analysis of the flow in a swirl atomizer [19] assumes a nonviscous liquid, which allows the spray cone angle to be expressed as a function of nozzle dimensions only. It leads to the following expression for the mean value of the spray cone half-angle θ:

$$\sin \theta = \frac{(\pi/2)C_D}{K(1 + \sqrt{X})} \qquad (7.10)$$

where $K = A_p/D_s d_o$ and $X = A_a/A_o$.

The same theory also provides an expression for discharge coefficient in terms of atomizer dimensions:

$$C_D = \left[\frac{(1 - X)^3}{1 + X}\right]^{0.5} \qquad (5.26)$$

Substitution of Eq. (5.26) into Eq. (7.10) gives

$$\sin \theta = \frac{(\pi/2)(1 - X)^{1.5}}{K(1 + \sqrt{X})(1 + X)^{0.5}} \qquad (7.11)$$

This equation gives a relationship between the atomizer dimensions, the size of the air core, and the mean spray cone angle. To eliminate one of these variables, Giffen and Muraszew applied the condition that the size of the air core in the orifice will always be such as to give maximum flow, i.e., the value of the discharge coefficient expressed as a function of X is a maximum, or $1/C_D^2$ is a minimum.

Putting $d(1/C_D^2)/dX = 0$ leads to the following expression for K in terms of X:

$$K^2 = \frac{\pi^2(1 - X)^3}{32X^2} \tag{7.12}$$

Equations (7.11) and (7.12) allow the spray cone angle to be expressed in terms of either X or K. Figure 7.5, which shows spray cone angle plotted against the relevant atomizer dimensions, is based on calculations carried out by Simmons [20].

Since the value of X is a function only of K, it is evident that the spray cone angle is, theoretically, a function only of this atomizer constant and is independent of liquid properties and injection pressure.

Starting from the assumption of a vortex flow pattern within the nozzle of the form $vr^n = $ constant, Babu et al. [21] derived the following empirical equation for the initial spray cone half-angle:

$$\tan \theta = \frac{(\pi/4)(1 - X)K_\theta}{B} \tag{7.13}$$

where

$$B = \frac{A_p}{D_m d_o} \left(\frac{D_m}{d_o}\right)^{1-n}$$

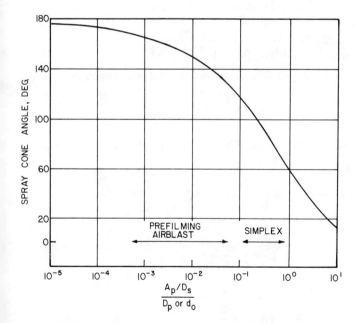

Figure 7.5 Theoretical relationship between cone angle and atomizer dimensions [19, 20].

For $\Delta P_L > 2.76$ MPa (400 psi)

$$n = 17.57 \frac{A_o^{0.1396} A_p^{0.2336}}{A_s^{0.1775}}$$

$$K_\theta = 0.00812 \frac{A_p^{0.034048}}{A_o^{0.17548} A_s^{0.24579}}$$

For $\Delta P_L < 2.76$ MPa (400 psi)

$$n = 28 A_o^{0.14176} \frac{A_p^{0.27033}}{A_s^{0.17634}}$$

$$K_\theta = 0.0831 \frac{A_p^{0.34873}}{A_o^{0.26326} A_s^{0.32742}}$$

The general validity of Eq. (7.13) is supported by a considerable body of experimental data obtained with several different simplex nozzles. According to Babu et al., the maximum deviation between the predicted and experimental results is less than 10%.

Rizk and Lefebvre [22] start from the notion that any given particle of liquid leaves a point on the lip of the discharge orifice in a plane that lies tangential to the lip cylinder, at an angle that is the resultant of the axial and tangential components of velocity at this point. The included angle between the tangents to the spray envelope gives the cone angle, 2θ.

The axial component of velocity in the orifice is given by

$$U_{ax} = \frac{\dot{m}_L}{\rho_L (A_o - A_a)} = \frac{\dot{m}_L}{\rho_L A_o (1 - X)} \qquad (7.14)$$

The average velocity of the liquid film at the nozzle exit is obtained as

$$\bar{U} = K_v \left(\frac{2\Delta P_L}{\rho_L} \right)^{0.5} \qquad (7.15)$$

where K_v is the velocity coefficient (see Chapter 5). The mass flow rate is given by

$$\dot{m}_L = C_D A_o \rho_L \left(\frac{2\Delta P_L}{\rho_L} \right)^{0.5} = C_D A_o \rho_L \frac{\bar{U}}{K_v}$$

or

$$\bar{U} = \frac{\dot{m}_L K_v}{C_D A_o \rho_L} \qquad (7.16)$$

Now

$$\cos \theta = \frac{U_{ax}}{\bar{U}}$$

Substituting for U_{ax} and $\bar{U}$ from Eqs. (7.14) and (7.16), respectively, gives

$$\cos\theta = \frac{C_D}{K_v\,(1 - X)} \qquad (7.17)$$

Rizk and Lefebvre [22] also derived the following equation for spray cone angle, solely in terms of air core size:

$$\cos^2\theta = \frac{1 - X}{1 + X} \qquad (7.18)$$

In this equation θ is the cone half-angle, as measured close to the nozzle. As the spray in this region has a small but definite thickness, the cone angle formed by the outer boundary of the spray is defined as 2θ, while $2\theta_m$ represents the mean cone angle in this near-nozzle region. Equation (7.18) was derived by regarding the average value of the tangential component of velocity as

$$v_m = \frac{v_s + v_o}{2} \qquad (7.19)$$

where v_s is the tangential velocity at the liquid surface in the air core and v_o is the tangential velocity at the orifice diameter d_o.

Thus, according to free-vortex theory

$$v_o = \frac{d}{d_o}\,v_s = \sqrt{X}\,v_s \qquad (7.20)$$

Hence, by assuming a constant axial velocity U_{ax} across the liquid film, we have

$$\tan\theta_m = \frac{v_s + v_o}{2U_{ax}} = 0.5\tan\theta\left(1 + \sqrt{X}\right) \qquad (7.21)$$

Of course, X is directly related to film thickness t, since the difference between the air core diameter and the discharge orifice diameter is determined solely by the thickness of the liquid film. We have

$$X = \frac{(d_o - 2t)^2}{d_o^2} \qquad (5.37)$$

Values of t for insertion into Eq. (5.37) may be obtained from Eq. (5.41) or (5.43). Equations (7.18), (7.21), and (5.37), (5.41), and (5.43) may be used to examine the effects of changes in atomizer dimensions, liquid properties, air properties, and injection pressure on spray cone angle.

Nozzle dimensions. For nonviscous liquids, the spray cone angle is governed mainly by the atomizer dimensions A_p, D_s, and d_o, as shown in Figs. 7.4 and 7.5. However, in practice, some modifications to the "basic" spray angle can be made by using different orifice edge configurations (radiusing, chamfers, lips, steps, etc.).

Rizk and Lefebvre [23] estimated values of cone angle for different nozzle dimensions, using Eqs. (7.18) and (7.21) along with corresponding values of t from Eq. (5.41). Some of the results obtained are shown in Fig. 7.6, which illustrates the effect of varying discharge orifice diameter on cone angle. It is seen that increase in orifice diameter produces a wider cone angle, while raising the injection pressure also widens the cone angle but to a lesser extent. The experimental results in [24] generally confirm the above trends, as illustrated in Fig. 7.6.

Similar calculations were conducted to determine the effects of variations in inlet port diameter, swirl chamber diameter, orifice length, swirl chamber length, and nozzle constant on spray cone angle. Figure 7.7 shows that cone angle diminishes slightly with increase in inlet port diameter, while increase in swirl chamber diameter has the opposite effect, as illustrated in Fig. 7.8. These effects may be attributed directly to changes in liquid flow rates and in the relative magnitudes of the tangential and axial flow velocities. Figures 7.9 and 7.10 indicate very small changes in cone angle with variations in orifice length and swirl chamber length, respectively. They also demonstrate good agreement between predicted and experimental values of cone angle.

Experimental data on the relationship between cone angle and the atomizer constant K are shown in Fig. 7.11. Also plotted in this figure, and showing the same general trend, are curves representing the predicted variation of θ_m with K for various values of ΔP_L.

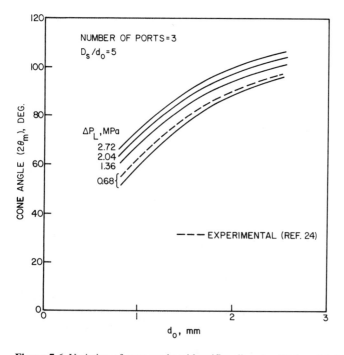

Figure 7.6 Variation of cone angle with orifice diameter *(Rizk and Lefebvre [23])*.

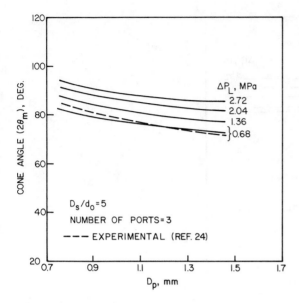

Figure 7.7 Variation of cone angle with inlet port diameter *(Rizk and Lefebvre [23])*.

Liquid properties. Based on numerous calculations of spray angle using Eqs. (5.41), (7.18), and (7.21), Rizk and Lefebvre [23] derived the following dimensionally correct equation for spray angle:

$$2\theta_m = 6K^{-0.15}\left(\frac{\Delta P_L d_o^2 \rho_L}{\mu_L^2}\right)^{0.11} \tag{7.22}$$

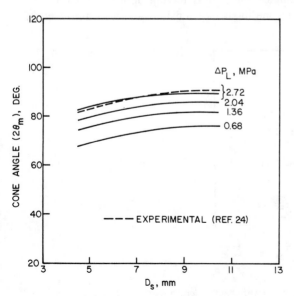

Figure 7.8 Variation of cone angle with swirl chamber diameter *(Rizk and Lefebvre [23])*.

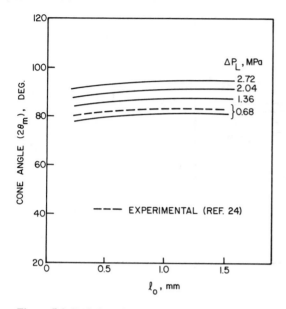

Figure 7.9 Variation of cone angle with orifice length *(Rizk and Lefebvre [23])*.

This equation provides a useful indication of the effects of liquid injection pressure and liquid properties on spray angle.

Surface tension According to Eq. (7.22), surface tension should have no effect on spray cone angle, and this is generally confirmed by experiment. Giffen and Massey [17] measured cone angles for liquids having approximately the same viscosity but a range of three to one in surface tension, and no significant effect

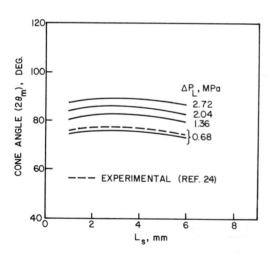

Figure 7.10 Variation of cone angle with swirl chamber length *(Rizk and Lefebvre [23])*.

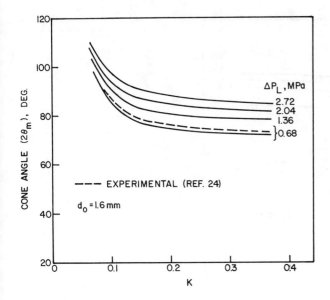

Figure 7.11 Variation of cone angle with atomizer constant *(Rizk and Lefebvre [23])*.

of surface tension was observed. More recently, Wang and Lefebvre [25] compared the equivalent cone angles exhibited by several simplex swirl nozzles when flowing water and diesel oil (DF-2). These liquids have surface tensions that differ by a factor of almost three. In all cases the spray cone angle was slightly less for diesel oil, and this was attributed to its slightly higher viscosity. Wang and Lefebvre concluded that surface tension has no appreciable effect on cone angle, thus confirming the conclusion reached by Giffen and Massey.

Density The influence of liquid density on spray cone angle is illustrated in Fig. 7.12, where it can be seen that the spray angle widens slightly with increase in density. It may be noted in this figure that increase in nozzle pressure differential also causes the spray cone angle to widen. This is one of the reasons why increase in ΔP_L improves atomization quality.

Viscosity The viscosity of the liquid is of great importance to cone angle. It modifies the flow of an ideal liquid in two ways: (1) by friction in the body of the liquid and (2) by friction at the boundary between the liquid and its containing walls. Both effects are caused by the friction force created by the velocity gradient in a viscous liquid. The friction force produced by this velocity gradient tends to reduce the tangential velocity, and this effect increases with decreasing value of atomizer radius, reaching a maximum at the air core. The higher the viscosity, the greater will be the divergence between the actual tangential velocity and the value calculated from the equation

$$vr = v_i R \qquad (7.23)$$

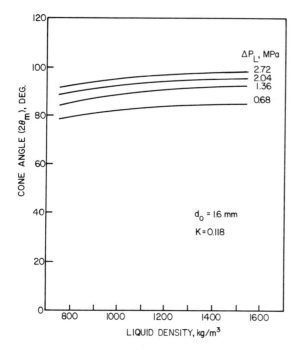

Figure 7.12 Effect of liquid density on initial spray cone angle *(Rizk and Lefebvre [23])*.

The velocity gradient at the boundaries between the liquid and the containing walls is another source of friction, and this causes the formation of a boundary layer of liquid in which the velocity of the liquid changes from zero at the boundary to the ordinary velocity of the nearby liquid. The liquid in the boundary layer does not obey the same laws as the main body of liquid away from the solid boundaries, and if the thickness of the boundary layer is appreciable, a significant modification of the conditions of flow will occur [19].

Giffen and Massey [17] examined the influence of viscosity on cone angle over a range of viscosities from 2×10^{-6} to 50×10^{-6} m²/s (2 to 50 cS). They found that the relation between the cone half-angle and viscosity for a swirl-plate atomizer could be expressed by the empirical equation

$$\tan \theta = 0.169 v_L^{-0.131} \tag{7.24}$$

This equation indicates a dependence of spray angle on viscosity that is roughly the same as the viscosity dependence shown in Eq. (7.22).

Wang and Lefebvre [25] used blends of diesel oil (DF-2) and polybutene to examine the influence of viscosity on equivalent spray angle for a simplex nozzle. Figure 7.13 is typical of the results obtained. It shows that cone angle decreases with increase in viscosity, but the effect is not as great as Eqs. (7.22) and (7.24) appear to suggest.

In a swirl atomizer of the helical-groove type the liquid has to flow through tortuous passages before it enters the final orifice; therefore viscosity might be expected to have a greater effect in this type of atomizer. This was confirmed by Giffen and Massey [17], who found that the decrease of cone angle with increased viscosity was more pronounced than in an atomizer of the swirl-plate type.

Air properties. De Corso and Kemeny [3] carried out the first investigation on the effects of ambient gas pressure on spray cone angle. Their results show that, over a range of gas pressures from 10 to 800 kPa, the equivalent spray angle is an inverse function of $P_A^{1.6}$. A similar study carried out by Neya and Sato [26] on water sprayed into stagnant air produced a similar result, namely that spray contraction $\Delta\phi$ increases with $P_A^{1.2}$.

These studies were limited to ambient air pressure below 1 MPa (10 atm). It should also be noted that the equivalent spray angles studied by these workers were not measured close to the nozzle but at some distance downstream. In consequence, they tend to be considerably smaller than angles measured in the near-nozzle region, because as the drops move away from the nozzle they are subjected to radial air currents generated by their own kinetic energy, as described by De Corso and Kemeny [3].

Ortman and Lefebvre [4] used a radial patternator of the type shown in Fig. 7.1 to examine, for four different simplex nozzles, the effects of variations in liquid injection pressure and ambient air pressure on equivalent spray angle. The liquid employed was aviation kerosine having the following properties: $\rho = 780$ kg/m³, $\sigma = 0.0275$ kg/s², $\mu = 0.0013$ kg/m s. Some of the results obtained are shown in Figs. 7.14. to 7.17. Initially, an increase in gas pressure above normal atmospheric causes the spray to contract sharply. However, with continuing in-

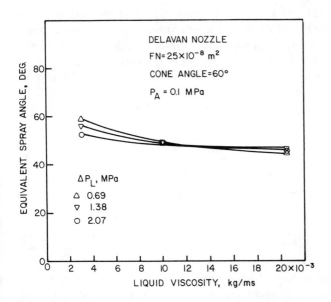

Figure 7.13 Influence of viscosity on equivalent spray angle *(Wang and Lefebvre [25])*.

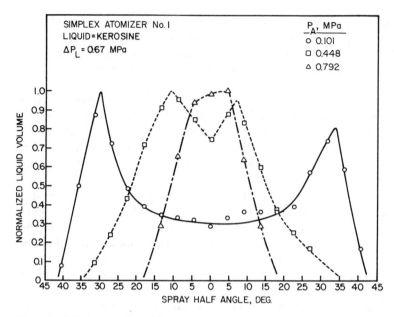

Figure 7.14 Influence of ambient gas pressure on radial liquid distribution *(Ortman and Lefebvre [4])*.

crease in gas pressure the rate of spray contraction decreases. Finally, a point is reached where further increase in ambient pressure has virtually no influence on the equivalent spray angle, as shown in Fig. 7.17.

Dodge and Biaglow [27] used a Hago simplex nozzle of nominal 80° cone angle to examine the influence of atomizing conditions on spray characteristics. The liquids employed were kerosine (Jet-A) and diesel oil (DF-2). Cone angles were measured 10 mm from the nozzle. They were defined as twice the half-angle, where the half-angle was calculated from half the total width of the spray at 10 mm divided by the distance from the nozzle plus the effective origin of the spray, which was 1.1 mm upstream of the nozzle tip. The half-cone angle θ is thus

$$\theta = \arctan \frac{\text{spray width, mm}}{2(10 + 1.1)} \tag{7.25}$$

The experimental data were correlated using an expression for cone angle of the form

$$2\theta = 79.8 - 0.918 \frac{\rho_A}{\rho_{A_0}} \tag{7.26}$$

where ρ_{A_0} is the air density at normal atmospheric pressure and temperature.

According to De Corso and Kemeny [3], the phenomenon of decreasing spray angle is caused by the aerodynamic effects due to the motion of the liquid spray

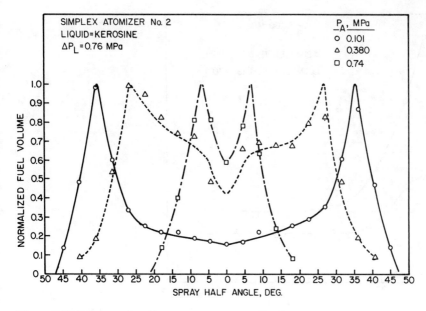

Figure 7.15 Influence of ambient gas pressure on radial liquid distribution *(Ortman and Lefebvre [4])*.

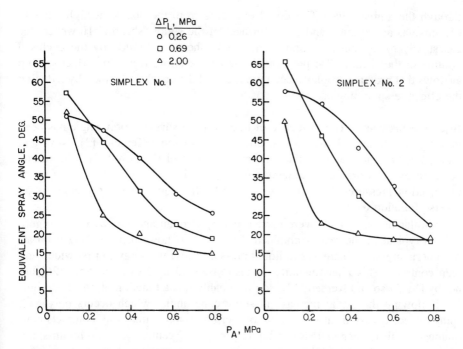

Figure 7.16 Influence of ambient gas pressure on equivalent spray cone angle *(Ortman and Lefebvre [4])*.

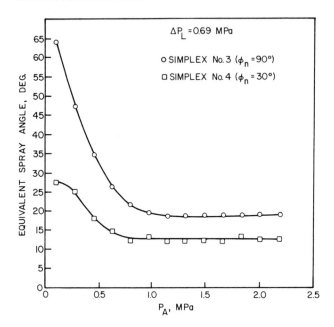

Figure 7.17 Influence of ambient gas pressure on equivalent spray angle *(Ortman and Lefebvre [4]).*

through the ambient gas. The liquid emerging from the nozzle at high velocity entrains gas at the inner and outer surfaces of the spray "sheath." However, the gas supply to the inner portion of the spray sheath is limited by the enclosed volume in the sheath. The pressure difference resulting from this effect sets up airflows that produce droplet acceleration toward the nozzle axis, thereby reducing the effective spray angle.

Injection pressure. The influence of injection pressure on spray angle has been investigated by several workers including De Corso and Kemeny [3], Neya and Sato [26], Ortman and Lefebvre [4], and Dodge and Biaglow [27]. The results obtained by De Corso and Kemeny and Neya and Sato show that, over a range of injection pressures from 0.17 to 2.7 MPa, the equivalent spray angle is an inverse function of ΔP_L.

Similar measurements were carried out by Ortman and Lefebvre [4]. Figure 7.18 is typical of the results obtained. Figure 7.19 shows that, starting from atmospheric pressure, increases in liquid pressure cause the spray to first widen and then contract. This phenomenon was also observed by Neya and Sato [26] but not by De Corso and Kemeny [3], so presumably it is a function of nozzle design.

Following the initial increase in spray cone angle (which occurs only with sprays injected into air at normal atmospheric pressure), the spray contracts continuously with further increase in liquid pressure. Eventually, a minimum spray angle is attained, and further increase in injection pressure has no discernible effect on spray angle.

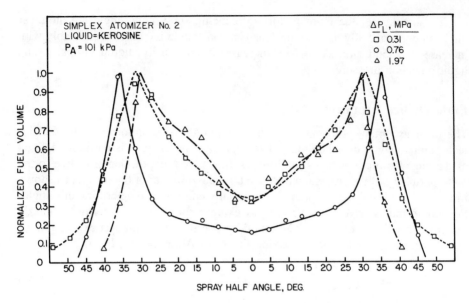

Figure 7.18 Influence of injection pressure on radial liquid distribution *(Ortman and Lefebvre [4])*.

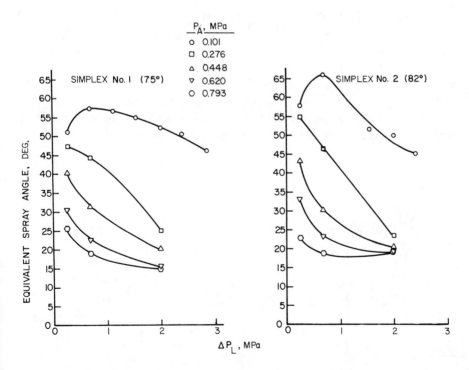

Figure 7.19 Influence of injection pressure on equivalent spray angle *(Ortman and Lefebvre [4])*.

To summarize, the main factors influencing spray cone angle are atomizer dimensions, especially d_o, and air density. Cone angle increases appreciably with increase in d_o and diminishes with increases in air density, liquid viscosity, and liquid injection pressure.

Plain-Orifice Atomizers

The most important practical example of a plain-orifice atomizer is the fuel injector employed on diesel engines. The angle of a diesel spray is normally defined as the angle formed by two straight lines drawn from the discharge orifice to the outer periphery of the spray at a distance $60d_o$ downstream of the nozzle, as shown in Fig. 7.20. Photographic methods are usually employed to determine this angle. Several formulas have been derived to express spray angle in terms of nozzle dimensions and the relevant air and liquid properties. The simplest expression for spray angle is given by the jet mixing theory of Abramovich [29] as

$$\tan \theta = 0.13\left(1 + \frac{\rho_A}{\rho_L}\right) \tag{7.27}$$

Yokota and Matsuoka [30] correlated their experimental data on spray angles obtained at high ambient air pressures using the expression

$$\theta = 0.0676 \text{Re}_L^{0.64}\left(\frac{l_o}{d_o}\right)^{-n}\left[1 - \exp-\left(0.023\frac{\rho_L}{\rho_A}\right)\right]^{-1} \tag{7.28}$$

where

$$n = 0.0284\left(\frac{\rho_L}{\rho_A}\right)^{0.39} \tag{7.29}$$

According to Reitz and Bracco [31], the spray angle can be determined by combining the radial velocity of the fastest growing of the unstable surface waves

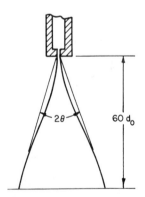

Figure 7.20 Definition of diesel spray angle [28].

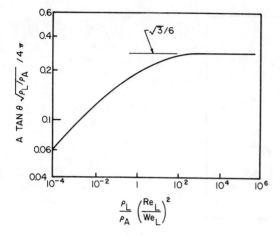

Figure 7.21 Plot of function in Eq. (7.30) [31].

with the axial injection velocity. This hypothesis results in the following expression for spray angle:

$$\tan \theta = \frac{4\pi}{A} \left(\frac{\rho_A}{\rho_L}\right)^{0.5} f \left(\frac{\rho_L Re_L}{\rho_A We_L}\right)^2 \tag{7.30}$$

where A is a function of l_o/d_o and must be determined experimentally. The function f is shown in Fig. 7.21.

In a more recent paper, Bracco et al. [32] quote Eq. (7.30) in a simplified form as

$$\tan \theta = \frac{2\pi}{\sqrt{3}A} \left(\frac{\rho_A}{\rho_L}\right)^{0.5} \tag{7.31}$$

Hiroyasu and Arai [33] applied dimensional analysis to their experimental data acquired at high pressures to derive the following equation for spray angle:

$$\theta = 0.025 \left(\frac{\rho_A \Delta P_L d_o^2}{\mu_A^2}\right)^{0.25} \tag{7.32}$$

Note that in Eqs. (7.28) and (7.32) the angle θ is expressed in radians.

The effects of Reynolds number and nozzle orifice length-to-diameter ratio on spray angle are illustrated in Fig. 7.22. The data shown in this figure were obtained by injecting water through a 0.3-mm orifice into quiescent air at a pressure of 3 MPa (30 atm). The actual experimental points are omitted from the figure to improve clarity.

According to Arai et al. [28], the shapes of the curves shown in Fig. 7.22 are the result of complex interactions between jet turbulence, jet breakup length, and orifice discharge coefficient, all of which are affected by changes in Reynolds number and nozzle l_o/d_o ratio. Figure 7.22 shows that increase in Reynolds num-

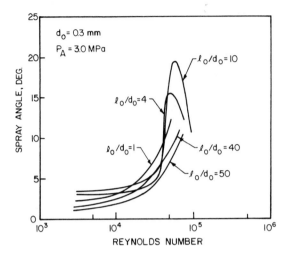

Figure 7.22 Influence of Reynolds number and nozzle l_o/d_o ratio on spray angle *(Hiroyasu and Arai [33])*.

ber above 30,000 causes the spray cone angle to widen appreciably. For the range of nozzle l_o/d_o ratios of most practical interest, i.e., between 4 and 10, the maximum spray angle occurs at the Reynolds number corresponding to minimum jet breakup length.

CIRCUMFERENTIAL LIQUID DISTRIBUTION

Very few studies appear to have been made on the circumferential distribution of liquid in the sprays produced by pressure-swirl atomizers. This is surprising, because in gas turbine combustors nonuniformities in circumferential fuel distribution can give rise to local pockets of fuel-lean mixture in which burning rates are low, thereby producing high concentrations of carbon monoxide and unburned hydrocarbons. By the same token, other regions of the spray in which the fuel concentration exceeds the mean value are characterized by high soot formation, leading to excessive flame radiation and exhaust smoke. Irregularities in circumferential fuel distribution can also adversely affect liner wall temperatures and turbine blade life.

Ortman and Lefebvre [4] examined the spray characteristics of several nozzles of different design and manufacture. They found that spray quality, expressed in terms of symmetry of radial liquid distribution and uniformity of circumferential liquid distribution, varied between different nozzles. These variables in performance were not due to physical damage or errors in manufacture but could be attributed directly to differences in nozzle design. The best nozzles exhibited excellent radial symmetry and circumferential maldistributions of less than 10%. Other nozzles were less satisfactory, as illustrated in Figs. 7.23 and 7.24, in which

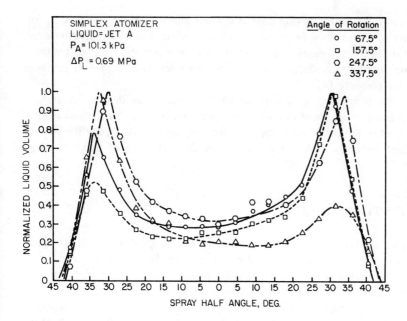

Figure 7.23 Measurements of radial liquid distribution plotted at different angles to show circumferential uniformity [4].

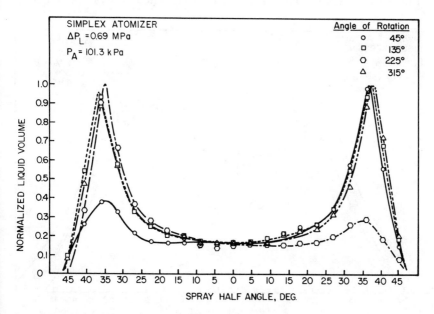

Figure 7.24 Measurements of radial liquid distribution plotted at different angles to show circumferential uniformity [4].

eight radial liquid distributions obtained at normal atmospheric pressure are plotted at intervals of 45° to show both radial and circumferential liquid distributions. From inspection of these figures it is apparent that some nonuniformities are present. These maldistributions were also found at other levels of injection pressure and ambient air pressure [4].

Further work is needed to determine which geometrical design features of pressure-swirl nozzles are most important to spray uniformity. This could lead to the establishment of simple procedures for achieving satisfactory spray uniformity of the kind that currently exist for securing any desired value of nominal spray cone angle.

AIRBLAST ATOMIZERS

With prefilming airblast atomizers, the distribution of drops throughout the spray is governed partly by the symmetry and uniformity of the liquid film created at the atomizing lip and also by the flow patterns generated in the high-velocity atomizing airstream during its passage through the atomizer.

Ortman et al. [34] have measured radial and circumferential liquid distributions for a number of airblast atomizers of the prefilming type. Figures 7.25 to 7.28 are typical of the results obtained. Figure 7.25 shows that increasing the air pressure differential across the nozzle causes the two lobes in the liquid distribution curve to move closer together, while the concentration of liquid in the middle of the spray increases. The effect of increasing liquid flow rate is to reduce

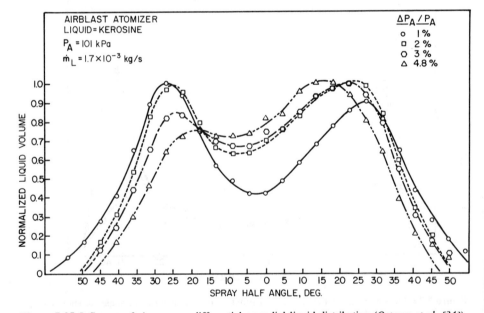

Figure 7.25 Influence of air pressure differential on radial liquid distribution *(Ortman et al. [34])*.

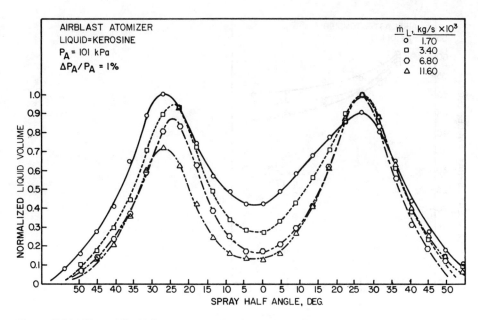

Figure 7.26 Effect of liquid flow rate on radial liquid distribution *(Ortman et al. [34])*.

the density of the spray in the center with no significant change in the position of the peaks of the curves, as shown in Fig. 7.26.

The variation of equivalent spray angle with air pressure differential is shown in Fig. 7.27. There is a continuous reduction in spray angle as ΔP_A increases. An increase in liquid flow rate at low air pressure differential causes the cone angle to first decrease slightly and then increase slightly, while at higher values of ΔP_A the equivalent spray angle decreases initially and then remains sensibly constant, as shown in Fig. 7.28.

Experimental data obtained with other prefilming airblast atomizers showed the same general trends as those observed in Figs. 7.25 to 7.28, but further work is needed before any general conclusions can be drawn.

DROP DRAG COEFFICIENTS

When a liquid is sprayed into a flowing gas stream, estimation of the trajectories of individual drops in the spray is usually fairly straightforward, provided the flow conditions are accurately defined and the drop drag coefficient is known.

Stokes [35] analyzed the steady flow of a real fluid past a solid sphere (or a solid sphere moving in a real fluid) and showed that the drag force on a sphere can be expressed as

$$F = 3\pi D \mu_L U_R \qquad (7.33)$$

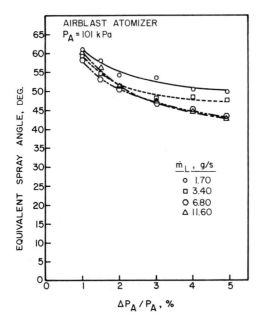

Figure 7.27 Effect of air pressure differential on equivalent spray angle *(Ortman et al. [34])*.

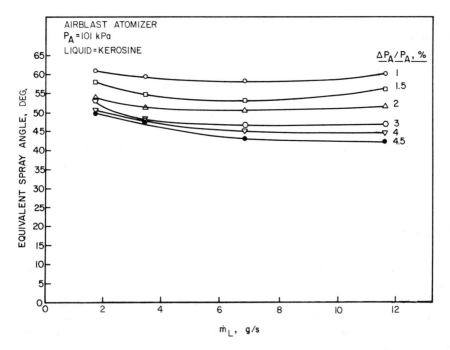

Figure 7.28 Influence of liquid flow rate on equivalent spray angle *(Ortman et al. [34])*.

If the drag coefficient C_D is defined as

$$C_D = \frac{F}{(\pi/4)D^2(\rho_A U_A^2/2)} \tag{7.34}$$

it can be expressed as

$$C_D = \frac{24}{\text{Re}} \tag{7.35}$$

where $\text{Re} = U_R D \rho_A / \mu_A$

Stokes' law holds only for flows that are entirely dominated by viscous forces, that is, for flows characterized by low Reynolds numbers. For higher Reynolds numbers, the inertial forces become significant and the drag force may be increased due to the formation of a wake and possible detachment of energy-consuming vortices. Another departure from Stokes' law is caused by the presence of a boundary layer, since the actual volume moving through the fluid consists of the sphere together with its boundary layer. These considerations have led to various theoretical correction factors to Stokes' drag law, but most of the correlations for drop drag coefficient in common use are based on the analysis of experimental data.

There are many empirically determined functions that approximate the experimental data obtained by different investigators. For example, Langmuir and Blodgett [36] suggested the following equation:

$$C_D \frac{\text{Re}}{24} = 1 + 0.197\text{Re}^{0.63} + 2.6 \times 10^{-4}\text{Re}^{1.38} \tag{7.36}$$

while according to Prandtl [37]

$$C_D = \frac{24}{\text{Re}} + 1 \qquad \text{(for Re} < 1000) \tag{7.37}$$

Mellor [38] employed the following equation for low values of Re for his prediction of drop trajectories:

$$C_D = \frac{1}{\text{Re}} [23 + (1 + 16\text{Re}^{0.33})^{0.5}] \tag{7.38}$$

All the above drag coefficient equations approach Stokes' law, i.e., $C_D = 24/\text{Re}$, as Reynolds number tends to zero. Some of these equations do not lend themselves readily to the mathematical integration that is necessary for trajectory predictions. Equations (7.37) and (7.38) are both suitable for integration, but the former is too simple and the latter is too complicated. This is why Putnam [39] tried to obtain an empirical equation that is suitable for integration and also reasonably accurate. He suggested the following equation for trajectory prediction, which fits the experimental data well and also is mathematically tractable.

$$C_D = \frac{24}{\text{Re}} \left(1 + \frac{1}{6}\text{Re}^{2/3} \right) \tag{7.39}$$

This equation is easily integrated to obtain drop velocities and trajectories and is recommended for Reynolds numbers less than 1000. It is convenient to use since it covers the Reynolds number range that is normally encountered in spray analysis. Since spray analysis involves drop trajectory predictions, an equation for drop drag coefficient that can be integrated easily is clearly very desirable.

Effect of Acceleration on Drag Coefficient

So far we have considered only drops moving steadily in a flow field with no acceleration or deceleration. For this situation the drag coefficients obtained theoretically or experimentally are called "standard" drag coefficients. However, the drops are accelerated by the gas flow if the gas velocity is higher than the drop velocity, or decelerated if the gas velocity is lower than the drop velocity. Thus the effect of acceleration on drag coefficient is always present, although for low accelerations the drop drag coefficient approaches the steady-state value.

Ingebo [40] conducted experiments in which clouds of liquid and solid spheres accelerating in airstreams were studied for various values of airstream pressure, temperature, and velocity. Diameter and velocity data for individual drops and solid spheres in the clouds were obtained with a high-speed camera. From these data the linear accelerations of spheres 20 to 120 μm in diameter were determined, and instantaneous drag coefficients for unsteady momentum transfer were calculated. The drag coefficients for drops of isooctane, water, and trichloroethylene and solid spheres (magnesium and calcium silicide) were found to correlate with Reynolds number according to the following equation, for $6 < \text{Re} < 400$:

$$C_D = \frac{27}{\text{Re}^{0.84}} \tag{7.40}$$

Drag Coefficients for Evaporating Drops

The influence of evaporation on drop drag may be due to two factors: (1) the effect of mass transfer on drag, known as the "blowing" effect, and (2) the temperature and concentration gradient near the drop surface (due to evaporation), which affects the dependence of drag coefficient on Reynolds number. This means that the physical properties chosen (variable properties or constant properties with different averaging methods) will affect the equation $C_D = f(\text{Re})$.

Yuen and Chen [41] have shown that the drag data for evaporating drops can be correlated with the drag data for solid spheres provided the average viscosity in the boundary layer is calculated at the one-third reference state [see Eqs. (8.7) and (8.8)]. A possible explanation for this result is that at high Reynolds numbers the decrease in friction drag due to evaporation is offset by the increase in pressure drag from flow separation due to blowing. At low Reynolds numbers, mass efflux has little effect on drag. Thus, at both high and low Reynolds numbers, the effect of mass efflux (due to evaporation) from a sphere has little effect on the drag coefficient. Yuen and Chen's basic conclusion is that the deviation of drag coefficient for an evaporating drop from that for a corresponding nonevaporating drop is due not to the blowing effect but to changes in the physical properties of the gas at the drop surface.

Eisenklam et al. [42] determined drag coefficients using experimental data on small, free-falling, burning and/or evaporating drops of fuel. They proposed the following correlation for evaporating or burning drops:

$$C_{D,m} = \frac{C_D}{1 + B_M} \qquad (7.41)$$

where $C_{D,m}$ is the drag coefficient evaluated at a mean condition for intense mass transfer, and B_M is the mass transfer number (see Chapter 8).

Equation (7.41) is considered valid over the following range of conditions: drop diameter = 25–500 μm, approach Re = 0.01–15, transfer number = 0.06–12.3.

According to Law [43,44], the drop drag coefficient can be written as

$$C_D = \frac{23}{Sc^{0.14}Re} (1 + 0.276Sc^{0.33}Re^{0.5})(1 + B_M)^{-1} \qquad (7.42)$$

This expression is valid only for Re < 200. Law argues that it is seldom necessary to investigate the behavior of the system for larger Reynolds numbers, since for most liquids that are not too viscous the drop would then become unstable and tend to break up. This statement is true only for low pressures; for high pressures and relative low temperatures, Re > 1000 is quite common. As Re approaches zero this correlation fails to approach the Stokes expression $C_D = 24/Re$, but the difference is small and is usually unimportant, since in such situations the drops follow the gas motion almost exactly.

The drop drag coefficient equations suggested by Lambiris and Combs [45] and used by many researchers are as follows:

$$C_D = 27 \times Re^{-0.84} \qquad \text{(for Re < 80)} \qquad (7.43)$$

$$= 0.271 \times Re^{0.217} \qquad \text{(for 80 < Re < 10,000)} \qquad (7.44)$$

$$= 2 \qquad \text{(for Re > 10,000)} \qquad (7.45)$$

Finally, it can be stated that at present there is no single expression for drop drag coefficient that applies to all conditions. Until there is, the following equations are recommended. For low temperatures and Re < 1000, for easy integration use

$$C_D = \frac{24}{Re} \left(1 + \frac{1}{6} Re^{2/3} \right) \qquad (7.39)$$

For high temperatures (with the drop evaporating) use Eisenklam's correlation, Eq. (7.41).

NOMENCLATURE

A_a	air core area, m^2
A_p	total inlet ports area, m^2
A_o	discharge orifice area, m^2
A_s	swirl chamber area, m^2
B_M	mass transfer number
C_D	discharge coefficient or drag coefficient
D_p	inlet port diameter, m
D_s	swirl chamber diameter, m
d_o	discharge orifice diameter, m
F	drag force on sphere
FN	flow number ($=m_L/\Delta P_L \rho_L^{0.5}$), m^2
K	atomizer constant ($=A_p/d_o D_s$)
K_v	velocity coefficient; ratio of actual to theoretical discharge velocity
L_s	length of parallel portion of swirl chamber, m
l_o	orifice length, m
$\dot{m}_L$	liquid flow rate, kg/s
P	total pressure, Pa
ΔP	pressure differential across nozzle, Pa
R	disk radius, m
R_s	radius of swirl chamber, m
r	local radius, m
S	spray tip penetration, m
t	film thickness, m, or time, s
T	temperature, K
U	velocity, m/s
$\bar{U}$	resultant velocity in orifice, m/s
U_{ax}	axial velocity in orifice, m/s
v	tangential velocity at radius r, m/s
v_i	inlet velocity to swirl chamber, m/s
v_m	mean tangential velocity, m/s
v_o	tangential velocity at orifice diameter d_o, m/s
v_r	radial velocity of liquid, m/s
v_s	tangential velocity at liquid surface in air core, m/s
We	Weber number
X	A_a/A_o
θ	maximum spray cone half-angle, degrees
θ_m	mean spray cone half-angle, degrees
μ	absolute viscosity, kg/m s
ν	kinematic viscosity, m^2/s
ρ	density, kg/m^3
σ	surface tension, kg/s^2
Φ_n	nominal spray angle, i.e., angle of spray silhouette measured at $\Delta P_L = 0.69$ MPA (100 psi) and $P_A = 0.101$ MPa (14.7 psia)

Φ equivalent spray angle, based on angle between centers of mass in left-and right-hand lobes of spray

Φ_o equivalent spray angle measured at $\Delta P_L = 0.69$ MPa (100 psi) and $P_A = 0.101$ MPa (14.7 psia)

Subscripts

A air
L liquid
R relative value between liquid and air

REFERENCES

1. Jones, R. V., Lehtinen, J. R., and Gaag, J. M., The Testing and Characterization of Spray Nozzles; the Manufacturer's Viewpoint, paper presented at the 1st National Conference on Liquid Atomization and Spray Systems, Madison, Wis., June 1987.
2. Tate, R. W., Spray Patternation, *Ind. Eng. Chem.,* Vol. 52, No. 10, 1960, pp. 49– 52.
3. De Corso, S. M., and Kemeny, G. A., Effect of Ambient and Fuel Pressure on Nozzle Spray Angle, *Trans. ASME,* Vol. 79, No. 3, 1957, pp. 607–615.
4. Ortman, J., and Lefebvre, A.H., Fuel Distributions from Pressure-Swirl Atomizers, *AIAA J. Propul. Power,* Vol. 1, No. 1, 1985, pp. 11–15.
5. Joyce, J. R., Report ICT 15, Shell Research Ltd., London, 1947.
6. Hiroyasu, H., Diesel Engine Combustion and Its Modelling, *Proceedings of the International Symposium on Diagnostics and Modelling of Combustion in Reciprocating Engines,* September 1985, pp. 53–75.
7. Sitkei, G., *Kraftstoffaufbereitung und Verbrennung bei Dieselmotoren,* Springer-Verlag, Berlin, 1964.
8. Taylor, D. H., and Walsham, B. E., Combustion Processes in a Medium Speed Diesel Engine, *Proc. Inst. Mech. Eng.,* Vol. 184, Part 3J, 1970, pp. 67–76.
9. Dent, J. C., A Basis for the Comparison of Various Experimental Methods for Studying Spray Penetration, *SAE Trans.,* Vol. 80, 1971, Paper 710571.
10. Hay, N., and Jones, P. L., Comparison of the Various Correlations for Spray Penetration, *SAE* Paper 720776, 1972.
11. Hiroyasu, H., Kadota, T., and Tasaka, S., Study of the Penetration of Diesel Spray, *Trans. JSME,* Vol. 34, No. 385, 1978, p. 3208.
12. Hiroyasu, H., and Arai, M., Fuel Spray Penetration and Spray Angle in Diesel Engines, *Trans. JSAE,* Vol. 21, 1980, pp. 5–11.
13. Levich, V. G., *Physicochemical Hydrodynamics,* Prentice-Hall, Englewood Cliffs, N.J., 1962, pp. 639–650.
14. Lefebvre, A. H., Factors Controlling Gas Turbine Combustion Performance at High Pressures, *Combustion in Advanced Gas Turbine Systems,* Cranfield International Symposium Series, Vol. 10, I. E. Smith, ed., 1968, pp. 211–226.
15. Taylor, G. I., The Mechanics of Swirl Atomizers, *Seventh International Congress of Applied Mechanics,* Vol. 2, Pt. 1, 1948, pp. 280–285.
16. Watson, E. A., unpublished report, Joseph Lucas Ltd., London, 1947.
17. Giffen, E., and Massey, B. S., Report 1950/5, Motor Industry Research Association, England, 1950.
18. Carlisle, D. R., Communication on The Performance of a Type of Swirl Atomizer, by A. Radcliffe, *Proc. Inst. Mech. Engs.,* Vol. 169, 1955, p. 101.
19. Giffen, E., and Muraszew, A., *Atomization of Liquid Fuels,* Chapman & Hall, London, 1953.
20. Simmons, H. C., Parker Hannifin Report, BTA 136, 1981.

21. Babu, K. R., Narasimhan, M. V., and Narayanaswamy, K., Correlations for Prediction of Discharge Rate, Cone Angle, and Air Core Diameter of Swirl Spray Atomizers, *Proceedings of the 2nd International Conference on Liquid Atomization and Spray Systems,* Madison, Wisc., 1982, pp. 91–97.

22. Rizk, N. K., and Lefebvre, A. H., Internal Flow Characteristics of Simplex Swirl Atomizers, *AIAA J. Propul. Power,* Vol. 1, No. 3, 1985, pp. 193–199.

23. Rizk, N. K., and Lefebvre, A. H., Prediction of Velocity Coefficient and Spray Cone Angle for Simplex Swirl Atomizers, *Proceedings of the 3rd International Conference on Liquid Atomization and Spray Systems,* London, 1985, pp. 111C/2/1–16.

24. Sankaran Kutty, P., Narasimhan, M. V., and Narayanaswamy, K., Design and Prediction of Discharge Rate, Cone Angle, and Air Core Diameter of Swirl Chamber Atomizers, *Proceedings of the 1st International Conference on Liquid Atomization and Spray Systems,* Tokyo, 1978, pp. 93–100.

25. Wang, X. F., and Lefebvre, A. H., unpublished work, 1986.

26. Neya, K., and Sato, S., Effect of Ambient Air Pressure on the Spray Characteristics and Swirl Atomizers, Ship. Res. Inst. Tokyo, Paper 27, 1968.

27. Dodge, L. G., and Biaglow, J. A., Effect of Elevated Temperature and Pressure on Sprays from Simplex Swirl Atomizers, ASME Paper 85-GT-58, presented at the 30th International Gas Turbine Conference, Houston, Texas, March 18–21, 1985.

28. Arai, M., Tabata, M., Hiroyasu, H., and Shimizu, M., Disintegrating Process and Spray Characterization of Fuel Jet Injected by a Diesel Nozzle, *SAE Trans.,* Vol. 93, 1984, Paper 840275.

29. Abramovich, G. N., *Theory of Turbulent Jets,* MIT Press, Cambridge, Mass., 1963.

30. Yokota, K., and Matsuoka, S., An Experimental Study of Fuel Spray in a Diesel Engine, *Trans. JSME,* Vol. 43, No. 373, 1977, pp. 3455–3464.

31. Reitz, R. D., and Bracco, F. V., On the Dependence of Spray Angle and Other Spray Parameters on Nozzle Design and Operating Conditions, *SAE* Paper 790494, 1979.

32. Bracco, F. V., Chehroudi, B., Chen, S. H., and Onuma, Y., On the Intact Core of Full Cone Sprays, *SAE Trans.,* Vol. 94, 1985, Paper 850126.

33. Hiroyasu, H., and Arai, M., Fuel Spray Penetration and Spray Angle in Diesel Engines, *Trans. JSAE,* Vol. 21, 1980, pp. 5–11.

34. Ortman, J., Rizk, N. K., and Lefebvre, A. H., unpublished report, School of Mechanical Engineering, Purdue University, 1984.

35. Stokes, G. G., *Scientific Papers,* University Press, Cambridge, 1901.

36. Langmuir, I., and Blodgett, K., A Mathematical Investigation of Water Droplet Trajectories, A.A.F. Tech. Rep. 5418, Air Material Command, Wright Patterson Air Force Base, 1946.

37. Prandtl, L., *Guide to the Theory of Flow,* 2nd ed., Braunschweig, 1944, p. 173.

38. Mellor, R., Ph.D. thesis, University of Sheffield, 1969.

39. Putnam, A., Integratable Form of Droplet Drag Coefficient, *J. Am. Rocket Soc.,* Vol. 31, 1961, pp. 1467–1468.

40. Ingebo, R. D., Drag Coefficients for Droplets and Solid Spheres in Clouds Accelerating in Airstreams, NACA TN 3762, 1956.

41. Yuen, M. C., and Chen, L. W., On Drag of Evaporating Liquid Droplets, *Combust. Sci. Technol.,* Vol. 14, 1976, pp. 147–154.

42. Eisenklam, P., Arunachlaman, S. A., and Weston, J. A., Evaporation Rates and Drag Resistance of Burning Drops, *11th Symposium (International) on Combustion,* The Combustion Institute, 1967, pp. 715–728.

43. Law, C. K., Motion of a Vaporizing Droplet in a Constant Cross Flow, *Int. J. Multiphase Flow,* Vol. 3, 1977, pp. 299–303.

44. Law, C. K., A Theory for Monodisperse Spray Vaporization in Adiabatic and Isothermal Systems, *Int. J. Heat Mass Transfer,* Vol. 18, 1975, pp. 1285–1292.

45. Lambiris, S., and Combs, L. P., Steady State Combustion Measurement in a LOX RP-1 Rocket Chamber and Related Spray Burning Analysis, *Detonation and Two Phase Flow, Prog. Astronaut. Rocketry,* Vol. 6, 1962, pp. 269–304.

EIGHT

DROP EVAPORATION

INTRODUCTION

The evaporation of drops in a spray involves simultaneous heat and mass transfer processes in which the heat for evaporation is transferred to the drop surface by conduction and convection from the surrounding hot gas, and vapor is transferred by convection and diffusion back into the gas stream. The overall rate of evaporation depends on the pressure, temperature, and transport properties of the gas; the temperature, volatility, and diameter of the drops in the spray; and the velocity of the drops relative to that of the surrounding gas.

Much of the material contained in this chapter is relevant to all types of liquid sprays, but the main emphasis is on the evaporation of liquid hydrocarbon fuels. Examples of combustion devices burning such fuels are diesel engines, spark ignition engines, gas turbines, liquid rocket engines, and industrial furnaces.

The most commonly considered case in the published literature is that of droplet combustion involving a volatile liquid fuel burning in a surrounding oxidizing atmosphere. The drop evaporates and acts as a source of fuel vapor, which burns with the surrounding oxidant (usually air) as a diffusion flame around the drop. However, this type of *heterogeneous* combustion represents only one extreme form of the combustion of liquid fuels. At the other extreme, the fuel is completely vaporized and mixed with air prior to combustion, and the combustion characteristics are essentially the same as those of fully mixed gaseous fuel-air mixtures. This type of combustion process may be described as *premixed* or *homogeneous*.

In most practical combustors the fuel leaves the nozzle in the form of ligaments or shreds, which rapidly disintegrate into drops of varying size. The primary purpose of atomization is to increase the surface area of the fuel and thereby enhance the rate of heat transfer from the surrounding gases to the fuel. As heat transfer takes place the drops heat up (or possibly cool down, depending on the ambient gas conditions and the initial fuel temperature) and, at the same time, lose part of their mass by vaporization and diffusion into the surrounding air or gas. The rates of heat and mass transfer are markedly affected by the drop Reynolds number, whose value varies throughout the lifetime of the drop, since neither the drop diameter nor the drop velocity remains constant. The history of the drop velocity is determined by the relative velocity between the drop and the surrounding gas and also by the drop drag coefficient. The latter again depends on the Reynolds number. After a certain time has elapsed each drop attains its steady-state or "wet-bulb" temperature corresponding to the prevailing conditions.

The larger drops take longer to attain equilibrium conditions, and their trajectories are different from those of the smaller drops since they are less influenced by aerodynamic drag forces. However, the smaller drops evaporate faster and produce a cloud of vapor that moves along with the air. At some point a combustible mixture of air and fuel vapor is formed that is ready for ignition. In many combustion systems a source of ignition is provided by the recirculation of burned products into the reaction zone. In other systems the mixture is heated by the air until it reaches its self-ignition temperature. In almost all cases the time required to produce a combustible mixture of air and fuel vapor represents a significant fraction of the total time available for completion of combustion. For this reason the subject of fuel drop evaporation merits discussion in some detail.

STEADY–STATE EVAPORATION

The term steady state is perhaps a misnomer when applied to drop evaporation, because a fuel drop may not attain steady-state evaporation during its lifetime. This is especially true for multicomponent fuel drops, which may contain several different petroleum compounds, each possessing its own physical and chemical properties. Nevertheless, for most light distillate fuel oils it is convenient to consider a quasi-steady gas phase that embodies the main features of the mass and thermal diffusion processes, which allows mass evaporation rates and drop lifetimes to be estimated to a reasonable level of accuracy.

Measurement of Evaporation Rate

Two general approaches have been taken in the experimental study of single-drop evaporation. In one, a drop is suspended from a silica fiber or thermocouple wire. (The latter method allows the liquid temperature to be measured.) The change of drop diameter with time is recorded using a cine camera operating at a speed of around 100 frames per second. The elliptical shape of the drop is corrected to a sphere of equal volume. After an initial transient period, steady-state evaporation is soon established, and the drop diameter diminishes with time according to the relationship

$$D_0^2 - D^2 = \lambda t \tag{8.1}$$

This is sometimes called the D^2 *law* of droplet evaporation; λ is known as the *evaporation constant*.

The alternative method of determining evaporation rates is by feeding the liquid into the inside of a hollow sphere constructed with a porous wall, as illustrated in Fig. 8.1. The liquid supply should be adjusted to maintain a wetted surface on the outside of the sphere. With this technique the diameter of the vaporizing spherical surface remains constant and the rate of vaporization is equal to the supply rate of the liquid. A typical result of this type of experiment is shown in Fig. 8.2. Values of λ for various fuels are listed in Table 8.1.

Theoretical Background

The development of drop evaporation theory has been largely motivated by the needs of the aero gas turbine and the liquid propellant rocket engine. Following Godsave [2] and Spalding [3], the approach generally used is to assume a spherically symmetric model of a vaporizing drop in which the rate-controlling process is that of molecular diffusion. The following assumptions are usually made:

1. The drop is spherical.
2. The fuel is a pure liquid having a well-defined boiling point.
3. Radiation heat transfer is negligible.

Except for conditions of very low pressure or for highly luminous flames, these assumptions are considered valid.

Consider the hypothetical case of a pure fuel drop that is suddenly immersed in gas at high temperature. According to Faeth [4], the ensuing evaporation process proceeds according to the sketch provided in Fig. 8.3. At normal fuel injec-

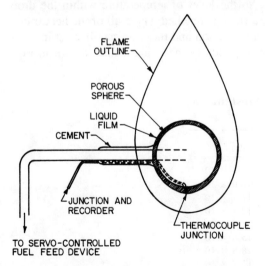

Figure 8.1 Schematic diagram of fuel-wetted porous sphere with thermocouple for measurement of liquid surface temperature.

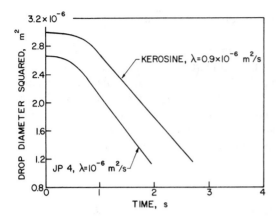

Figure 8.2 Burning rate curves for kerosine and JP 4 *(Wood et al. [1]).*

tion temperatures the concentration of fuel vapor at the liquid surface is low, and there is little mass transfer from the drop in this initial stage, which corresponds to the first part of the plots in Fig. 8.2, where the slope is quite small. Under these conditions the fuel drop heats up exactly like any other cold body when placed in a hot environment. Because of the limited heat conductivity of the fuel, the temperature inside the drop is not uniform but is cooler at the center of the drop than at the liquid surface.

Initially, almost all of the heat supplied to the drop serves to raise its temperature. As the liquid temperature rises, the fuel vapor formed at the drop surface has two effects: (1) part of the heat transferred to the drop is needed to furnish the heat of vaporization of the liquid, and (2) the outward flow of fuel vapor impedes the rate of heat transfer to the drop. This tends to diminish the rate of increase of the surface temperature, so the level of temperature within the drop becomes more uniform. Eventually, a stage is reached where all of the heat transferred to the drop is used as heat of vaporization and the drop stabilizes at its wet-bulb temperature. This condition corresponds to the straight lines drawn in Fig. 8.2.

Table 8.1 Values of evaporation constant for various stagnant fuel-air mixtures

Fuel	λ (m^2/s)	λ (ft^2/s)
Gasoline	1.06×10^{-6}	11.4×10^{-6}
Gasoline	1.49×10^{-6}	16.0×10^{-6}
Kerosine	1.03×10^{-6}	11.1×10^{-6}
Kerosine	1.12×10^{-6}	12.1×10^{-6}
Kerosine	1.28×10^{-6}	13.8×10^{-6}
Kerosine	1.47×10^{-6}	15.8×10^{-6}
Diesel oil	0.79×10^{-6}	8.5×10^{-6}
Diesel oil	1.09×10^{-6}	11.7×10^{-6}

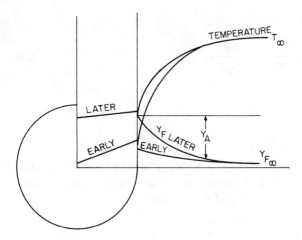

Figure 8.3 Variation of temperature and gas concentration during droplet evaporation *(Faeth [4])*.

Mass transfer number. An expression for the rate of evaporation of a fuel drop may be derived as follows. More comprehensive and rigorous treatments are provided in the publications of Faeth [4], Goldsmith and Penner [5], Kanury [6], Spalding [7], and Williams [8].

Neglecting thermal diffusion and assuming that the driving force for species diffusion is a concentration gradient in the direction of the diffusion path, the following expression is obtained for an evaporating drop:

$$\frac{dY_F}{dr} = -\frac{RT}{D_c P}(m_F Y_A) \tag{8.2}$$

where
Y_F = fuel mass fraction
Y_A = air mass fraction
m_F = mass rate of diffusion per unit area
D_c = diffusion coefficient
P = gas pressure
r = radius ($r = 0$ at center of drop and $r = r_s$ at drop surface)

From continuity considerations the mass rate of diffusion at the drop surface is given by

$$m_{F_s} 4\pi r_s^2 = m_F 4\pi r^2$$

$$\frac{dY_F}{dr} = -\frac{RT}{D_c P} Y_A \dot{m}_{F_s}\left(\frac{r_s^2}{r^2}\right)$$

or, since $Y_A = 1 - Y_F$,

$$\frac{dY_F}{dr} = -\frac{RT}{D_c P}(1 - Y_F)\dot{m}_{F_s}\left(\frac{r_s^2}{r^2}\right)$$

Rearranging gives

$$\frac{dY_F}{1 - Y_F} = -\frac{RT}{D_c P} \dot{m}_{F_s} r_s^2 \frac{dr}{r^2} \tag{8.3}$$

Now the boundary conditions are

$$r = r_s; \quad T = T_s; \quad Y_F = Y_{F_s}$$

$$r = \infty; \quad T = T_\infty; \quad Y_F = Y_{F_\infty} = 0$$

A final boundary condition, implied by the insolubility assumption first invoked by Spalding [3], is that the mass flux of ambient gas is zero at the liquid surface.

Integrating Eq. (8.3) between $r = 0$ and $r = \infty$ yields

$$[\ln(1 - Y_F)]_{Y_{F_s}}^0 = \left[\frac{RT}{D_c P} \dot{m}_{F_s} r_s^2 \left(-\frac{1}{r}\right)\right]_{r_s}^\infty$$

or

$$0 - \ln(1 - Y_{F_s}) = -\frac{RT}{D_c P} \dot{m}_{F_s} r_s^2 \left(-\frac{1}{r_s}\right)$$

$$= \frac{RT}{D_c P} \dot{m}_{F_s} r_s$$

Hence,

$$\dot{m}_{F_s} = -\frac{PD_c}{RT} \frac{\ln(1 - Y_{F_s})}{r_s}$$

$$= \frac{\rho D_c}{r_s} \ln(1 - Y_{F_s})$$

and

$$\dot{m}_F = 4\pi r_s^2 \dot{m}_{F_s} = 4\pi r_s \rho D_c \ln(1 - Y_{F_s}) \tag{8.4}$$

The quantity ρD_c can be replaced with $(k/c_p)_g$, assuming a Lewis number of unity, where k and c_p are the mean thermal conductivity and specific heat, respectively.

If we now define

$$B_M = \frac{Y_{F_s}}{1 - Y_{F_s}} \tag{8.5}$$

then

$$\ln(1 - Y_{F_s}) = -\ln(1 + B_M)$$

Substituting for $\ln(1 - Y_{F_s})$ and for $2r_s = D_s$ in Eq. (8.4) gives

$$\dot{m}_F = 2\pi D_s \left(\frac{k}{c_p}\right)_g \ln(1 + B_M) \tag{8.6}$$

This is the basic equation for the rate of evaporation of a fuel drop of diameter D_s. Its accuracy is very dependent on the choice of values of k_g and c_{p_g}. According to Hubbard et al. [9], best results are obtained using the one-third rule of Sparrow and Gregg [10], where average properties are evaluated at the following reference temperature and composition:

$$T_r = T_s + \frac{T_\infty - T_s}{3} \tag{8.7}$$

$$Y_{F_r} = Y_{F_s} + \frac{Y_{F_\infty} - Y_{F_s}}{3} \tag{8.8}$$

where T is temperature, Y_F is mass fraction of fuel vapor, and subscripts r, s, and ∞ refer to reference, surface, and ambient conditions. If the fuel concentration at an infinite distance from the drop is assumed to be zero, Eq. (8.8) becomes

$$Y_{F_r} = \tfrac{2}{3} Y_{F_s} \tag{8.9}$$

and

$$Y_{A_r} = 1 - Y_{F_r} = 1 - \tfrac{2}{3} Y_{F_s} \tag{8.10}$$

Equations (8.7)–(8.10) are used to calculate the reference values of the relevant physical properties of the vapor-air mixture that constitutes the environment of an evaporating drop. For example, the reference specific heat at constant pressure is obtained as

$$c_{p_g} = Y_{A_r}(c_{p_A} \text{ at } T_r) + Y_{F_r}(c_{p_v} \text{ at } T_r) \tag{8.11}$$

The reference value of thermal conductivity is estimated in a similar manner as

$$k_g = Y_{A_r}(k_A \text{ at } T_r) + Y_{F_r}(k_v \text{ at } T_r) \tag{8.12}$$

Values for the specific heat of air as a function of pressure and temperature are shown in Fig. 8.4. The variation of specific heat of hydrocarbon fuel vapor with temperature is described by the relationship

$$c_{p_v} = (363 + 0.467T)(5 - 0.001\rho_{F_0}) \text{ J/kg K} \tag{8.13}$$

where ρ_{F_0} is the fuel density at a temperature of 288.6 K. Equation (8.13) is plotted in Figure 8.5 along with experimental data for gasoline, diesel oil, and kerosine from [11].

Values for the thermal conductivity of air at various temperature levels are provided in Fig. 8.6. The thermal conductivity of hydrocarbon fuel vapor is given to a reasonable level of accuracy by the expression

$$k_{v_T} = 10^{-3}[13.2 - 0.0313(T_{bn} - 273)]\left(\frac{T}{273}\right)^n \tag{8.14}$$

where

$$n = 2 - 0.0372\left(\frac{T}{T_{bn}}\right)^2 \tag{8.15}$$

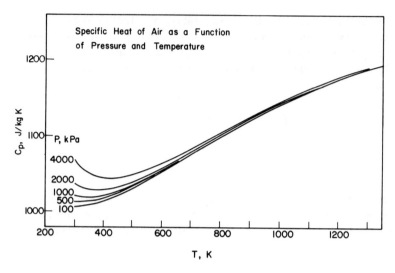

Figure 8.4 Specific heat of air.

The influence of temperature on the thermal conductivity of fuel vapor is illustrated in Fig. 8.7. This figure shows experimental data from [11] and [12] plotted with Eq. (8.14).

Heat transfer number. Arguments similar to those employed above, but based on considerations of conductive and convective heat fluxes across a thin shell

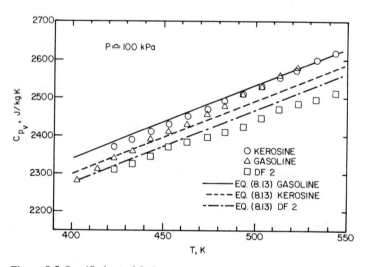

Figure 8.5 Specific heat of fuel vapor.

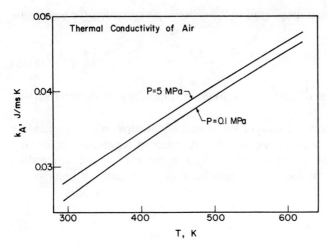

Figure 8.6 Thermal conductivity of air. Data from [11, 12].

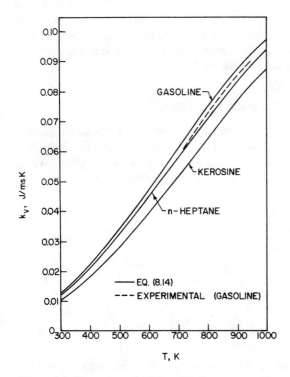

Figure 8.7 Thermal conductivity of fuel vapor.

surrounding the evaporating drop, lead to the following expression for heat transfer number:

$$B_T = \frac{c_{p_g}(T_\infty - T_s)}{L} \tag{8.16}$$

where L is the latent heat of fuel vaporization corresponding to the fuel surface temperature T_s.

The number B_T denotes the ratio of the available enthalpy in the surrounding gas to the heat required to evaporate the fuel. As such, it represents the driving force for the evaporation process. Where heat transfer rates are controlling for evaporation, the rate of fuel evaporation for a Lewis number of unity is obtained as

$$\dot{m}_F = 2\pi D \left(\frac{k}{c_p}\right)_g \ln(1 + B_T) \tag{8.17}$$

Under steady-state conditions $B_M = B_T$ and either Eq. (8.6) or Eq. (8.17) may be used to calculate the rate of fuel evaporation. The advantage of Eq. (8.6) is that it applies under all conditions, including the transient state of droplet heat-up, whereas Eq. (8.17) can only be used for steady-state evaporation. However, Eq. (8.17) is usually easier to evaluate, since the magnitudes of the various terms are either contained in the data of the problem or readily available in the literature. This is especially true when the ambient gas temperature is significantly higher than the fuel surface temperature T_s, in which case it is sufficiently accurate to replace T_s by the boiling temperature of the fuel.

Calculation of Steady-State Evaporation Rates

The term steady-state is used to describe the stage in the drop evaporation process where the drop surface has attained its wet-bulb temperature and all of the heat reaching the surface is utilized in providing the latent heat of vaporization. Where $T_{s_{st}}$ is known, the transfer number B is easy to evaluate. We have

$$B_M = \frac{Y_{F_s}}{1 - Y_{F_s}} \tag{8.5}$$

Now

$$Y_{F_s} = \frac{P_{F_s}M_F}{P_{F_s}M_F + (P - P_{F_s})M_A} \tag{8.18}$$

$$= \left[1 + \left(\frac{P}{P_{F_s}} - 1\right)\frac{M_A}{M_F}\right]^{-1} \tag{8.19}$$

where P_{F_s} is the fuel vapor pressure at the drop surface, P is the ambient pressure, which is the sum of the fuel vapor pressure and the air partial pressure at the drop surface, and M_F and M_A are the molecular weights of fuel and air, respectively.

For any given value of surface temperature, the vapor pressure is readily estimated from the Clausius-Clapeyron equation as

$$P_{F_s} = \exp\left(a - \frac{b}{T_s - 43}\right) \qquad (8.20)$$

Values of a and b for several hydrocarbon fuels are listed in Table 8.2.

At the steady-state condition, $B_M = B_T = B$ and the mass rate of fuel evaporation is given by

$$\dot{m}_F = 2\pi D \left(\frac{k}{c_p}\right)_g \ln(1 + B) \qquad (8.21)$$

Example A drop of n-heptane fuel, 200 μm in diameter, is undergoing steady-state evaporation in air at normal atmospheric pressure and a temperature of 773 K. Its surface temperature is 341.8 K. Find the rate of fuel evaporation.

For $T_s = 341.8$ the vapor pressure is obtained using Table 8.2 and Eq. (8.20) as 38.42 kPa. $M_A = 28.97$ and $M_F = 100.2$. At normal atmospheric pressure, $P = 101.33$ kPa. Substituting these values into Eq. (8.19) gives

$$Y_{F_s} = \left[1 + \left(\frac{101.33}{38.42} - 1\right)\frac{28.97}{100.2}\right]^{-1}$$

$$= 0.679$$

Hence, from Eq. (8.5)

$$B_M = \frac{Y_{F_s}}{1 - Y_{F_s}} = 2.12$$

To obtain B_T it is necessary first to derive the reference values of T, Y_A, and Y_F.

Table 8.2 Relevant thermophysical properties

Fuel	n-Heptane	Aviation gasoline	JP 4	JP 5	DF-2
T_{cr}, K	540.17	548.0	612.0	684.8	725.9
$L_{T_{bn}}$, J/kg	371,800	346,000	292,000	266,500	254,000
Molecular mass	100.16	108.0	125.0	169.0	198.0
Density, kg/m³	687.8	724	773	827	846
T_{bn}, K	371.4	333	420	495.3	536.4
a in Eq. (8.20) for $T > T_{bn}$	14.2146	14.1964	15.2323	15.1600	15.5274
a in Eq. (8.20) for $T < T_{bn}$	14.3896	13.7600	15.2323	15.1600	15.5274
b in Eq. (8.20) for $T > T_{bn}$, K	3151.68	2777.65	3999.66	4768.77	5383.59
b in Eq. (8.20) for $T < T_{bn}$, K	3209.45	2651.13	3999.66	4768.77	5383.59

From Eq. (8.7) $T_r = 341.8 + \frac{1}{3}(773 - 341.8) = 485.5$

From Eq. (8.9) $Y_{F_r} = \frac{2}{3}(0.679) = 0.453$

From Eq. (8.10) $Y_{A_r} = 1 - Y_{F_r} = 0.547$

From Figs. 8.4 and 8.5, c_{P_A} and c_{p_v} at the reference temperature of 485.5 are 1026 and 2450 J/kg K, respectively.
From Eq. (8.11)

$$c_{P_g} = 0.547 \times 1026 + 0.453 \times 2450$$

$$= 1671 \text{ J/kg K}$$

The latent heat of vaporization is given by Watson [13] as

$$L = L_{T_{bn}}\left(\frac{T_{cr} - T_s}{T_{cr} - T_{bn}}\right)^{-0.38} \qquad (8.22)$$

Values of $L_{T_{bn}}$, T_{cr}, and T_{bn} for some representative fuels are listed in Table 8.2. Substitution of appropriate values for *n*-heptane into Eq. (8.22) gives, for $T_s = 341.8$, $L = 339,000$ J/kg. Hence

$$B_T = \frac{1671(773 - 341.8)}{339,000} = 2.13$$

Thus the calculated values of B_M and B_T are the same, which is to be expected.

The rate of fuel evaporation may be evaluated using either Eq. (8.6) or Eq. (8.17). At this point in the calculation the only unknown term in these equations is k_g. This is estimated using Eq. (8.12). For air and fuel vapor at the reference temperature of 485.5 the values of k_A and k_v are obtained from Figs. 8.6 and 8.7, respectively, as 0.0384 and 0.0307 J/m s K, respectively. Hence

$$k_g = 0.547 \times 0.0384 + 0.453 \times 0.0307 = 0.0349 \text{ J/m s K}$$

From Eq. (8.21)

$$\dot{m}_F = \frac{2\pi \times 0.0002 \times 0.0349 \times \ln(1 + 2.13)}{1671} = 3 \times 10^{-8} \text{ kg/s}$$

Evaporation Constant

It will be shown later that during the steady-state period of an evaporating drop its diameter at any instant may be related to its initial diameter by the expression

$$D_0^2 - D^2 = \lambda_{st}t \qquad (8.1)$$

where λ is the evaporation constant as defined by Godsave [2]. If Eq. (8.1) is rewritten as

$$\lambda_{st} = \frac{d(D)^2}{dt} \qquad (8.23)$$

it is apparent that λ_{st} represents the slope of the lines drawn in Fig. 8.2.

Values of λ_{st} may be used to determine the transfer number B. From Eq. (8.23) it can be shown that

$$\dot{m}_F = \frac{\pi}{4} \rho_F \lambda_{st} D \qquad (8.24)$$

while from Eq. (8.21) we have

$$\dot{m}_F = 2\pi D \left(\frac{k}{c_p}\right)_g \ln(1 + B)$$

Equating (8.21) and (8.24) gives

$$\lambda_{st} = \frac{8k_g \ln(1 + B)}{c_{p_g} \rho_F} \qquad (8.25)$$

Calculation of evaporation constant. In the previous example the calculation of B, the transfer number for steady-state evaporation, was greatly simplified because the steady-state value of surface temeprature, $T_{s_{st}}$, was given. Usually it is necessary to evaluate $T_{s_{st}}$ in order to derive B and λ. A graphical method for determining T_{st} and B has been advocated by Spalding [7] and Kanury [6]. The basic idea is to plot graphs of B_M and B_T against T_s until the two curves intersect. The point of intersection of the two lines defines B and $T_{s_{st}}$. The corresponding value of λ_{st} is then obtained from Eq. (8.25).

> **Example** A spherical drop of n-heptane fuel is immersed in air at normal atmospheric pressure and a temperature of 773 K. It is required to calculate the steady-state values of B and λ_{st}.
>
> Since $T_{s_{st}}$ must always be less than the normal boiling temperature T_{bn}, which for n-heptane at 1 atmosphere is 371.4 K, it is proposed to calculate B_M and B_T for a range of values of T_s below 371.4 K, starting at 320 K and then increasing T_s in small steps until the graphs of B_M and B_T versus T_s converge.
>
> Calculation of B_M:
>
> 1. From Eq. (8.20), P_{F_s} at 320 K = 16.5 kPa.
> 2. From Eq. (8.19)
>
> $$Y_{F_s} = \left[1 + \left(\frac{P}{P_{F_s}} - 1\right) \frac{M_A}{M_F}\right]^{-1}$$
>
> Now $M_A = 28.97$ and, for n-heptane, $M_F = 100.2$. Also, we have $P = 1$ atm = 101.33 kPa. Hence $Y_{F_s} = 0.4022$.
> 3. $B_M = Y_{F_s}/(1 - Y_{F_s}) = 0.6727$.
>
> Calculation of B_T:
>
> 1. From Eq. (8.9), $Y_{F_r} = \frac{2}{3}Y_{F_s}$. Hence $Y_{F_r} = \frac{2}{3}(0.4022) = 0.2681$.
> 2. From Eq. (8.10), $Y_{A_r} = 1 - Y_{F_r} = 0.7319$.

3. From Eq. (8.7), $T_r = T_s + \frac{1}{3}(T_\infty - T_s) = 320 + \frac{1}{3}(773 - 320) = 471$ K.
4. Read off c_{p_A} and c_{p_v} at 471 K from Figs. 8.4 and 8.5, respectively. This gives

$$c_{p_A} = 1024 \text{ J/kg K}$$

$$c_{p_v} = 2400 \text{ J/kg K}$$

5. From Eq. (8.11)

$$c_{p_g} = 0.7319 \times 1024 + 0.2681 \times 2400$$

$$= 1393 \text{ J/kg K}$$

6. From Eq. (8.22) the value of L at 320 K is 350,000 J/kg.
7. We have

$$B_T = \frac{c_{p_g}(T_\infty - T_s)}{L}$$

$$= \frac{1393(773 - 320)}{350,000}$$

$$= 1.803$$

Thus for $T_s = 320$ K, B_M is smaller than B_T. This means that the selected value of T_s is too low. By repeating the above calculations at increasing values of T_s and plotting the results as graphs of B_M and B_T versus T_s, the required values of $T_{s_{st}}$ and B are obtained as the values of T_s and B corresponding to the point of intersection of the two lines. For this example, this procedure yields the results shown plotted in Fig. 8.8, which indicate that $T_{s_{st}} = 341.8$ K and $B = 2.13$.

To calculate the steady-state value of λ, the additional information needed is c_{p_g}, k_g, and ρ_F. Fuel density may be calculated from the expression

$$\rho_{F_T} = \rho_{F_{288.6}} \left[1 - 1.8 C_{ex}(T - 288.6) - 0.090 \frac{(T - 288.6)^2}{(T_{cr} - 288.6)^2} \right] \quad (8.26)$$

in which C_{ex} is a coefficient of thermal expansion. Values of C_{ex} are plotted as a function of fuel density at 288.6 K in Fig. 8.9. The close agreement between the predictions of Eq. (8.26) and the experimental data given in reference [11] is shown in Fig. 8.10.

For n-heptane at a temperature of 341.8 K, ρ_F is obtained from Eq. (8.26) as 638 kg/m³. From the previous example for identical experimental conditions, $c_{p_g} = 1671$ J/kg K and $k_g = 0.0349$ J/m s K. Substituting these values into Eq. (8.25) gives

$$\lambda_{st} = \frac{8 \ln(1 + 2.13)}{638(1671/0.0349)}$$

$$= 0.30 \times 10^{-6} \text{ m}^2/\text{s} = 0.30 \text{ mm}^2/\text{s}$$

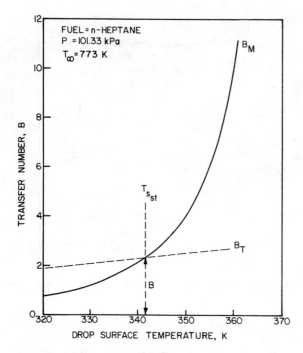

Figure 8.8 Variation of B_M and B_T with drop surface temperature.

The main drawback of the graphical method, as outlined above, is that it is tedious and time-consuming. Furthermore, it is not very accurate if the B_M and B_T lines intersect each other at a narrow angle. Much greater accuracy is possible with the numerical method described by Chin and Lefebvre [14].

Values of T_{st} and λ_{st}, calculated using this method, are shown in Figs. 8.11 and 8.12, respectively, for aviation gasoline, JP 4, JP 5, and DF-2 at

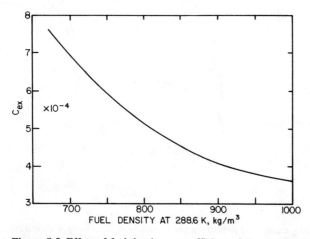

Figure 8.9 Effect of fuel density on coefficient of thermal expansion.

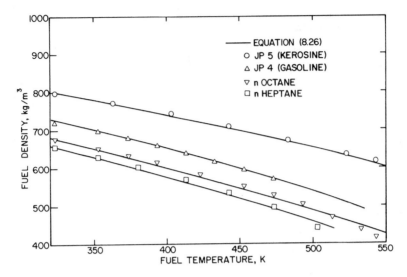

Figure 8.10 Comparison of predicted values of fuel density with measured values from Vargaftik [11].

ambient temperatures up to 2000 K and pressures up to 2 MPa (approximately 20 atm). Unfortunately, the paucity of experimental data on λ_{st} for commercial fuels at pressures other than normal atmospheric virtually prohibits any worthwhile comparison between experiment and calculation. However, where comparison can be made the result is encouraging, as illustrated in Table 8.3.

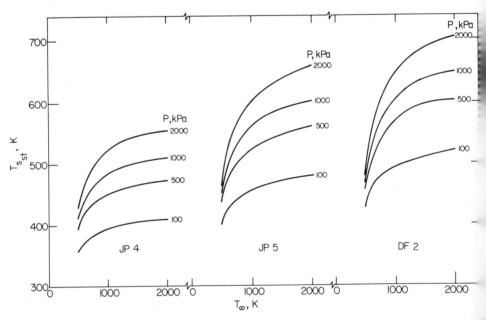

Figure 8.11 Influence of pressure and temperature on steady-state drop surface temperature for gasoline (JP 4), kerosine (JP 5), and diesel oil (DF-2) *(Chin and Lefebvre [14])*.

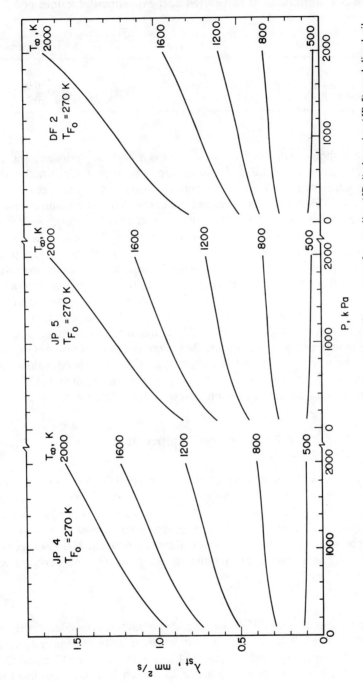

Figure 8.12 Influence of pressure and temperature on steady-state evaporation constant for gasoline (JP 4), kerosine (JP 5), and diesel oil (DF-2) (*Chin and Lefebvre [14]*).

Table 8.3 Comparison of calculated and experimental values of λ_{st}

| Fuel | λ_{st} (mm^2/s) | |
	Calculated	Experimental
n-Heptane	0.979	0.97
Gasoline	0.948	0.99
Diesel oil (DF-2)	0.802	0.79

This table shows calculated values of λ_{st} for *n*-heptane, gasoline, and DF-2 at a temperature of 2000 K and normal atmospheric pressure, listed with the corresponding measured values for burning drops at the same pressure from Godsave [2]. The agreement between the calculated and measured values of λ_{st} is clearly very satisfactory. The calculated values of λ_{st} shown in Fig. 8.12 represent a small number of important engine fuels. However, the data presented cannot readily be applied to other fuels or fuel blends. To provide similar information, but in a more general form, it is necessary to select a correlative fuel property that will define to a sufficient degree of accuracy the evaporation characteristics of any fuel. It is recognized that no single chemical or physical property is completely satisfactory for this purpose. However, the average boiling point (50% recovered) has much to commend it, since it is directly related to fuel volatility and fuel vapor pressure. It also has the virtue of being easy to measure and is usually quoted in fuel specifications. Plots of λ_{st} versus T_{bn} are shown in Fig. 8.13 for ambient temperatures of 500, 800, 1200, 1600, and 2000 K at pressure levels of 100, 500, 1000, and 2000 kPa.

Influence of Ambient Pressure and Temperature on Evaporation Rates

The graphs drawn in Fig. 8.12 show clearly that evaporation rates increase markedly with increase in ambient temperature. The effect of pressure on evaporation rates is more complex. The figures show that λ_{st} increases with pressure when the ambient temperature is high (>800 K) and declines with increase in pressure when the ambient temperature is low (<600 K). Between 600 and 800 K evaporation rates are sensibly independent of pressure. If the pressure dependence of λ_{st} is expressed in the form

$$\lambda_{st} \propto P^n \tag{8.27}$$

it is found that, over the range of pressures and temperatures considered, the value of n varies between ±0.25. At the highest levels of ambient temperature, corresponding to burning conditions, Figs. 8.14 and 8.15 show for JP 5 and DF-2, respectively, that n lies between 0.15 and 0.25, which is in fair agreement with the value of 0.25 determined experimentally by Hall and Diederichsen [15] for suspended burning fuel drops over a pressure range from 1 to 20 atm.

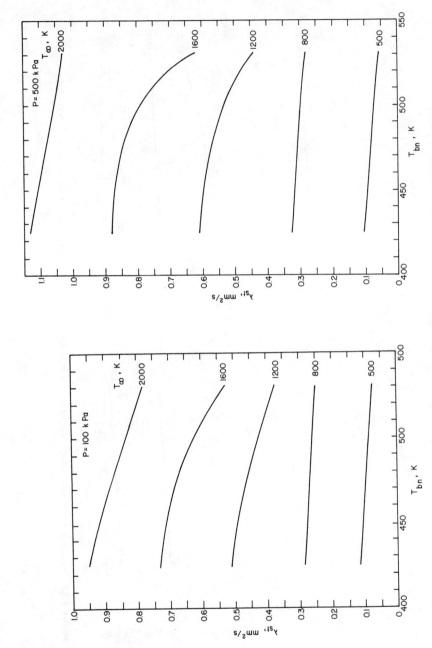

Figure 8.13 Relationship between normal boiling temperature and evaporation constant (*Chin and Lefebvre [14]*).

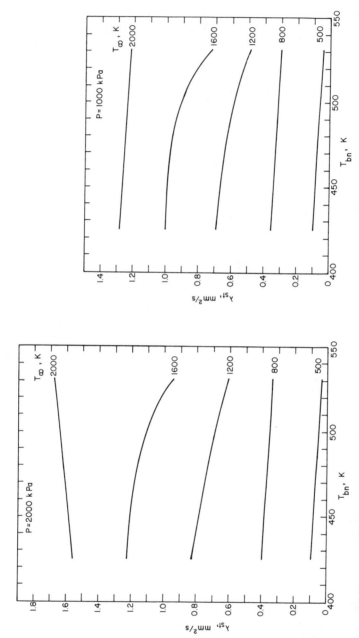

Figure 8.13 Relationship between normal boiling temperature and evaporation constant (*Chin and Lefebvre [14]*) (*Continued*).

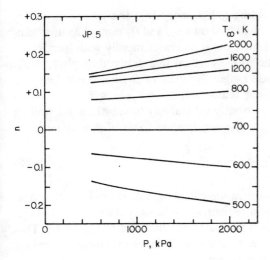

Figure 8.14 Effect of ambient pressure and temperature on the pressure dependence of evaporation rates for JP 5 *(Chin and Lefebvre [14])*.

Evaporation at High Temperatures

In many practical combustion systems, the liquid fuel is sprayed directly into the flame zone, so evaporation occurs under conditions of high ambient temperature, usually around 2000 K. At such elevated temperatures the wet-bulb temperature is only slightly lower than the boiling temperature at that pressure. This obser-

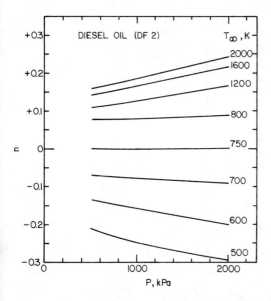

Figure 8.15 Effect of ambient pressure and temperature on the pressure dependence of evaporation rates for DF-2 *(Chin and Lefebvre [14])*.

vation was made previously by Spalding [7], Kanury [6], and others. It can be illustrated by reference to Fig. 8.16, which shows B_M and B_T plotted against temperature. At low temperatures B_M is low, but it increases rapidly with increase in temperature and becomes asymptotic to the vertical line drawn through $T_s = T_b$. The figure shows that there is a wide range of values of B_T, corresponding to a wide range of high values of ambient temperature, over which $T_{s_{st}} \simeq T_b$. This means that for high temperatures it is usually satisfactory to substitute the boiling temperature T_b for T_s in Eq. (8.16), so the steady-state value of B becomes

$$B = B_{T_{st}} = \frac{c_{p_g}(T_\infty - T_b)}{L} \tag{8.28}$$

Clearly this substitution greatly simplifies the calculation of B_T and λ_{st}. It is also of interest to note that for high ambient temperatures the pressure exponent in Eq. (8.27) correlates well with critical fuel pressure, as illustrated in Fig. 8.17. This figure also shows that the dependence of evaporation rates on ambient pressure declines with increase in critical fuel pressure.

UNSTEADY–STATE ANALYSIS

Figure 8.2 shows a straight-line relationship between (drop diameter)^2 and time during most of the evaporation period. However, inspection of the graphs reveals

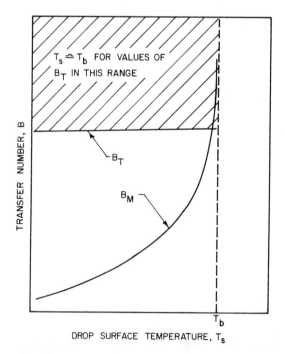

Figure 8.16 Graphs illustrating that $T_s \simeq T_b$ for high values of B_T corresponding to high ambient temperatures.

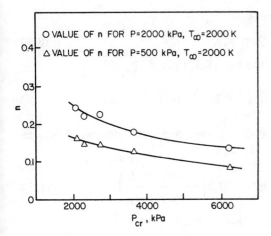

Figure 8.17 Relationship between pressure exponent and critical fuel pressure. See Eq. (8.27).

that the slope of the D^2/t line is almost zero in the first stage of evaporation and then gradually increases with time until the drop attains its wet-bulb temperature, after which the slope of the D^2/t line remains fairly constant throughout the remainder of the drop lifetime.

Although fuel drops reach their wet-bulb temperature asymptotically with time, for the purpose of analysis the vaporization process can roughly be separated into the *transient* or *unsteady* state and the *steady* state. For low-volatility fuels and low ambient temperatures, the drop spends only a very small portion of its time in the unsteady state. However, for many fuels at high ambient pressures and temperatures, the unsteady state is relatively much larger in magnitude and cannot be neglected.

Calculation of Heat–Up Period

Chin and Lefebvre [16] have discussed the role of the heat-up period in drop evaporation in some detail. For calculation purposes, a quasi-steady gas phase is assumed in which the boundary layer around the drop has the same characteristics as a steady boundary layer for the same conditions of drop size, velocity, and surface and ambient temperatures. Thus the heat transfer coefficient is determined by

$$\mathrm{Nu} = \frac{hD}{k_g} = 2\,\frac{\ln(1 + B_M)}{B_M} \tag{8.29}$$

The heat transferred from the gas to the drop is

$$Q = \pi D^2 h(T_\infty - T_s) \tag{8.30}$$

Substituting for h from Eq. (8.29) into Eq. (8.30) gives

$$Q = 2\pi D k_g(T_\infty - T_s)\,\frac{\ln(1 + B_M)}{B_M} \tag{8.31}$$

The heat used in vaporizing the fuel is

$$Q_e = \dot{m}_F L \tag{8.32}$$

Substituting for $\dot{m}_F$ from Eq. (8.6) into Eq. (8.32) gives

$$Q_e = 2\pi D \left(\frac{k}{c_p}\right)_g L \ln(1 + B_M) \tag{8.33}$$

The heat available for heating up the drop is obtained as the difference between Q and Q_e. From Eqs. (8.31) and (8.33) we have

$$Q - Q_e = 2\pi D k_g \ln(1 + B_M)\left(\frac{T_\infty - T_s}{B_M} - \frac{L}{c_{p_g}}\right) \tag{8.34}$$

Substituting for $B_T = c_{p_g}(T_\infty - T_s)/L$ yields

$$Q - Q_e = 2\pi D \left(\frac{k}{c_p}\right)_g L \ln(1 + B_M)\left(\frac{B_T}{B_M} - 1\right) \tag{8.35}$$

or

$$Q - Q_e = \dot{m}_F L\left(\frac{B_T}{B_M} - 1\right) \tag{8.36}$$

It will be noted in Eq. (8.36) that when $B_T = B_M$ the value of $Q - Q_e$ becomes zero, denoting the end of the heat-up period.

The rate of change of drop surface temperature is given by

$$\frac{dT_s}{dt} = \frac{Q - Q_e}{c_{p_F} m} \tag{8.37}$$

or

$$\frac{dT_s}{dt} = \frac{\dot{m}_F L}{c_{p_F} m}\left(\frac{B_T}{B_M} - 1\right) \tag{8.38}$$

where

$$m = \text{drop mass} = \frac{\pi}{6} \rho_F D^3 \tag{8.39}$$

Also, we have

$$\dot{m}_F = 2\pi D \left(\frac{k}{c_p}\right)_g \ln(1 + B) = \frac{d}{dt}\left(\frac{\pi}{6} \rho_F D^3\right) \tag{8.6}$$

hence

$$\frac{dD}{dt} = \frac{4 k_g \ln(1 + B_M)}{\rho_F c_{p_g} D} \tag{8.40}$$

Iterative method. Calculations of temperatures and evaporation rates for the heat-up period may be simplified using the iterative procedure proposed by Chin and Lefebvre [16]. Starting from $t = 0$, $D = D_0$, $m = m_0$, $T_s = T_{s_0}$, all the terms in the right-hand side of Eq. (8.38) can be calculated for any given fuel and ambient conditions using Eqs. (8.5), (8.6), (8.16), (8.22), (8.39), and (8.40). Thus, for any time increment Δt_1, we obtain from Eq. (8.38) a temperature increment ΔT_{s_1}, since

$$\Delta T_{s_1} = \frac{dT_s}{dt} \Delta t_1 \qquad (8.41)$$

Alternatively, for any specified temperature increment ΔT_{s_1}, Eq. (8.41) may be used to estimate the corresponding time increment Δt_1.

At the end of the first increment we have $t_1 = t_0 + \Delta t_1$; from Eq. (8.41) $T_{s_1} = T_{s_0} + \Delta T_{s_1}$; from Eq. (8.40) $D_1 = D_0 - \Delta D_1$; and from Eq. (8.39) $m_1 = (\pi/6)\rho_F(D_1)^3$. Note that in the calculation of m the variation of ρ_F with T_s should be accounted for (see Fig. 8.10).

The summation of $\Delta t_1 + \Delta t_2 + \Delta t_3 + \Delta t_n$ until $B_M = B_T$ and dT_s/dt becomes zero represents the heat-up period, and $D_0 - \Delta D_1 - \Delta D_2 - \Delta D_3 - \Delta D_n$ represents the drop diameter at the end of the heat-up period. The results obtained by this procedure for an n-heptane droplet 200 μm in diameter, in air at normal atmospheric pressure and a temperature of 773 K, are shown in Fig. 8.18. The initial drop temeprature is 288 K. Figure 8.18 illustrates the variation of T_s, D, and λ (plotted in nondimensional form as λ/λ_{st}) with time.

Approximate method. The iterative procedure described above for calculating the duration of the heat-up period and the size and surface temperature of the drop at the end of this period is fairly precise but time-consuming. A less accurate method, but one that is satisfactory for most practical purposes and is much less demanding in time, is described below.

We start by defining a nondimensional sensible heat parameter for the heat-up period as

$$\bar{Q} = \frac{Q - Q_e}{Q_0 - Q_{e_0}} \qquad (8.42)$$

where Q_0 is the total initial rate of heat transfer to the drop, and Q_{e_0} represents the portion of the initial heat supply that is utilized as latent heat of vaporization.

Thus when $T = T_{s_0}$, $\bar{Q} = 1$, and when $T = T_{s_{st}}$, $Q = Q_e$ and $\bar{Q} = 0$. The parameter $\bar{Q}$ is a function of surface temperature and also a function of time. Its value changes from unity to zero from the beginning to the end of the heat-up period. The manner in which $\bar{Q}$ varies with $\bar{t}$ is illustrated in Fig. 8.19. The parameter $\bar{t}$ is defined as the ratio of any given time during the heat-up period to the time required for the drop to attain 95% of its total temperature rise during the heat-up period. This artifice is necessary because T_s approaches $T_{s_{st}}$ asymptotically with time, which makes it impossible to define precisely the length of

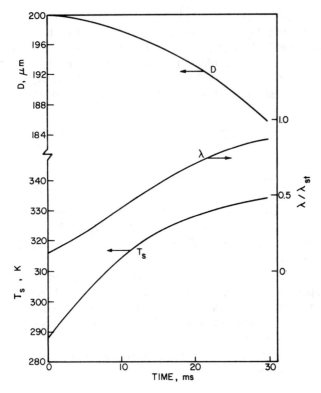

Figure 8.18 Graphs illustrating the variation of drop size, evaporation constant, and surface temperature during heat-up period *(Chin and Lefebvre [16])*.

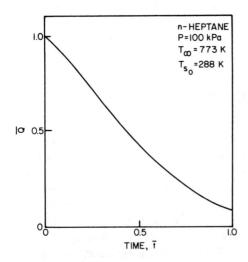

Figure 8.19 Variation of sensible heat input to drop with time *(Chin and Lefebvre [16])*.

the heat-up period. The line drawn in Fig. 8.19 approximates a straight line, so it is reasonable to regard $0.5\bar{Q}$ as the average value.

We now define a nondimensional surface temperature $\bar{T}_s$ as

$$\bar{T}_s = \frac{T_s - T_{s_0}}{T_{s_{st}} - T_{s_0}} \tag{8.43}$$

or

$$T_s = T_{s_0} + \bar{T}_s(T_{s_{st}} - T_{s_0}) \tag{8.44}$$

Note that when $t = 0$, $T_s = T_{s_0}$ and $\bar{T}_s = 0$. Also, at the end of the heat-up period $T_s = T_{s_{st}}$ and $\bar{T}_s = 1$. Thus the value of $\bar{T}_s$ increases from zero to unity during the heat-up period.

Chin and Lefebvre [16] calculated the variation of $\bar{Q}$ with T_s for different fuels at different levels of pressure and temperature and different initial fuel temperatures. From these calculations it was found possible to express $\bar{T}_s$ (at $\bar{Q} = 0.5$) in terms of ambient air temperature T_∞ and the ratio T_{s_0}/T_b, as illustrated in Fig. 8.20. Note that T_b is not the normal boiling point at atmospheric pressure, but the boiling temperature at the pressure under consideration.

To estimate the heat-up period, Eqs. (8.6), (8.38), and (8.39) are combined to give

$$\Delta t_{hu} = \frac{c_{p_F}\rho_F c_{p_g}D_{hu}^2(T_{s_{st}} - T_{s_0})}{12k_g \ln(1 + B_M)L(B_T/B_M - 1)} \tag{8.45}$$

where D_{hu} is the effective mean diameter during the heat-up period. It is related to the initial drop diameter D_0 by the equation

$$D_{hu} = D_0\left[1 + \frac{c_{p_F}(T_{s_{st}} - T_{s_0})}{2L(B_T/B_M - 1)}\right]^{-0.5} \tag{8.46}$$

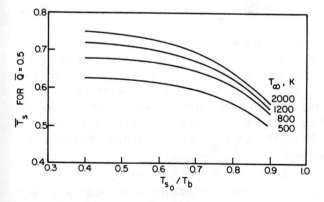

Figure 8.20 Correlation of nondimensional surface temperature for various temperatures and fuels *(Chin and Lefebvre [16]).*

For any stipulated conditions of air pressure and temperature, a value of $\bar{T}_s$ is read off Fig. 8.20. An effective mean value of drop surface temperature during the heat-up period is then defined by Eq. (8.44) as

$$T_{s_{hu}} = T_{s_0} + \bar{T}_s(T_{s_{st}} - T_{s_0}) \qquad (8.47)$$

This temperature is used to calculate values of L, B_M, B_T, and ρ_F for insertion into Eqs. (8.45) and (8.46). For c_{p_F} the appropriate temperature is $0.5(T_{s_0} + T_{s_{st}})$. Equations (8.11) and (8.12) may be used to calculate c_{p_g} and k_g, respectively, with the provision that the reference temperature now becomes

$$T_r = T_{s_{hu}} + \frac{T_\infty - T_{s_{hu}}}{3} \qquad (8.48)$$

Values of specific heat for liquid hydrocarbon fuels may be calculated from the expression

$$c_{p_F} = \frac{760 + 3.35T}{(0.001\rho_F)^{0.5}} \text{ J/kg K} \qquad (8.49)$$

where ρ_F is the fuel density at 289 K.

The mean or effective value of evaporation constant for the heat-up period, λ_{hu}, is readily obtained by inserting these calculated values of c_{p_g}, k_g, ρ_F, and B_M into Eq. (8.25).

The drop diameter at the end of the heat-up period is given by

$$D_1^2 = D_0^2 - \lambda_{hu}\Delta t_{hu} \qquad (8.50)$$

while the drop lifetime is obtained as the sum of the heat-up and steady-state periods, i.e.,

$$t_e = \Delta t_{hu} + \Delta t_{st} \qquad (8.51)$$

$$= \Delta t_{hu} + \frac{D_1^2}{\lambda_{st}} \qquad (8.52)$$

The subdivision of the drop evaporation time into two components, one for the heat-up period and another for the steady-state phase, leads to a representation of drop size variation with time as sketched in Fig. 8.21. Similar plots based on the results of calculations on λ_{hu}, λ_{st}, and D_1 for an n-heptane fuel drop 100 μm in diameter are shown in Fig. 8.22. It is of interest to note in these figures that change in T_{s_0} affects only the heat-up period and has no influence on the subsequent evaporation process. However, increase in ambient temperature reduces both the heat-up and steady-state periods.

Example Consider an n-heptane drop 200 μm in diameter in air at a temperature of 773 K and a pressure of 101.33 kPa. The initial drop temperature is 288 K. It is required to estimate the duration of the heat-up period, the drop diameter at the end of the heat-up period, and the total drop lifetime.

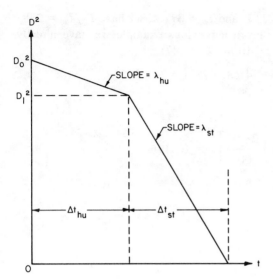

Figure 8.21 Representation of drop size variation with time, including heat-up period.

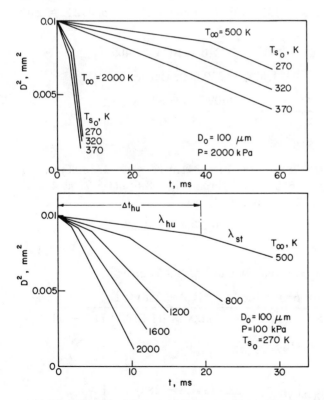

Figure 8.22 Graphs illustrating the effects of ambient air pressure and temperature and initial drop temperature on drop evaporation history; fuel is *n*-heptane *(Chin and Lefebvre [16])*.

From Fig. 8.20, for $T_\infty = 773$ K and $T_b = 371.4$, we have $T_{s0}/T_b = 288/371.4 = 0.776$ and $\bar{T}_s = 0.62$. From a previous example we have already calculated $T_{s_{st}} = 341.8$ K. Hence, from Eq. (8.47)

$$T_{s_{hu}} = 288 + 0.62(341.8 - 288) = 321.3 \text{ K}$$

From Eq. (8.20), $P_{F_s} = 17.4$.
From Eq. (8.19)

$$Y_{F_s} = \left[1 + \left(\frac{101.33}{17.4} - 1\right)\frac{28.97}{100.2}\right]^{-1} = 0.4176$$

$$B_M = \frac{Y_{F_s}}{1 - Y_{F_s}} = 0.717$$

$$Y_{F_r} = \tfrac{2}{3}Y_{F_s} = 0.2784$$

$$Y_{A_r} = 1 - Y_{F_r} = 0.7216$$

$$T_r = T_{s_{hu}} + \tfrac{1}{3}(T_\infty - T_{s_{hu}}) = 321.3 + \tfrac{1}{3}(773 - 321.3) = 472 \text{ K}$$

From Fig. 8.4, c_{p_A} at 472 K = 1024 J/kg K. From Fig. 8.5, c_{p_v} at 472 K = 2406 J/kg K.

$$c_{p_g} = 0.2784 \times 2406 + 0.7216 \times 1024 = 1409$$

From Eq. (8.22), for $T_s = 321.3$, $L = 350,000$. Hence

$$B_T = \frac{c_{p_g}(T_\infty - T_{s_{hu}})}{L} = \frac{1409(773 - 321.3)}{350,000}$$

$$B_T = 1.818$$

From Eq. (8.49), at $0.5(288 + 341.8) = 341.9$ K, $c_{p_F} = 2330$. From Eq. (8.26), at 321.3 K, $\rho_F = 650$ kg/m³. From Figs. 8.6 and 8.7, at 472 K, $k_A = 0.0376$ and $k_V = 0.0288$ J/m s K. Hence

$$k_g = 0.7216 \times 0.0376 + 0.2784 \times 0.0288$$

$$= 0.03515$$

Substitution of these property values into Eq. (8.45) gives

$$\Delta t_{hu} = \frac{2330 \times 650 \times 1409 \times (341.8 - 288) \times D_{hu}^2}{12 \times 0.03515 \times \ln(1.717) \times 350,000 \times [(1.818/0.717) - 1]}$$

$$= 0.9375 \times 10^6 \times D_{hu}^2 \text{ s}$$

From Eq. (8.46)

$$D_{hu} = 200 \times 10^{-6}\left[1 + \frac{2330(341.8 - 288)}{700,000[(1.818/0.717) - 1]}\right]^{-0.5}$$

$$= 187 \times 10^{-6} \text{ m}$$

Hence,

$$\Delta t_{hu} = 0.9375 \times 10^6 \times (187)^2 \times 10^{-12}$$

$$= 0.0328 \text{ s}$$

$$= 32.8 \text{ ms}$$

Thus the length of the heat-up period is 32.8 ms, during which the drop diameter is reduced from 200 to 187 μm.

The variation of drop diameter with time as obtained by the approximate method is indicated by the dashed line in Fig. 8.23. Also shown in this figure is the variation in drop diameter with time as calculated by the iterative method. This approach results in a value of Δt_{hu} of 31 ms.

The average value of λ during the heat-up period is obtained as

$$\lambda_{hu} = \frac{D_0^2 - D_1^2}{\Delta t_{hu}} = \frac{200^2 - 187^2}{0.0328 \times 10^{12}}$$

$$= 0.1534 \times 10^{-6} \text{ m}^2/\text{s} = 0.1534 \text{ mm}^2/\text{s}$$

From the previous example $\lambda_{st} = 0.30 \times 10^{-6}$ m^2/s. Hence, from Eq. (8.51), the total drop lifetime is given by

$$t_e = \Delta t_{hu} + \frac{D_1^2}{\lambda_{st}}$$

$$= 0.0328 + (187^2 \times 10^{-12})(0.30 \times 10^{-6})$$

$$= 0.0328 + 0.1049$$

$$= 0.138 \text{ s} = 138 \text{ ms}$$

Influence of Pressure and Temperature on Heat-Up Period

Inspection of Eq. (8.45) shows that the heat-up period is proportional to the square of the fuel drop diameter. Less obvious in this equation are the effects of ambient pressure and temperature. Calculations by Chin and Lefebvre [16], using their simplified method of estimating the heat-up period as outlined above, show that its duration increases with increase in pressure and decreases with increase in temperature. For example, Fig. 8.24 shows that for an n-heptane drop 100 μm in diameter, in air at a temperature of 500 K, an increase in pressure from 100 to 2000 kPa extends the heat-up period from 19.3 to 41.6 ms. At the same time the proportion of the total drop lifetime occupied by the heat-up period increases from 24.5 to 34%. If, instead of raising the pressure, the air temperature is raised from 500 to 2000 K, the heat-up period is reduced from 19.3 to 1.9 ms, while the portion occupied by the heat-up period in the total drop lifetime decreases from 24.5 to 16.8%. Similar data illustrating the effects of ambient pressure and temperature on Δt_{hu} for gasoline (JP 4), kerosine (JP 5), and diesel oil (DF-2) are shown in Fig. 8.25.

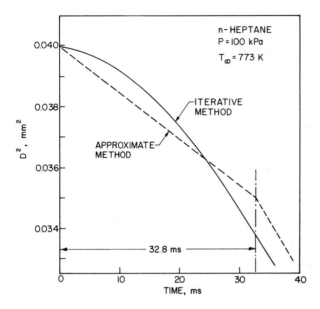

Figure 8.23 Variation of drop diameter with time during heat-up period *(Chin and Lefebvre [16])*.

The initial fuel temperature has a significant effect on the heat-up period, as shown in Fig. 8.26 for an *n*-heptane drop 100 μm in diameter at an ambient pressure of 2000 kPa. This figure reveals that increase in T_{s_0} from 270 to 370 K reduces Δt_{hu} by about 20% at all levels of ambient temperature. Thus fuel preheating provides a useful practical means of reducing the evaporation time of fuel drops. Incidentally, a further reduction in evaporation time accrues from the low-

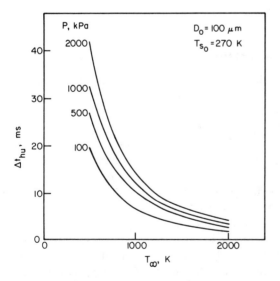

Figure 8.24 Effect of ambient air pressure and temperature on duration of heat-up period for *n*-heptane *(Chin and Lefebvre [16])*.

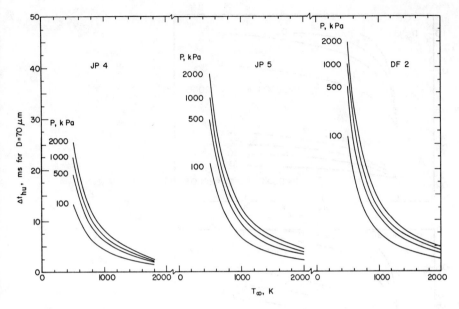

Figure 8.25 Effect of ambient pressure and temperature on duration of heat-up period; $D_0 = 70$ μm *(Chin and Lefebvre [16])*.

ering of viscosity and surface tension with increase in fuel temperature, which results in better atomization. As $\Delta t_{hu} \propto D_0^2$, any change in initial drop size has a pronounced effect on Δt_{hu}, as shown in Fig. 8.27.

The influence of ambient pressure and temperature on the average value of evaporation constant during the heat-up period is illustrated for JP 4, JP 5, and

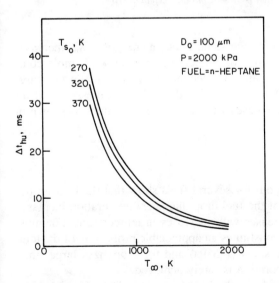

Figure 8.26 Graphs illustrating the effect of initial fuel temperature on heat-up period *(Chin and Lefebvre [16])*.

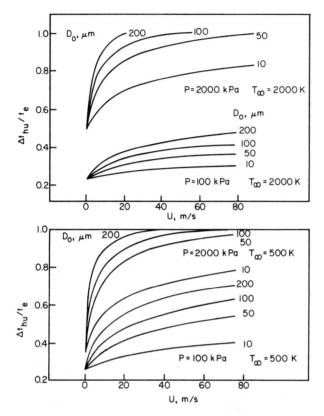

Figure 8.27 Influence of air pressure, temperature, and velocity and of drop size on proportion of drop lifetime occupied by heat-up period; fuel is JP 4 *(Chin and Lefebvre [16]).*

DF-2 in Fig. 8.28. The data all show the same trend, namely that λ_{hu} increases markedly with increase in T_∞. The effect of ambient pressure on λ_{hu} is more complex. At the lowest temperatures studied λ_{hu} declines with increase in pressure. At temperatures around 750 K, λ_{hu} is sensibly independent of pressure. At all higher levels of temperature, λ_{hu} increases with increase in pressure, as shown in Fig. 8.28.

Summary of Main Points

The results of calculations carried out for several fuels show that the heat-up period is proportional to the square of the fuel drop diameter. Its duration increases with increase in pressure and decreases with increases in temperature. For most practical fuels, the heat-up period constitutes an appreciable proportion of the total drop lifetime. In fact, at high pressures (2 MPa) and for drop sizes larger than around 200 μm, steady-state evaporation is rarely achieved.

Raising the fuel temperature reduces the heat-up period. This effect, in con-

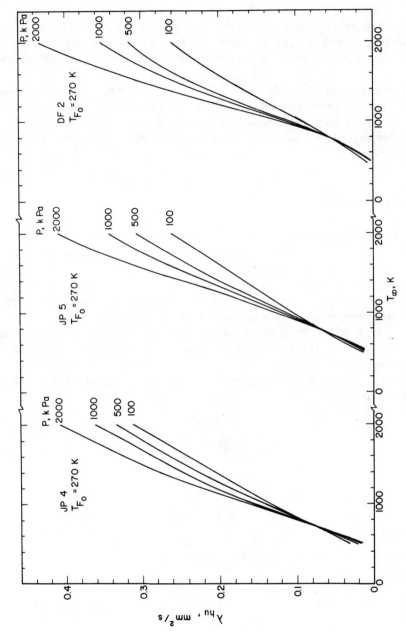

Figure 8.28 Influence of ambient air pressure and temperature on average evaporation constant during heat-up period (*Chin and Lefebvre [16]*).

junction with the improved atomization that is normally associated with higher fuel temperatures, provides a useful practical means of enhancing fuel spray evaporation.

Where relative velocity exists between a fuel drop and the surrounding medium, the drop lifetime is reduced. However, forced convection does not affect either the steady-state evaporation temperature or the duration of the heat-up period.

Since the heat-up period normally occupies a significant proportion of the total drop lifetime, calculations of drop lifetime and evaporation rates that ignore the heat-up period are prone to serious error. This is especially true at high pressures, where neglect of the heat-up period results in three- to fourfold underestimations of drop lifetimes.

DROP LIFETIME

In many practical combustion systems, the rates of chemical reaction are so high that the burning rate is controlled mainly by the rate of fuel vaporization. In such situations the evaluation of drop lifetime could be important, since it determines the residence time needed to ensure completion of combustion. Drop evaporation times are also important in the design of premix-prevaporize combustors for gas turbines, since they govern the length of the prevaporization passage.

One of the first theoretical attacks on the problem of droplet evaporation was made by Godsave [2]. Using the usual assumptions of a spherical system with a steady temperature distribution, where both the flow and temperature distribution are spherically symmetrical, he derived the rate of vaporization of a single drop as follows. We have,

$$m = \rho_F \left(\frac{\pi D^3}{6} \right) \tag{8.39}$$

Hence,

$$\frac{\dot{m}_F}{\rho_F} = \frac{d}{dt} \left(\frac{\pi D^3}{6} \right)$$

$$= \frac{\pi D^2}{2} \left(\frac{dD}{dt} \right)$$

$$= - \left(\frac{\pi}{4} \right) \lambda D$$

where

$$\lambda = - \frac{d(D^2)}{dt} \tag{8.23}$$

This definition of the evaporation constant λ by Godsave and his co-worker Probert [17] greatly simplifies many engineering calculations. For example, the rate of evaporation of a liquid drop is given by

$$\dot{m}_F = \left(\frac{\pi}{4}\right)\rho_F \lambda D \tag{8.24}$$

Also, the drop lifetime is readily obtained by assuming λ is constant and integrating (8.23) to give

$$t_e = \frac{D_0^2}{\lambda} \tag{8.53}$$

Substituting λ from Eq. (8.25) into Eq. (8.53) gives the following alternative expression for t_e:

$$t_e = \frac{\rho_F D_0^2}{8(k/c_p)_g \ln(1 + B)} \tag{8.54}$$

Equations (8.53) and (8.54) provide a ready means of calculating the time required for a drop to evaporate completely. In some applications, however, it is useful to know the time required to evaporate a certain fraction of the total mass contained within a drop.

If the heat-up period is neglected, the variation of D^2 with time can be represented by a straight-line relationship, as illustrated in Fig. 8.29. If f is used to denote the fraction of mass (or volume) evaporated in a time interval Δt_e, then

$$f = \frac{D_0^3 - D_1^3}{D_0^3} = 1 - \left(\frac{D_1}{D_0}\right)^3$$

or

$$\left(\frac{D_1}{D_0}\right)^3 = 1 - f$$

and

$$\left(\frac{D_1}{D_0}\right)^2 = (1 - f)^{2/3} \tag{8.55}$$

where D_1 is the drop diameter after a time Δt_e.

Now

$$\Delta t_e = \frac{D_0^2 - D_1^2}{\lambda} = \frac{D_0^2}{\lambda}\left[1 - \left(\frac{D_1}{D_0}\right)^2\right] \tag{8.56}$$

Substituting for $(D_1/D_0)^2$ from Eq. (8.55) into Eq. (8.56) gives

$$\Delta t_e = \frac{D_0^2}{\lambda}[1 - (1 - f)^{2/3}] \tag{8.57}$$

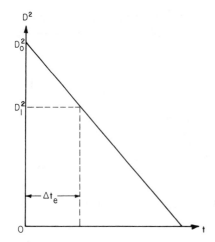

Figure 8.29 Representation of drop size variation with time, neglecting heat-up period.

This equation gives the time required to evaporate a fraction f of the initial mass (or volume) of a drop having an initial diameter D_0.

Effect of Heat-Up Phase on Drop Lifetime

Where the heat-up period is too long to be neglected, the time required to fully evaporate the drop is obtained as the sum of the unsteady-state period and the steady-state period, i.e.,

$$t_e = \frac{D_0^2 - D_1^2}{\lambda_{hu}} + \frac{D_1^2}{\lambda_{st}} \qquad (8.58)$$

where D_1 is the diameter of the drop at the end of the heat-up period, as illustrated in Fig. 8.21. Both D_1 and λ_{hu} may be estimated using the methods described in the previous section.

For the purpose of calculating the time required to evaporate a fraction f of the initial mass (or volume) of the drop it is necessary to consider two separate cases.

Case 1. When f is small, say less than 0.1, only evaporation occurring during the heat-up period need be considered, and the time required is given by

$$\Delta t_e = \frac{D_0^2}{\lambda_{hu}} [1 - (1 - f)^{2/3}] \qquad (8.59)$$

Case 2. If f is large enough to involve evaporation in both the heat-up and steady-state phases, the evaporation time is obtained as

$$\Delta t_e = \frac{D_0^2}{\lambda_{hu}} [1 - (1 - f_1)^{2/3}] + \frac{D_0^2}{\lambda_{st}} [1 - (1 + f_1 - f)^{2/3}](1 - f_1)^{2/3} \qquad (8.60)$$

where f_1 is the fraction of the initial liquid mass that is vaporized during the heat-up period. An appropriate expression for f_1 is

$$f_1 = 1 - \left[\frac{(D_0^2 - \lambda_{hu}\Delta t_{hu})^{0.5}}{D_0} \right]^3 \tag{8.61}$$

where Δt_{hu} is the duration of the heat-up period as given by Eq. (8.45).

An alternative expression for Δt_e is

$$\Delta t_e = \Delta t_{hu} + \frac{D_0^2}{\lambda_{st}} [1 - (1 + f_1 - f)^{2/3}](1 - f_1)^{2/3} \tag{8.62}$$

Effect of Prevaporization on Drop Lifetime

We now examine for steady-state conditions the effect of prevaporization on drop evaporation time and on the time to produce any given level of vapor concentration.

Case 1. Consider first a control volume into which flows a mixture of drops, vapor, and air. The vapor constitutes a fraction Ω of the total mass (liquid plus vapor). It is required to derive an expression for the time Δt_e that the mixture must spend in the control volume to raise the vapor content from its initial value Ω to a higher value f. Let

D_i = initial drop diameter
D_0 = diameter of drops entering control volume
D_1 = diameter of drops after time Δt_e
Ω = fraction of total mass (liquid plus vapor) entering control volume in vapor form

It is assumed that steady-state evaporation has been achieved by the time the drops enter the control volume.

We have, by definition,

$$\frac{D_i^3 - D_0^3}{D_i^3} = \Omega$$

or

$$\left(\frac{D_0}{D_i} \right)^3 = 1 - \Omega$$

or

$$D_0^2 = D_i^2 (1 - \Omega)^{2/3} \tag{8.63}$$

Also, by definition

$$\frac{D_i^3 - D_1^3}{D_i^3} = f$$

and

$$D_1^2 = D_i^2(1 - f)^{2/3} \tag{8.64}$$

Now

$$\Delta t_e = \frac{D_0^2 - D_1^2}{\lambda_{st}} \tag{8.65}$$

Substituting for D_0^2 from Eq. (8.63) and D_1^2 from Eq. (8.64) into Eq. (8.65) gives

$$\Delta t_e = \frac{D_i^2}{\lambda_{st}}[(1 - \Omega)^{2/3} - (1 - f)^{2/3}] \tag{8.66}$$

or, from Eq. (8.59),

$$\Delta t_e = \frac{D_0^2}{\lambda_{st}}\left[1 - \left(\frac{1 - f}{1 - \Omega}\right)^{2/3}\right] \tag{8.67}$$

Case 2. We now consider the situation where a mixture of drops, vapor, and air enters a control volume, as in case 1, except that no assumption is made concerning the origin of the vapor. It could arise from partial prevaporization of the existing drops, from a separate injection of vapor, or from a multiplicity of vapor sources.

This situation is most readily dealt with by assuming that all the incoming vapor, regardless of its origin, is due to partial prevaporization of the drops present in the system. The time required in the control volume to raise the concentration of vapor to a fraction f of the total mass (vapor plus liquid) is again given by Eq. (8.67), in which D_0 is the diameter of the drops entering the control volume, Ω is the fraction of the total mass (liquid plus vapor) that enters the control volume as vapor, and f denotes the fraction of the total mass (liquid plus vapor) that is present in vapor form after a residence time Δt_e in the control volume.

CONVECTIVE EFFECTS ON EVAPORATION

For drop evaporation under quiescent conditions, the principal mode of heat transfer is thermal conduction. Where relative motion exists between the drop and the surrounding air or gas, the rate of evaporation is enhanced. Surprisingly, perhaps, convective effects do not change the steady-state evaporation temperature $T_{s_{st}}$. This is because forced convection affects both Q, the rate of heat transfer to the drop, and Q_e, the rate of evaporation, to the same extent, so that under steady-state conditions, when $Q = Q_e$, the two effects cancel each other. Also, from inspection of Eq. (8.45) it is clear that the heat-up period Δt_{hu} is unaffected by convective effects. However, since rates of evaporation are enhanced by convection, the total drop lifetime must diminish. Thus the effect of convection is to reduce the duration of the steady-state portion of the drop evaporation process.

If the rate of heat transfer to a drop per unit surface area per unit time is denoted by h, then

$$h = \frac{\dot{m}_F L}{\pi D^2} \tag{8.68}$$

Substituting for $\dot{m}_F$ from Eq. (8.17) gives

$$h = \frac{2(k/c_p)_g L \ln(1 + B_T)}{D} \tag{8.69}$$

For constant fluid properties, in the absence of convection, the Nusselt number is

$$Nu = \frac{hD}{k(T_\infty - T_s)}$$

Substituting for h from Eq. (8.69) gives

$$Nu = 2 \left[\frac{L}{c_{p_g}(T_\infty - T_s)} \right] \ln(1 + B_T)$$

or, since $c_{p_g}(T_\infty - T_s)/L = B_T$,

$$Nu = 2 \frac{\ln(1 + B_T)}{B_T} \tag{8.70}$$

As B_T tends toward zero, $\ln(1 + B_T)/B_T$ approaches unity and $Nu = 2$, which is the normal value for low heat transfer rates to a sphere.

In a comprehensive theoretical and experimental study, Frossling [18] showed that the effects of convection on heat and mass transfer rates could be accommodated by a correction factor that is a function of Reynolds number and Schmidt (or Prandtl) number. Where diffusion rates are controlling, the correction factor is

$$1 + 0.276 Re_D^{0.5} Sc_g^{0.33} \tag{8.71}$$

For the more usual case where heat transfer rates are controlling, it becomes

$$1 + 0.276 Re_D^{0.5} Pr_g^{0.33} \tag{8.72}$$

The velocity term in Re_D should be the relative velocity between the drop and the surrounding gas, i.e., $Re_D = UD\rho_g/\mu_g$. However, both calculations and experimental observations suggest that small drops rapidly attain the same velocity as the surrounding gas, after which they are susceptible only to the fluctuating component of velocity, u'. The appropriate value of Reynolds number then becomes $u' D\rho_g/\mu_g$.

Another important convective heat transfer correlation is the following, due to Ranz and Marshall [19]:

$$Nu = 2 + 0.6 Re_D^{0.5} Pr_g^{0.33} \tag{8.73}$$

which corresponds to a correction factor of

$$1 + 0.3\text{Re}_D^{0.5}\text{Pr}_g^{0.33} \qquad (8.74)$$

Note that the physical properties ρ_g, μ_g, c_{p_g}, and k_g embodied in Re_D and Pr_g should be evaluated at the reference temperature T_r. Values of c_{p_g} and k_g are given by Eqs. (8.11) and (8.12), in which the appropriate values of c_{p_A}, c_{p_v}, k_A, and k_V may be obtained from Figs. 8.4 to 8.6. Corresponding equations for ρ_g and μ_g are

$$\rho_{g_r} = \left(\frac{Y_{A_r}}{\rho_A} + \frac{Y_{F_r}}{\rho_F}\right)^{-1} \qquad (8.75)$$

$$\mu_g = Y_{A_r}(\mu_A \text{ at } T_r) + Y_{F_r}(\mu_V \text{ at } T_r) \qquad (8.76)$$

Plots of ρ_V, μ_A, and μ_V versus temperature are provided for some hydrocarbon fuels in Figs. 8.30 and 8.31.

Faeth [4] has analyzed the available data on convective effects and proposed a synthesized correlation for Nu that approaches the correct limiting values at low and high Reynolds numbers ($\text{Re}_D < 1800$). This correlation yields the following correction factor to account for the augmentation of evaporation due to forced convection:

$$\frac{1 + 0.276\text{Re}_D^{0.5}\text{Pr}_g^{0.33}}{[1 + 1.232/(\text{Re}_D\text{Pr}_g^{1.33})]^{0.5}} \qquad (8.77)$$

Thus the primary factors affecting the lifetime of a drop are

1. The physical properties of the ambient air or gas, i.e., pressure, temperature, thermal conductivity, specific heat, and viscosity.

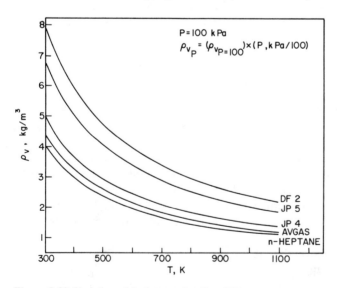

Figure 8.30 Variation of fuel vapor density with temperature.

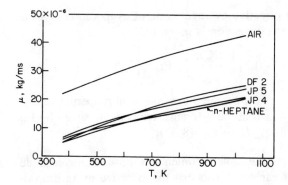

Figure 8.31 Variation of viscosity with temperature for air and various fuel vapors.

2. The relative velocity between the drop and the surrounding medium.
3. The properties of the liquid and its vapor, including density, vapor pressure, thermal conductivity, and specific heat.
4. The initial condition of the drop, especially size and temperature.

The combination of Eqs. (8.21) and (8.74) yields the following equation for the rate of fuel evaporation with forced convection:

$$\dot{m}_F = 2\pi D \left(\frac{k}{c_p}\right)_g \ln(1 + B)(1 + 0.30 \mathrm{Re}_d^{0.5} \mathrm{Pr}_g^{0.33}) \tag{8.78}$$

This equation gives the instantaneous rate of evaporation for a drop of diameter D. To obtain the *average* rate of evaporation of the drop during its lifetime, the constant should be reduced from 2 to 1.33. Substituting also $\bar{D} = 0.667 D_0$ and $\mathrm{Pr}_g = 0.7$ gives

$$\bar{\dot{m}}_F = 1.33\pi D_0 \left(\frac{k}{c_p}\right)_g \ln(1 + B)(1 + 0.22 \mathrm{Re}_{D_0}^{0.5}) \tag{8.79}$$

and

$$t_e = \frac{\rho_F D_0^2}{8(k/c_p)_g \ln(1 + B)(1 + 0.22 \mathrm{Re}_{D_0}^{0.5})} \tag{8.80}$$

Determination of Evaporation Constant and Drop Lifetime

Heat-up period. It can be shown [16] that during the heat-up period the average drop diameter with forced convection, D'_{hu}, is given by

$$D'_{\mathrm{hu}} = 0.25 \left(D_0^{0.5} + \left\{ D_0^2 - \lambda_{\mathrm{hu}} \Delta t_{\mathrm{hu}} \left[1 + 0.3 \left(\frac{\rho_g D'_{\mathrm{hu}} U}{\mu_g}\right)^{0.5} \mathrm{Pr}_g^{0.33} \right] \right\}^{0.25} \right)^2 \tag{8.81}$$

As D'_{hu} appears on both sides of the above equation, some trial and error is involved in its solution.

Having determined D'_{hu}, the effective average value of evaporation constant for the heat-up period with forced convection is given by

$$\lambda'_{hu} = \lambda_{hu}\left[1 + 0.3\left(\frac{\rho_g D'_{hu} u'}{\mu_g}\right)^{0.5} Pr_g^{0.33}\right] \tag{8.82}$$

As noted earlier for quiescent mixtures, the thermophysical properties in Eqs. (8.81) and (8.82) should be evaluated at the reference temperature given by Eq. (8.7) in which T_s is replaced with $T_{s_{hu}}$ from Eq. (8.47).

Steady-state period. By following the same arguments as those employed in deriving Eqs. (8.81) and (8.82), it can be shown that the effective mean drop diameter during the steady-state period is given by

$$D'_{st} = 0.25D_1 = 0.25(D_0^2 - \lambda'_{hu}\Delta t_{hu})^{0.5} \tag{8.83}$$

and

$$\lambda' = \lambda_{st}\left[1 + 0.3\left(\frac{\rho_g D'_{st} u'}{\mu_g}\right)^{0.5} Pr_g^{0.33}\right] \tag{8.84}$$

For the steady-state condition, $T_s = T_{st}$, and the reference temperature to be used in determining ρ_g, μ_g, c_{p_g}, and k_g is obtained from Eq. (8.7) as

$$T_r = T_{st} + \frac{T_\infty - T_{st}}{3}$$

Drop Lifetime

The drop lifetime is given by

$$t_e = \Delta t_{hu} + \Delta t_{st} \tag{8.51}$$

$$= \Delta t_{hu} + \frac{D_0^2 - \lambda'_{hu}\Delta t_{hu}}{\lambda'_{st}}$$

$$= \Delta t_{hu}\left(1 - \frac{\lambda'_{hu}}{\lambda'_{st}}\right) + \frac{D_0^2}{\lambda'_{st}} \tag{8.85}$$

The fraction of the total evaporation time occupied by the heat-up period is obtained from Eqs. (8.45) and (8.85) as the ratio $\Delta t_{hu}/t_e$. Graphs of $\Delta t_{hu}/t_e$ versus velocity for JP 4 and DF-2 are shown in Figs. 8.27 and 8.32, respectively. These figures demonstrate that over wide ranges of operating conditions the heat-up period represents an appreciable portion of the total drop evaporation time, as first noted by El Wakil, Priem, et al. [20–22]. This is especially true for high air pressures and temperatures, where all but the smallest drops fail to attain steady-state evaporation during their lifetime.

Figures 8.27 and 8.32 show that $\Delta t_{hu}/t_e$ increases with increase in fuel drop size. It also increases markedly with increase in velocity. This is because higher

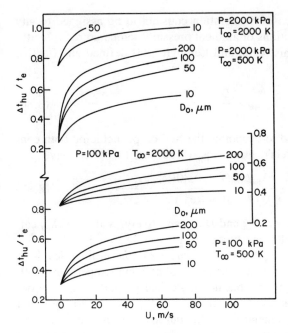

Figure 8.32 Influence of air pressure, temperature, and velocity and of drop size on proportion of drop lifetime occupied by heat-up period; fuel is DF-2 *(Chin and Lefebvre [16]).*

velocities enhance evaporation rates and thereby reduce t_e, while Δt_{hu} remains sensibly constant. The effect of pressure on $\Delta t_{hu}/t_e$ is brought out most clearly in Fig. 8.27, where it is shown that increase in pressure always raises $\Delta t_{hu}/t_e$. This is because Δt_{hu} is strongly affected by changes in pressure. The effect of temperature on $\Delta t_{hu}/t_e$ is less marked but more complex. Figure 8.32 shows that at low pressure (100 kPa) $\Delta t_{hu}/t_e$ diminishes with increase in temperature, whereas at higher pressure (2000 kPa) the opposite trend is observed.

CALCULATION OF EFFECTIVE EVAPORATION CONSTANT

During steady-state evaporation, all the heat transferred to the drop is employed in fuel evaporation and consequently evaporation rates are relatively high. However, during the heat-up period much of the heat transferred to the drop is absorbed in heating it up, so the amount of heat available for fuel vaporization is correspondingly less. This lower rate of vaporization, when considered in conjunction with the significant proportion of the total drop lifetime occupied by the heat-up period, means that overall evaporation rates can be appreciably lower and drop lifetimes much longer than the corresponding values calculated on the assumption of zero heat-up time.

From a practical viewpoint it would be very convenient if the effect of the heat-up period could be combined with that of forced convection in a manner that would allow an "effective" value of evaporation constant to be assigned to any given fuel at any stipulated conditions of ambient pressure, temperature, velocity, and drop size. To accomplish this, Chin and Lefebvre [23] defined an effective evaporation constant as

$$\lambda_{\text{eff}} = \frac{D_0^2}{t_e} \tag{8.86}$$

where t_e is the total time required to evaporate the fuel drop, including both convective and transient heat-up effects. From Eqs. (8.50) and (8.71) we have

$$t_e = \Delta t_{\text{hu}} + \frac{[D_0^2 - \lambda_{\text{hu}}\Delta t_{\text{hu}}(1 + 0.30\text{Re}_{\text{hu}}^{0.5}\text{Pr}_g^{0.33})]}{\lambda_{\text{st}}(1 + 0.30\text{Re}_{\text{st}}^{0.5}\text{Pr}_g^{0.33})} \tag{8.87}$$

in which Re_{hu} and Re_{st} are based on D_{hu} and D_{st}, respectively. Values of λ_{eff} calculated from Eqs. (8.86) and (8.87) are plotted in Figs. 8.33 to 8.35. These figures represent plots of λ_{eff} versus T_{bn} for various values of UD_0 at three levels of pressure, 100, 1000, and 2000 kPa, and three levels of ambient temperature, 500, 1200, and 2000 K. While recognizing that no single fuel property can fully describe the evaporation characteristics of any given fuel, the average boiling point (50% recovered) has much to commend it for this purpose, because it is directly related to fuel volatility and vapor pressure. It also has the virtue of being easy to measure and is usually quoted in fuel specifications.

The figures show that λ_{eff} increases with increase in ambient temperature, pressure, velocity, and drop size and diminishes with increase in normal boiling temperature.

The concept of an effective value of evaporation constant considerably simplifies calculations of the evaporation characteristics of fuel drops. For example, for any given conditions of pressure, temperature, and relative velocity, the lifetime of a fuel drop of any given size is obtained from Eq. (8.86) as

$$t_e = \frac{D_0^2}{\lambda_{\text{eff}}} \tag{8.88}$$

while the average rate of fuel evaporation is readily determined by dividing m_F from Eq. (8.39) by t_e from Eq. (8.88) as

$$\dot{m}_F = \frac{\pi}{6}\rho_F\lambda_{\text{eff}}D_0 \tag{8.89}$$

Example It is required to calculate the lifetime of a fuel drop 65 μm in diameter in a flowing stream where the ambient pressure and temperature are 1400 kPa and 1100 K, respectively, and the relative velocity between the air and the drop is 20 m/s. The normal boiling temperature of the fuel is 485 K, and its initial density is 810 kg/m³.

From Fig. 8.34 we obtain by interpolation that the value of λ_{eff} for $P = 1000$ kPa and $UD_0 = 20 \times 65 = 1300$ is 0.54 mm²/s. Similarly, we find from Fig. 8.35 that for $P = 2000$ kPa and $UD_0 = 1300$, λ_{eff} is 0.59 mm²/s. Hence,

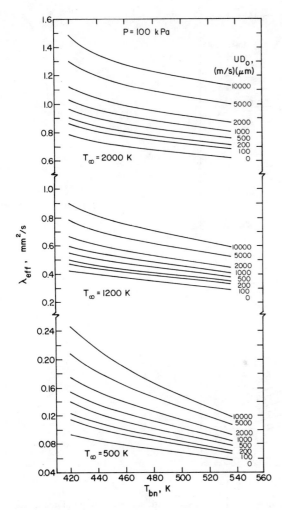

Figure 8.33 Variation of effective evaporation constant with normal boiling point at a pressure of 100 kPa *(Chin and Lefebvre [23])*.

by interpolation, λ_{eff} for $P = 1400$ kPa is 0.56 mm^2/s. From Eq. (8.86)

$$t_e = \frac{D_0^2}{\lambda_{\text{eff}}} = \frac{(65 \times 10^{-6})^2}{0.56 \times 10^{-6}}$$
$$= 7.55 \times 10^{-3} \text{ s} = 7.55 \text{ ms}$$

From Eq. (8.89)

$$\dot{m}_F = \frac{\pi}{6} \rho_F \lambda_{\text{eff}} D_0 = \frac{\pi}{6} \times 810 \times 0.56 \times 10^{-6} \times 65 \times 10^{-6}$$
$$= 1.54 \times 10^{-8} \text{ kg/s}$$

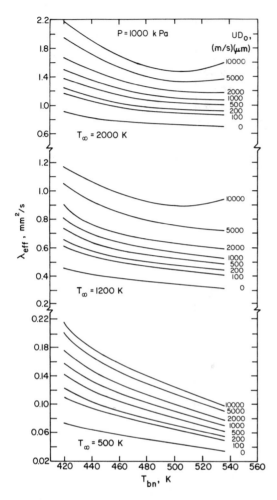

Figure 8.34 Variation of effective evaporation constant with normal boiling point at a pressure of 1000 kPa *(Chin and Lefebvre [23])*.

INFLUENCE OF EVAPORATION ON DROP SIZE DISTRIBUTION

Although the evaporation characteristics of a single drop provide useful guidance on the evaporation history of a complete spray, there are certain key features of an evaporating spray that can be explored only by considering the spray as a whole. For example, for the ignition of a fuel spray it is the initial rate of evaporation that is important, whereas for high combustion efficiency the time required for complete evaporation of the fuel spray is the dominant factor. Another example is provided by the lean premix-prevaporize combustor. A key feature of this concept is the attainment of complete evaporation of the fuel and complete

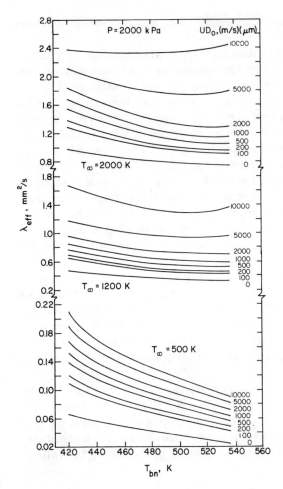

Figure 8.35 Variation of effective evaporation constant with normal boiling point at a pressure of 2000 kPa *(Chin and Lefebvre [23])*.

mixing of the fuel vapor with air prior to combustion. Failure to fully vaporize the fuel results in higher NO_x emissions. It is important, therefore, to know how the fraction or percentage of fuel evaporated varies with time, so that the best compromise can be made between the conflicting requirements of minimum NO_x emissions and minimum length of premixing chamber.

It is also of interest to ascertain how the initial mean drop size and the drop size distribution (as defined by the Rosin-Rammler distribution parameter q) affect the subsequent evaporation history.

Chin et al. [24] have studied in detail the evaporation histories of JP 5 fuel sprays. The results of their calculations are presented in a form that allows the time required to vaporize any given percentage of the initial spray mass to be estimated for other fuels at any stipulated conditions of ambient pressure and temperature.

The manner in which the mass median diameter MMD changes as evaporation proceeds is shown in Fig. 9.1 for a JP 5 fuel spray in air at a temperature of 2000 K and a pressure of 2 MPa. This figure contains several different curves to illustrate the influence of the initial drop size distribution parameter, q_0, on the variation in mean drop size with time. Over the range of values of q_0 of most practical interest, Fig. 9.1 shows that evaporation causes MMD to increase with time, the effect being especially pronounced for sprays having low values of q_0. If time is expressed in dimensionless form as $t_{90\%}$, the time required to vaporize 90% by mass of the spray, and MMD is divided by its initial value MMD_0, we can plot the nondimensional mean diameter MMD/MMD_0 against nondimensional time $t/t_{90\%}$ as shown in Fig. 8.36, where the results for different values of MMD_0 combine to form a single curve. Since the evaporation time is now nondimensionalized, Fig. 8.36 can be seen as characteristic for the spray evaporation of most hydrocarbon fuels. It is perfectly general and applies equally well for all practical values of pressure, temperature, and MMD_0. In a similar manner, if the relative change of q, i.e., q/q_0, is plotted against nondimensional evaporation time, $t/t_{90\%}$, then the results for different MMD_0 values fall on a single curve as shown in Fig. 8.37. The general trend is for q to increase with evaporation time, and the effect is more significant for sprays having low values of q_0. As Fig. 8.37 is plotted in dimensionless form, it is valid for all practical values of MMD_0, pressure, and temperature.

The time required to vaporize a certain percentage of the fuel spray is proportional to the square of MMD_0, as shown in Fig. 8.38. The curve drawn in this figure represents calculated results for the time to vaporize 90% of the spray mass for $MMD_0 = 70$ μm. Other data points are converted from the calculated data for $MMD_0 = 30, 50, 90,$ and 120 μm by multiplying by the ratio $(70/MMD_0)^2$.

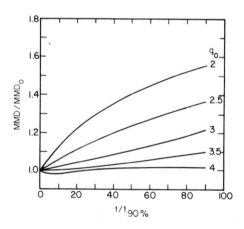

Figure 8.36 Nondimensional diameter plotted against nondimensional evaporation time for various values of q_0 (*Chin et al. [24]*).

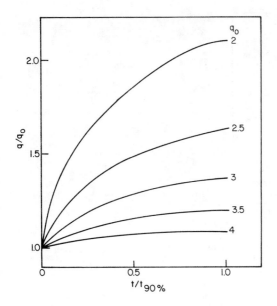

Figure 8.37 Relative change of drop size distribution parameter with nondimensional evaporation time for various values of q_0 *(Chin et al. [24]).*

The time required to vaporize 20% of the spray mass for $MMD_0 = 70$ μm is shown in Fig. 8.39. From inspection of Figs. 8.38 and 8.39 it is clear that a spray of large q_0 will have a low 90% evaporation time and a high 20% evaporation time. Thus, from a combustion efficiency viewpoint, it is desirable to have a fuel spray with a high value of q_0, but for good ignition performance it is better to have a low q_0 value.

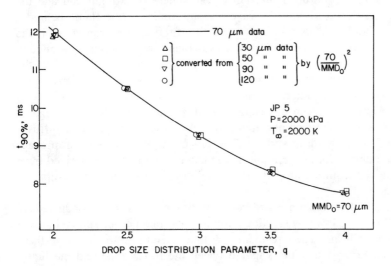

Figure 8.38 Data illustrating that evaporation time is proportional to MMD^2 *(Chin et al. [24]).*

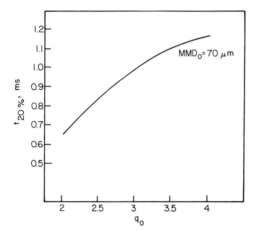

Figure 8.39 Time to vaporize 20% of spray mass (volume) for different values of q_0; fuel = JP 5, $P = 2$ MPa, $T = 2000$ K *(Chin et al. [24])*.

DROP BURNING

Most experimental studies on the burning of single fuel drops have either used drops suspended at the end of wires or simulated the drop-burning process by the use of a porous sphere covered with a liquid film, as discussed earlier. Spalding's early studies [3] on the influence of relative air velocity on the combustion of liquid fuel spheres revealed a critical velocity above which the flame could not be supported at the upstream portion of the sphere. At air velocities below the critical level the sphere is entirely enveloped in flame, whereas at higher relative velocities the flame on the upstream half of the sphere is extinguished and the burning zone is confined to a small wake region, similar to the flame behind a bluff-body flameholder. Thus within a liquid fuel-fired combustion zone, if the temperature and oxygen concentration of the ambient gas are both high enough to effect rapid ignition and the relative air velocity is below the critical value, the fuel drop will become surrounded by a thin spherical flame. This "envelope" flame is sustained by the exothermic reaction of fuel vapor and oxygen in the thin combustion zone, with oxygen diffusing radially inward from the outer regions (against the outward flow of combustion products) while the fuel vapor diffuses radially outward from the drop surface. The process is analogous to drop evaporation without combustion, except that in the latter case the heat is derived from regions far removed from the drop, while in the burning case the heat is supplied from the flame.

Measurements of temperature and concentration profiles were conducted by Aldred et al. [25] at the lower stagnation point of a 9.2-mm porous sphere burning n-heptane fuel in still air. Their results show that in addition to oxidation there is significant cracking of the fuel vapor between the flame zone and the liquid surface. This cracking produces the carbon particles that are largely responsible for the characteristic yellow flames of fuel sprays.

As the burning rates of individual fuel drops are limited by evaporation rates, the equations derived in the previous sections for the mass evaporation rate, $\dot{m}_F$ and drop lifetime t_e are directly applicable. Since heat transfer processes are dominant, it is appropriate to use B_T for the transfer number. At the beginning of drop combustion, while the temperature at the center of the drop still retains its initial value, a suitable expression for B is

$$B = \frac{c_{p_g}(T_f - T_b)}{L + c_{p_F}(T_b - T_{F_0})} \tag{8.90}$$

where T_f = flame temperature
T_b = boiling point of fuel
T_{F_0} = initial fuel temperature
L = latent heat of vaporization at T_b
c_{p_g} = mean specific heat of gas between T_b and T_f
c_{p_F} = mean specific heat of fuel between T_b and T_{F_0}.

As burning proceeds, the flow recirculations induced within the drop soon create a fairly uniform temperature distribution, so that throughout its volume the drop temperature is everywhere fairly close to the surface value T_b. Thus, under steady-state conditions, the transfer number simplifies to

$$B = \frac{c_{p_g}(T_f - T_b)}{L} \tag{8.91}$$

Substituting this value of B into Eq. (8.78) gives the following expression for the instantaneous burning rate of a fuel drop:

$$\dot{m}_F = 2\pi D \left(\frac{k}{c_p}\right)_g \ln\left[1 + \frac{c_{p_g}(T_f - T_b)}{L}\right](1 + 0.30\mathrm{Re}_D^{0.5}\mathrm{Pr}_g^{0.33}) \tag{8.92}$$

while the lifetime of a burning drop is obtained from Eqs. (8.53), (8.54), and (8.80) in conjunction with Eq. (8.91). For example, Eq. (8.80), which defines the lifetime of a drop under conditions where the heat-up period is effectively zero, becomes

$$t_e = \frac{\rho_F D_0^2}{8(k/c_p)_g \ln[1 + c_{p_g}(T_f - T_b)/L](1 + 0.22\mathrm{Re}_{D_0}^{0.5})} \tag{8.93}$$

MULTICOMPONENT FUEL DROPS

Much of the available information on drop evaporation and combustion has been gained using pure fuels such as n-heptane. For such fuels the boiling temperature is clearly defined and the drop cross-sectional area decreases linearly with time during the steady-state evaporation (or combustion) period. Commercial fuels, however, are generally multicomponent mixtures of various petroleum compounds, possessing individually different physical and chemical properties. According to Wood et al. [1], as combustion proceeds the composition of a multicomponent fuel drop changes by a process of simple batch distillation; that is, the vapors produced are removed continuously without further contact with the residual liquid mixture. Hence, as burning proceeds, the more volatile constituents of the liquid drop vaporize first and the concentration of the higher-boiling fractions in the liquid phase increases. Thus, the drop-sectional area decreases in a nonlinear manner with time. This characteristic, incidentally, offers a suitable means of analyzing the vaporization process occurring during the combustion of a multicomponent fuel.

There have been many experimental studies on multicomponent fuels and numerous analyses have been made to compare the predictions of diffusion flame models of drop combustion with measurements at moderate pressures. A review of these studies is beyond the scope of the present work, but further information is available in the various publications of Faeth, El Wakil, Law, and Sirignano (see, for example, references [26–40]).

The available experimental evidence on light distillate petroleum fuels suggests that λ remains sensibly constant during steady-state evaporation, as illustrated in Fig. 8.2 for JP 4 and kerosine burning in air. However, as pointed out by Faeth [26], while this may be the case for combustion, where flame temperatures are much higher than drop temperatures, it should not be expected for drop evaporation in a low-temperature environment. At low gas temperatures, variation of drop surface temperature due to composition changes exerts a stronger influence on the temperature difference between the drop and its surroundings, causing significant variations in B_M. More experimental work is needed on the vaporization characteristics of multicomponent fuels in low-temperature gas streams.

The combustion of medium and heavy fuel oils is complicated by the disruptive boiling and swelling of the burning drops and the formation of a carbonaceous residue [8,27]. The D^2 law no longer holds, although apparently it is still possible to define equivalent burning rate constants.

According to Williams [8], the course of combustion of a medium to a heavy fuel oil drop can be summarized as follows:

1. Heating up of the drop and vaporization of the low boiling point components.
2. Self-ignition and combustion with slight thermal decomposition and continued vaporization of the volatile components.
3. Extensive disruptive boiling and swelling of the drop together with considerable thermal decomposition to give a heavy tar.
4. Combustion of remaining volatile liquids and gases from the decomposition of the heavy tar, which collapses and forms a carbonaceous residue having an open structure called a cenosphere.
5. Slow heterogeneous combustion of the carbonaceous residue at a rate of about one-tenth that of the initial drop (in terms of burning rate constants).

It is claimed that when disruptive swelling is augmented by emulsification of water with the fuel, extensive swelling of the drop can produce secondary atomization by smaller drops being explosively ejected from the parent drop.

NOMENCLATURE

a, b	constants in Eq. (8.20)
B	transfer number
B_M	mass transfer number
B_T	thermal transfer number
C_{ex}	coefficient of thermal expansion
c_p	specific heat at constant pressure, J/kg K
D	drop diameter, m
D_0	initial drop diameter, m
D_1	drop diameter at end of heat-up period, m
D_c	diffusion coefficient, m²/s
D_{hu}	effective mean drop diameter during heat-up period, m
D'_{hu}	effective mean drop diameter during heat-up period with forced convection, m
D'_{st}	effective mean drop diameter during steady-state period with forced convection, m
f	fraction of initial liquid mass in vapor form after time Δt_e
h	heat transfer coefficient, J/m² s K
k	thermal conductivity, J/m s K
L	latent heat of fuel vaporization, J/kg
L_{T_b}	latent heat of fuel vaporization at boiling temperature T_b, J/kg
$L_{T_{bn}}$	latent heat of vaporization at normal boiling temperature T_{bn}, J/kg
m	mass of fuel drop, kg
m_F	rate of fuel evaporation per unit surface area, kg/m² s
$\dot{m}_F$	rate of fuel evaporation, kg/s
M	molecular weight
MMD	mass median diameter
Nu	Nusselt number
n	pressure exponent of evaporation rate
P	ambient air pressure, kPa
P_{cr}	critical pressure, kPa
P_{F_s}	fuel vapor pressure at drop surface, kPa
Pr	Prandtl number
Q	rate of heat transfer to drop from surrounding gas, J/s
Q_e	rate of heat utilization in fuel vaporization, J/s
$\bar{Q}$	nondimensional sensible heat parameter
q	Rosin-Rammler drop size distribution parameter [see Eq. (3.14)]
R	gas constant
r	drop radius, m
r_s	radius at drop surface, m
Re_D	drop Reynolds number
Sc	Schmidt number
T	temperature, K

T_b	boiling temperature, K
T_{bn}	boiling temperature at normal atmospheric pressure, K. For multicomponent fuels T_{bn} = average boiling point (50% recovered).
T_{cr}	critical temperature, K
T_f	flame temperature, K
T_r	reference temperature, K [see Eq. (8.7)]
T_s	drop surface temperature, K
$\bar{T}_s$	nondimensional surface temperature, K
T_∞	ambient temperature, K
t	time, s
t_e	drop evaporation time, s
Δt_{hu}	duration of heat-up period, s
Δt_{st}	duration of steady-state period, s
U	relative velocity between fuel drop and surrounding gas, m/s
u'	fluctuating component of turbulence velocity, m/s
Y_A	mass fraction of air
Y_F	mass fraction of fuel vapor
Y_i	mass fraction of i
λ	evaporation constant, m^2/s
λ'	evaporation constant for forced convection, m^2/s
λ_{eff}	effective average value of λ during drop lifetime, m^2/s
μ	dynamic viscosity, kg/m s
ρ	density, kg/m^3

Subscripts

A	air
F	fuel
V	vapor
g	gas
0	initial value
r	reference value
s	value at drop surface
st	steady-state value
hu	mean or effective value during heat-up period
∞	ambient value

REFERENCES

1. Wood, B. J., Wise, H., and Inami, S. H., Heterogeneous Combustion of Multicomponent Fuels, NASA TN D-206, 1959.
2. Godsave, G. A. E., Studies of the Combustion of Drops in a Fuel Spray—the Burning of Single Drops of Fuel, *Fourth Symposium (International) on Combustion,* Williams & Wilkins, Baltimore, 1953, pp. 818–830.
3. Spalding, D. B., The Combustion of Liquid Fuels, *Fourth Symposium (International) on Combustion,* Williams & Wilkins, Baltimore, 1953, pp. 847–864.

4. Faeth, G. M., Current Status of Droplet and Liquid Combustion, *Prog. Energy Combust. Sci.,* Vol. 3, 1977, pp. 191–224.

5. Goldsmith, M., and Penner, S. S., On the Burning of Single Drops of Fuel in an Oxidizing Atmosphere, *Jet Propul.,* Vol. 24, 1954, pp. 245–251.

6. Kanury, A. M., *Introduction to Combustion Phenomena,* Gordon & Breach, New York, 1975.

7. Spalding, D. B., *Some Fundamentals of Combustion,* Academic Press, New York; Butterworths Scientific Publications, London, 1955.

8. Williams, A., Fundamentals of Oil Combustion, *Prog. Energy Combust. Sci.,* Vol. 2, 1976, pp. 167–179.

9. Hubbard, G. L., Denny, V. E., and Mills, A. F., *Int. J. Heat Mass Transfer,* Vol. 16, 1973, pp. 1003–1008.

10. Sparrow, E. M., and Gregg, J. L., *Trans ASME,* Vol. 80, 1958, pp. 879–886.

11. Vargaftik, N. B., *Tables on the Thermophysical Properties of Liquids and Gases,* Halsted Press, New York, 1975.

12. Touloukian, Y., *Thermal-Physical Properties of Matter,* Plenum Press, New York, 1970.

13. Watson, K. M., Prediction of Critical Temperatures and Heats of Vaporization, *Ind. Eng. Chem.,* Vol. 23, No. 4, 1931, pp. 360–364.

14. Chin, J. S., and Lefebvre, A. H., Steady-State Evaporation Characteristics of Hydrocarbon Fuel Drops, *AIAA J.,* Vol. 21, No. 10, 1983, pp. 1437–1443.

15. Hall, A. R., and Diederichsen, J., An Experimental Study of the Burning of Single Drops of Fuel in Air at Pressures up to 20 Atmospheres, *Fourth Symposium (International) on Combustion,* Williams & Wilkins, Baltimore, 1953, pp. 837–846.

16. Chin, J. S., and Lefebvre, A. H., The Role of the Heat-up Period in Fuel Drop Evaporation, *Int. J. Turbo Jet Engines,* Vol. 2, 1985, pp. 315–325.

17. Probert, R. P., *Philos. Mag.,* Vol. 37, 1946, p. 94.

18. Frossling, N., On the Evaporation of Falling Droplets, *Gerlands Beitr. Geophys.,* Vol. 52, 1938, pp. 170–216.

19. Ranz, W. E., and Marshall, W. R., Evaporation from Drops, *Chem. Eng. Prog.,* Vol. 48, 1952, Part I, pp. 141–146; Part II, pp. 173–180.

20. El Wakil, M. M., Uyehara, O. A., and Myers, P. S., A Theoretical Investigation of the Heating-Up Period of Injected Fuel Drops Vaporizing in Air, NACA TN 3179, 1954.

21. El Wakil, M. M., Priem, R. J., Brikowski, H. J., Myers, P. S., and Uyehara, O. A., Experimental and Calculated Temperature and Mass Histories of Vaporizing Fuel Drops, NACA TN 3490, 1956.

22. Priem, R. J., Borman, G. L., El Wakil, M. M., Uyehara, O. A., and Myers, P. S., Experimental and Calculated Histories of Vaporizing Fuel Drops, NACA TN 3988, 1957.

23. Chin, J. S., and Lefebvre A. H., Effective Values of Evaporation Constant for Hydrocarbon Fuel Drops, *Proceedings of the 20th Automotive Technology Development Contractor Coordination Meeting,* 1982, pp. 325–331.

24. Chin, J. S., Durrett, R., and Lefebvre, A. H., The Interdependence of Spray Characteristics and Evaporation History of Fuel Sprays, *ASME J. Eng. Gas Turbines Power,* Vol. 106, July 1984, pp. 639–644.

25. Aldred, J. W., Patel, J. C., and Williams, A., The Mechanism of Combustion of Droplets and Spheres of Liquid *n*-Heptane, *Combust. Flame,* Vol. 17, 1971, pp. 139–149.

26. Faeth, G. M., Evaporation and Combustion of Sprays, *Prog. Energy Combust. Sci.,* Vol. 9, 1983, pp. 1–76.

27. Faeth, G. M., Spray Combustion Models—A Review, AIAA Paper No. 79-0293, 1979.

28. Law, C. K., Multicomponent Droplet Combustion with Rapid Internal Mixing, *Combust. Flame,* Vol. 26, 1976, pp. 219–233.

29. Law, C. K., Unsteady Droplet Combustion with Droplet Heating, *Combust. Flame,* Vol. 26, 1976, pp. 17–22.

30. Law, C. K., Recent Advances in Multicomponent and Propellant Droplet Vaporization and Combustion, ASME Paper 86-WA/HT-14, presented at Winter Annual Meeting, Anaheim, Calif., December 1986.

31. Law, C. K., Recent Advances in Droplet Vaporization and Combustion, *Prog. Energy Combust. Sci.*, Vol. 8, 1982, pp. 171–201.
32. Law, C. K., Prakash, S., and Sirignano, W. A., Theory of Convective, Transient, Multi-component Droplet Vaporization, *Sixteenth Symposium (International) on Combustion*, The Combustion Institute, 1977, pp. 605–617.
33. Sirignano, W. A., and Law, C. K., Transient Heating and Liquid Phase Mass Diffusion in Droplet Vaporization, Adv. Chem. Ser. 166, *Evaporation-Combustion of Fuels*, J. T. Zung, ed., 1978, pp. 1–26.
34. Law, C. K., and Sirignano, W. A., Unsteady Droplet Combustion with Droplet Heating—II: Conduction Limit, *Combust. Flame*, Vol. 28, 1977, pp. 175–186.
35. Sirignano, W. A., Theory of Multicomponent Fuel Droplet Vaporization, *Arch. Thermodyn. Combust.*, Vol. 9, 1979, pp. 235–251.
36. Lara-Urbaneja, P., and Sirignano, W. A., Theory of Transient Multicomponent Droplet Vaporization in a Convective Field, *Eighteenth Symposium (International) on Combustion*, The Combustion Institute, 1981, pp. 1365–1374.
37. Sirignano, W. A., Fuel Droplet Vaporization and Spray Combustion, *Prog. Energy Combust. Sci.*, Vol. 9, No. 4, 1983, pp. 291–322.
38. Aggarwal, S. K., and Sirignano, W. A., Ignition of Fuel Sprays: Deterministic Calculations for Idealized Droplet Arrays, *Twentieth Symposium (International) on Combustion*, The Combustion Institute, 1984, pp. 1773–1780.
39. Tong, A. Y., and Sirignano, W. A., Multicomponent Droplet Vaporization in a High Temperature Gas, *Combust. Flame*, Vol. 66, 1986, pp. 221–235.
40. Tong, A. Y., and Sirignano, W. A., Multicomponent Transient Droplet Vaporization with Internal Circulation: Integral Equation Formulation and Approximate Solution, *Numer. Heat Transfer*, Vol. 10, 1986, pp. 253–278.

DROP SIZING METHODS

INTRODUCTION

The problem of measuring the size of very small particles has been encountered in many branches of engineering science, and many different methods have been employed with varying degrees of success. However, special difficulties arise in the application of these methods to the measurement of drop sizes in a spray and are well known to those who have attempted such measurements. These difficulties include (1) the very large number of drops in a spray, (2) the high and varying velocity of the drops, (3) the wide range of drop sizes encountered in most practical sprays (the ratio of the largest to the smallest drop diameter often exceeds 100:1), and (4) the changes in drop size with time through evaporation and coalescence. All these factors must be considered when choosing a drop size measuring technique for any given application.

Drop sizing techniques for sprays should ideally have the following characteristics.

1. Create no disturbance to the atomization process or spray pattern. Caution is required when using intrusive devices to measure drop sizes in a flowing airstream since small drops generally follow streamlines whereas large drops tend to migrate across streamlines. This phenomenon can cause a size bias in the measurements.
2. Have a size range capability of at least 10, although a factor closer to 30 is desirable in some applications. Generally, probe methods such as the hot-wire technique have a size range from 1 to 600 μm and imaging techniques a size range above about 5 μm.

3. Have the ability to measure both spatial and temporal distributions. The latter is generally preferred, but many important optical techniques provide only spatial distributions. Conversion from one type of distribution to the other is possible if drop velocity data are also available.
4. Where appropriate, provide large representative samples. To achieve a reasonably accurate measurement of size distribution, the sample should contain at least 5000 drops.
5. Tolerate wide variations in the liquid and ambient gas properties. Many combustion systems operate at elevated pressures and temperatures. Successful development of mathematical models for fuel evaporation and combustion at these arduous conditions is heavily dependent on accurate and reliable spray diagnostics.
6. Provide rapid means of sampling and counting. Automated data acquisition and processing is needed to accumulate results within seconds. When the number density (drops/cm^3) is required, the system must be fast enough to measure and record every drop passing through the probe volume. The data management system should be able to reproduce promptly the drop size distribution in histogram form along with the various mean diameters and standard deviations.

As it is virtually impossible to fulfill all these criteria in a single instrument, the capabilities and limitations of the various available measuring techniques must be recognized.

Drop Sizing Methods

The various methods employed in drop size measurement may be grouped conveniently into three broad categories—mechanical, electrical, and optical—as outlined below. The third category includes optical systems that have been developed in recent years and are finding an increasing range of applications. Some of these instruments have the capability of measuring drop velocity and number density as well as size distribution.

1. Mechanical
    a. Drop collection on slides or in cells
    b. Molten-wax and frozen-drop techniques
    c. Cascade impactors
2. Electrical
    a. Charged-wire and hot-wire techniques
3. Optical
    a. Imaging—photography, holography
    b. Nonimaging—single-particle counters, light scattering, Malvern particle analyzer

The growing importance of spray diagnostics, especially in the combustion area, has prompted several instrument manufacturers to intensify their research and development activities to satisfy this expanding market. In this chapter no attempt is made to encompass all the techniques of relevance or potential application to the measurement of size distributions in sprays. Instead, attention is focused on three main categories: methods that have been widely used in the past, those still in widespread use, and methods that appear to have exceptional promise for enhanced diagnostic capabilities in the future.

Excellent reviews of drop size measuring techniques have been provided by Chigier [1,2], Bachalo [3], Ferrenberg [4], Hirleman [5], and Jones [6].

FACTORS INFLUENCING DROP SIZE MEASUREMENT

All measuring techniques, whether simple or sophisticated, are susceptible to various errors and ambiguities, the nature and importance of which depend on the particular method employed. However, a number of potential sources of error in drop size measurement are common to almost all methods. They include the method of sampling (spatial or temporal), sample size, drop saturation, drop evaporation, drop coalescence, and sampling location. In the design of instruments for drop size measurement every effort is made to minimize these types of errors, but the fact remains that all techniques are subject to errors and inaccuracies associated with the basic assumptions employed, their manner of use, and the quantity of data obtained.

Spatial and Temporal Sampling

Two types of sampling are employed to determine drop size distribution. One is *spatial sampling,* which describes the observation or measurement of drops contained within a volume during such short intervals of time that the contents of the volume do not change during any single observation. Examples of spatial sampling are single-flash, high-speed photography and laser holography. Any sum of such photographs would also constitute spatial sampling. The second type is *temporal sampling,* which describes the observation or measurement of drops that pass through a fixed area during a specific time interval, with each drop individually counted. Temporal distributions are generally produced by collection techniques and by optical instruments which are capable of sensing individual drops. Under certain circumstances, samples may be obtained that are neither spatial nor temporal, an example being the collection of drops on a slide that is moving, so that drops settle on the slide in addition to the drops collected by the motion of the slide through the swept volume.

If all the drops in a spray travel at the same velocity, the results obtained by spatial and temporal sampling are identical. If not, the spatial drop size distribution may be converted into the temporal distribution by multiplying the number of drops of a given velocity by that velocity.

With pressure atomizers, the smaller drops in the spray usually decelerate more rapidly than the larger drops, which leads to a high concentration of small drops just downstream of the atomizer. Thus, spatial sampling in this region yields mean diameters smaller than those indicated by temporal sampling [7,8]. Farther downstream, where all the drops are moving at the same velocity, the results obtained by both methods should be the same.

The relative merits of spatial and temporal sampling are often debated. It is sometimes argued that spatial sampling is incorrect because it gives results that do not correspond to the spectrum of drop sizes actually discharged by the atomizer during a given time interval. However, as Tate [9] has pointed out, neither method is basically inferior to the other. In practice, which method is best depends on the application. For example, when herbicides are sprayed on the ground, temporal sampling would be the most informative. However, spatial sampling might be advantageous in combustion applications, where ignition and burning rates are dependent on the instantaneous droplet population within a given volume or zone [9].

Sample Size

Although a spray contains a far larger proportion of small drops than large ones, it is the relatively few large drops that predominate in determining the average drop diameter of the spray. Thus if a sample of drops is to be truly representative of the spray as a whole, it is vitally important that the large drops be included. As Lewis et al. [10] have pointed out, the presence or absence of one large drop in a sample of 1000 drops may affect the average diameter of the sample by as much as 100%.

To achieve a reasonably accurate estimate of spray quality, it is necessary to measure about 5500 drops. The accuracy of the mean diameter obtained for various sample sizes, as estimated by Bowen and Davies [11] for 95% confidence limits, is shown in Table 9.1.

Drop Saturation

This occurs when the drop flux or population exceeds the capability of the sizing instrument or method. The problem is most evident when drops are collected on coated slides or in immiscible solvents. If the sample is too large, the probability of error due to overlapping of drops (or their impressions) is high.

Table 9.1 Influence of sample size on accuracy of drop size measurements [11]

No. drops in sample	Accuracy (%)
500	±17
1,500	±10
5,500	±5
35,000	±2

Saturation is also encountered in optical systems [9]. Certain instruments are designed to sense one particle at a time. The presence of more than one particle in the optical path will be recorded as a single drop of large apparent diameter unless the instrument is capable of recognizing and rejecting coincident particles. Saturation is also a problem when dense sprays are studied using high-speed photographic or video imaging techniques.

Drop Evaporation

As the lifetime of small drops is extremely short, evaporation effect are very important in the measurement of fine sprays. Whether evaporation leads to an increase or decrease in mean drop size depends on the initial drop size distribution. For a monodisperse spray, evaporation always reduces the mean drop size, but if the spray contains a wide range of drop sizes initially, then evaporation may produce an increase in mass median diameter (MMD). The effects of evaporation on mean drop size have been discussed by Chin et al. [12]. The changes in MMD with evaporation time for different initial values of the Rosin-Rammler size distribution parameter q are shown in Fig. 9.1 for a kerosine (JP 5) fuel spray evaporating in air at a temperature of 2000 K and pressure of 2 MPa. It is apparent in this figure that MMD generally increases with evaporation time, the change of MMD being more significant for sprays having smaller q_o values. For values of q_o higher than 4, Fig. 9.1 indicates that MMD declines as evaporation proceeds.

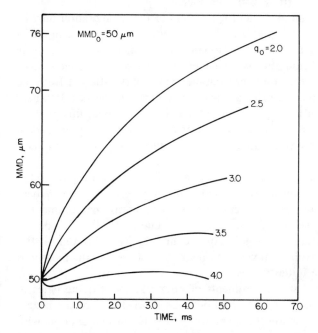

Figure 9.1 Change of mass median diameter with evaporation time for different values of q_0. Fuel = JP 5, P_A = 2000 kPa, T_A = 2000 K *(Chin et al. [12]).*

Drop Coalescence

The collision of drops sometimes results in drop coalescence. Whether or not the collision of two drops leads to coalescence depends on the diameter of the drops, their relative velocity, and the collision angle [13]. It also depends on the number density of the spray and the time available for collisions. Thus coalescence is most likely to occur in dense sprays when sampled at large distances from the atomizer.

Sampling Location

With some measuring techniques it is possible to sample the entire spray or, at least, a representative portion of it. In other methods the data are drawn from a very small volume, typically around 1 mm^3. With such methods the acquisition of sufficient data to characterize the total spray could prove tedious and time-consuming.

Most pressure-swirl atomizers produce a hollow-cone spray in which the largest drops, due to their higher inertia, are located at the outer periphery. In fan sprays also, the largest drops are usually found at the outer edges of the spray pattern. For such sprays local measurements of drop size distribution could clearly be misleading, but it should also be noted that even instruments that measure a line-of-sight average through the complete spray (for example, those that sample drops contained within a volume defined by the intersection of a laser beam and the spray) are not immune to errors from this source.

For low ambient temperatures, where the effects of fuel evaporation may be regarded as negligibly small, Chin et al. [8] have shown that when a liquid is injected from a pressure atomizer into a coflowing airstream, the values of mean drop size and drop size distribution, as measured by light-scattering techniques in the region downstream of the atomizer, may be appreciably different from the true values that would be measured in a plane fairly close to the nozzle where the atomization process is just complete. Unfortunately, in practice, this plane is very difficult to determine with any degree of precision, since it is very dependent on nozzle design features as well as liquid flow rate, injection pressure, and ambient air properties.

Generally it is found that a more uniform spray (higher q) or a higher initial MMD will exhibit less variation of MMD with changing axial location. Where drop acceleration or deceleration is present, the effect on spray measurement is to increase or decrease, respectively, the measured value of MMD. Far downstream of the nozzle, the distance depending on the air velocity and initial MMD, all the drops attain a sensibly uniform velocity, regardless of size, so the measured MMD again approaches the initial value.

These findings suggest that measurements of nozzle spray characteristics in a flowing airstream should ideally be carried out well downstream of the nozzle (200 to 300 mm), using air at low temperature (<300 K) in conjunction with liquids of low volatility.

The problems outlined above with pressure atomizers are far less pronounced with airblast atomizers, partly because of the turbulent mixing of droplets and air but mainly because the airborne drops are transported radially by the flowing air to an extent that prohibits the formation of steep drop size gradients. However, errors arising from the combined effects of coalescence and evaporation still apply to this type of atomizer.

In general, the spray patterns produced by most practical atomizers are so complex that fairly precise measurements of drop size distributions can be obtained only if accurate and reliable instrumentation and data reduction procedures are combined with a sound appreciation of their useful limits of application. As emphasized by Tate [9], the ability to recognize and challenge questionable data is especially important. Thus, experience and judgment are valuable attributes for those who measure drop sizes and use the information in the design and application of atomizing systems.

MECHANICAL METHODS

These usually involve the capture of a sample of the spray on a solid surface or in a cell containing a special liquid. The drops are then observed, or photographed, with the aid of a microscope. The method is fairly simple and has many variations.

Collection of Drops on Slides

With this technique a solid surface, usually a glass slide, is covered with a suitable coating, which must be of very fine grain structure to make distinct the impressions created by very small drops [9,14]. A very fine surface may be obtained by burning a kerosine-soaked wick under the glass to produce a thin coating of soot, or by burning a magnesium ribbon to deposit a thin layer of magnesium oxide. With such coatings, drop sizes down to 3 μm in diameter can be observed and measured [15].

After the slide is exposed to the spray, the sizes of the drop impressions are measured, sometimes by means of a microscope fitted with a traversing scale, but more usually with a Quantimet image analyzer computer. These sizes are then converted to actual drop sizes using a correction factor derived by May [15].

More recently, Elkotb et al. [16] used soot-coated slides to measure both the radial and overall drop size distributions in a spray. The sampling apparatus consists of a small-diameter cylinder with 10 holes of 5 mm diameter at different radial distances; the soot-coated glass slide is mounted on an axle fitted inside the cylinder and can be rotated around the cylinder axis by a wheel. The apparatus is placed a distance of 60 mm from the atomizer.

Sprays injected from the atomizer and passing through the holes on the rim of the apparatus leave impressions on the soot layer deposited on the glass slide. The impressions on the slide are photographed under a microscope with a total magnification of 350. Each group of impressions represents part of the spray at a radial distance equal to that of the hole responsible for these impressions. The diameters of the impressions are measured with a scale graduated down to 0.5 μm, and the drops that have the same diameter are counted and classified into size intervals of 10-μm width. The number and diameters of the impressions obtained for each radius of the spray define the spatial and total size distributions.

One problem associated with slides is that of determining what fraction of the slide area should be covered by drops. If too many drops are collected, the probability of error due to overlap is high and drop counting becomes tedious. On the other hand, if too few drops are collected, the sample may not be representative of the spray. From the standpoint of ease of measurement and counting, 0.2% coverage is satisfactory, but a coverage as high as 1.0% can be tolerated before the problem of drop overlap becomes significant.

Other important considerations are drop evaporation and collection efficiency. The lifetime of small drops is extremely short. For example, a water drop that is 10 μm in diameter has a lifetime of about 1 s in a 90% relative humidity atmosphere [17]. Thus, evaporation effects are very significant in the measurement of fine sprays. Collection efficiency is especially important with airblast atomizers, owing to the flow field created around the collecting surface. The large drops have enough inertia to hit the surface, but the small drops tend to follow the streamlines. For these reasons, drop size data for airblast atomizers obtained by direct drop collection tend to indicate larger than actual sizes.

Yet another problem that arises with the collection of drops on coated slides is the determination of the correction factor by which the diameter of the flattened drop must be multiplied to obtain the original diameter of the spherical volume. Its value depends on liquid properties, notably surface tension, and on the nature of the coating employed. For example, with oil drops the correction factor is around 0.5 for a clean glass surface and 0.86 for a magnesium-coated slide [18].

Most of the work using the impression method was performed more than 20 years ago, and the technique has now been largely supplanted by photographic and optical methods.

Collection of Drops in Cells

An improvement on the coated-slide technique is one in which the drops are caught on a target where they are held suspended while they are counted and measured. Hausser and Strobl [19] were among the first investigators to use as the target microscope slides coated with a special liquid in which the drop would not dissolve but would remain stable and suspended. An alternative approach is to collect the drops in a cell containing a suitable immersion liquid. This method has three advantages over collection on a slide: (1) the drops remain almost perfectly spherical, provided the density of the immersion liquid is only slightly less than that of the sprayed liquid, (2) evaporation is prevented, and (3) provided no splitting of the drop occurs on hitting the immersion liquid, the true sizes of the drops are obtained and can be measured directly. The method cannot be applied satisfactorily to coarse sprays due to the risk of disintegration of the largest drops on impact with the immersion liquid. The problem of drop breakup can be alleviated to some extent by choosing immersion liquids of low viscosity and surface tension. Such liquids facilitate the penetration of the liquid surface by the drops. The drops then settle to the bottom of the cell, where they remain suspended and stable for an appreciable time, provided, of course, the immersion liquid is absolutely immiscible with the sprayed liquid.

Another problem with the immersion sampling method is that of drop coalescence. According to Karasawa and Kurabayashi [20], this phenomenon is often overlooked because the occurrence of coalescence is difficult to detect. The remedy is the same as that for drop breakup mentioned above, namely to use an immersion liquid of low viscosity [20]. Rupe [21] used this method in his investigation of sprays. The liquids employed were water and water-alcohol mixtures, and the immersion liquids were Stoddard solvent and white kerosine. The Delavan Corporation in the United States has also used this technique extensively to accumulate considerable data on drop size distributions in sprays [22]. The experimental procedures employed follow closely those of Rupe. Dyed water is sprayed into a cell filled with Stoddard solvent. The cell is located beneath an air-operated shutter, whose speed is adjusted to obtain a representative number of drops. After settling at the bottom of the cell, the drops are photographed at high magnification (usually 50 ×). The magnified drop images are then processed and the final computer printouts show the number of sampled drops and the spray volume associated with each size class. The computer also determines mean and median drop diameters as well as parameters indicating the uniformity of the distribution [22].

Molten Wax Technique

Joyce [23] developed this technique to a high state of perfection during the period 1940 to 1946. The basic idea is that paraffin wax, when heated to a suitable temperature above its melting point, has physical properties close to those of aviation kerosine (density, 780 kg/m^3; surface tension, 0.027 kg/s^2; kinematic viscosity, 1.5×10^{-6} m^2/s). The molten wax is injected into the atmosphere of a large pressure vessel, where the drops rapidly cool and solidify. The sample is then subjected to a sieving operation in which the wax drops are separated into size groups. Each size group is weighed to obtain the volume (or mass) fraction in each size range. Thus the cumulative volume distribution and mass median diameter are measured directly without the large expenditure of time and personnel associated with the sizing and counting of multitudinous individual drops. Furthermore, the number of drops in a sample runs into millions, so clearly the technique does not suffer from a sample size that is too small to be statistically accurate.

One serious disadvantage of the hot wax technique is the limited choice of materials that can be used conveniently. If the properties of any given liquid are different from those of the simulant, it is necessary to establish the effects of these properties (notably surface tension and viscosity) on the atomization process. Other drawbacks of the method are the practical problems associated with preheating the wax and the errors incurred because of changes in the physical properties of the wax droplet as it rapidly cools after leaving the atomizer, so that the processes of formation and secondary recombination may not be accurately reproduced. For this reason the air near the nozzle, i.e., in the region where the key atomization processes are taking place, should be heated to the same temperature as the molten wax.

Drop Freezing Techniques

A natural extension of the molten wax technique is to solidify the drops by freezing as soon as they emerge from the nozzle. In an early investigation, Longwell [24] developed a technique for collecting the liquid drops from a fuel spray into a stream of flowing fluid at room temperature. The fluid carrying the drops was then fed into an alcohol bath kept at approximately the temperature of dry ice, which was cold enough to freeze the drops into solid spheres. The drops were then sieved while still frozen to separate them into different size groups.

In the early 1950s Taylor and Harmon [25] froze water drops by collecting them in a hexane pan, packed around with dry ice and containing hexane at $-20°C$. Their method was dependent on the time taken for the drops to fall through hexane. The drops froze during their descent through the fluid and came to rest on a shutter. When the sample had all settled, the shutter was opened and the time for the drops to fall another 30 cm was measured. After traveling the 30 cm, the drops landed on a scale pan. The variation of scale pan weight with time could then be converted to equivalent drop diameters, the time taken for the fall being a function of the density difference between the drops and the supporting liquid, the liquid viscosity, the depth of fall, and the drop diameter squared. The process was, however, a very lengthy one, with 5-μm drops taking up to 27 h to fall, 100-μm drops taking 4 min.

In 1957 Choudhury et al. [26] described a method for freezing the entire spray in a bath of liquid nitrogen. The atomizer was mounted with its discharge nozzle pointing downward at a height of approximately 0.4 m above the liquid-nitrogen surface. After sufficient drops had been collected, the liquid nitrogen was decanted off and the frozen drops were passed through a series of screens ranging in size from 53 to 5660 μm. It was found that to avoid any agglomeration of drops on the liquid-nitrogen surface, the density of the liquid being atomized should exceed 1200 kg/m^3. Thus, the method is unsuitable for kerosine, fuel oils, and water, which clearly represents a serious disadvantage—especially coupled with a minimum measurable drop size of 53 μm.

Various refinements to the nitrogen-freezing technique have been made by Nelson and Stevens [27], Street and Danaford [28], Rao [18], and Kurabayashi et al. [29]. Rao's method employs a specially designed isokinetic probe that is mounted with its inlet facing the atomizer. Gaseous or liquid nitrogen is conveyed to the tip of the probe and injected into the incoming spray through a narrow annular slot. The nitrogen, at a temperature of around 140 K, rapidly freezes the drops, which are then collected in a nitrogen-cooled Perspex pot and photographed through a microscope. The final prints, showing the drops enlarged by a factor of 50 or more, are subsequently analyzed to determine the number and sizes of the drops in the sample. The method is simple, elegant, and convenient, but it still involves a correction factor to account for the change in drop size that occurs during freezing. A typical frozen drops photograph is shown in Fig. 9.2.

Cascade Impactors

This method is based on the principle that a large drop moving at high velocity will, because of its momentum, hit an obstacle placed in its path, whereas small

Figure 9.2 Typical photograph of frozen drops.

drops traveling below certain velocities will follow the airflow around the obstacle.

A single-stage impactor is shown in Fig. 9.3. A sample of the spray under investigation is drawn through a nozzle, where it encounters a slide. The slide is usually coated with a mixture of carbon and magnesium oxide to retain the drops. The large drops hit the slide, whereas the smaller drops follow the airstream around the slide. As the velocity of the jet containing the drops is increased by decreasing the area of the nozzle through which the stream flows, smaller drops will hit the slide by virtue of their increased momentum. May [30] has designed a four-stage cascade impactor in which the four jets are made progressively smaller, so that the speed and therefore the efficiency of impaction of drops increases from slide to slide when air is drawn through at a steady rate. Deposition of liquid on the walls of the first orifice becomes a problem for drop sizes larger than 50 μm, so the greatest efficiency of sampling is achieved in the size range from 1.5 to 50 μm in diameter. To improve the sampling of larger drops, the dimensions of the impactor should be increased.

One advantage of the cascade impactor is that isokinetic sampling can be used to avoid discrimination against either very large or very small drops in the spray. Another advantage is that, after calibration, the quantity of liquid associated with each drop size range, collected in each stage of the cascade impactor, can readily be assessed by gravimetric or chemical means. In theory, any number of stages may be used to provide the desired drop size groupings.

The cascade impactor clearly lacks the sophistication of modern optical methods, but it can be made mechanically strong to withstand the rigors of routine handling. Moreover, it can operate satisfactorily under quite arduous conditions; for example, it has been used in flight tests to measure drop size distributions in clouds [14].

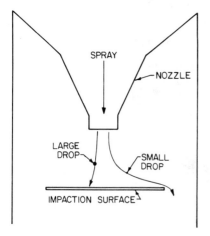

Figure 9.3 Schematic diagram illustrating principle of cascade impactor.

ELECTRICAL METHODS

Electrical methods generally rely on the detection and analysis of electronic pulses produced by drops for calculating size distributions.

Wicks-Dukler Technique

The method proposed by Wicks and Dukler [31] is based on the counting of pulses produced by drops momentarily bridging the gap between two sharp needles across which there is a potential difference. By adjusting the width of this gap, frequency counts can be obtained in terms of gap size and then converted into drop size distributions.

Some of the problems that can arise with this type of instrument have been described by Jones [6]. It has been found that the apparent resistance of a drop bridging the gap is a function of both drop velocity and the depth of immersion of the needles into the drop. Unless a very sensitive detection and counting system is employed, a large proportion of the pulses will be below the threshold that has to be set to avoid counting pulses due to noise. This could lead to large errors in drop size distribution [6]. Furthermore, changes in drop velocities will have the same effect on count rate as changes in drop sizes, so that apparent changes in size distribution could occur when, in fact, only drop velocities have changed. Despite these limitations, the system might find useful application for certain types of spray, bearing in mind its ease of operation. Comparison with a proven drop sizing method would be necessary to establish its useful operating range and to identify and quantify the principal sources of error.

Charged Wire Technique

This technique operates on the principle that when a drop impinges on an electrically charged wire it removes an amount of charge that depends on its size. This allows drop sizes to be obtained by converting the charge transfer into a measurable voltage pulse.

According to Gardiner [32], the limitations of the charged wire probe appear to depend on a combination of the electrical conductivity of the liquid and the flux of drops. Liquids of low conductivity produce pulses of long duration, which increases the probability of counting errors due to successive impingements being superimposed on each other. This problem appears to limit the useful application of this technique to dilute sprays of high-conductivity liquids.

Hot Wire Technique

When a liquid drop becomes attached to a heated wire it causes local cooling of the wire as it evaporates. This phenomenon can be used to obtain the sizes and concentrations of liquid drops present in a gas stream. Essentially, the device employed is a constant-temperature hot wire anemometer. When no drops are present, the electrical resistance of the wire is high and sensibly uniform along its length. When a drop attaches to the wire, local cooling by the drop reduces the resistance in proportion to the drop size. This reduction in resistance is manifested as a voltage drop across the wire supports. The constant-current electrical energy supplied to the wire subsequently evaporates the drop, leaving the device ready to receive another drop. The entire process takes place within a time of 2 ms, depending on the drop size [33].

Mahler and Magnus [33] have described the operating principles and calibration procedures of a commercial hot wire instrument in some detail. Under the right conditions it can yield accurate measurements of Sauter mean diameter, mass mean diameter, volume flow rate, and liquid concentration. Although it is an intrusive technique, the platinum wire employed is only 5 μm in diameter and 1 mm long and causes very little disturbance to the flow. However, when measuring large drops the device can operate only at flow velocities below around 10 m/s because of drop shattering on the wire [3]. Moreover, the technique cannot be used for liquids that leave a residue on the wire, as this affects the calibration.

For a more complete account of the hot wire technique and other electrical methods for drop size measurements, reference should be made to the review by Jones [6].

OPTICAL METHODS

Optical methods can be broadly divided into *imaging* and *nonimaging* types. The former, including flash photography and holography, are limited in practice to drop sizes larger than 5 μm. Imaging methods have the advantage of allowing the drops to be "seen" as they exist at the point and time where knowledge of their size is required. Another advantage is that errors that might arise from coalescence or evaporation of drops after sampling are eliminated [14].

Nonimaging methods can be subdivided into two classes—those that count and size individual drops, one at a time, and those that measure a large number of drops simultaneously. For accurate results it is important to know both drop size and velocity, and some nonimaging instruments can provide both sets of information.

A variety of optical methods have been employed for spray analysis. Each method has its own advantages and limitations, but they all have the important attribute of allowing size measurements to be made without the insertion of a physical probe into the spray.

High-Speed Photography

At present, photography is one of the most accurate and least expensive techniques for the measurement of drop sizes and velocities in sprays. Usually it involves taking a photograph with a light pulse of sufficient intensity and sufficiently short duration to produce a sharp image and then counting and sizing the images on the processed film. Much of the development work in the application of photography to spray analysis was carried out by Dombrowski and co-workers [34–36] and by Chigier and co-workers at Sheffield University [37–40].

Mercury vapor lamps, electrical sparks, flash lights, and laser pulses are generally used to create a high-intensity light source of short duration. Flash light sources have time durations of the order of 1 μs, while lasers have pulse durations of the order of nanoseconds.

According to Jones [6], photography is probably the only technique with the potential to provide drop size information from the dense, fast-moving sprays of interest in power generation. However, the method is not without problems, of which the most difficult is analysis of the photographic images, which always entails at least some degree of human involvement. Manual sizing is tedious and time-consuming and invariably subject to operator fatigue and bias. Automatic image analyzers, such as the Quantimet, also call for considerable judgment on the part of the operator, who must ultimately decide which drops are in focus. With high-density sprays, the drop images may become very closely spaced on the negative, or even overlapping, which detracts appreciably from the accuracy of automatic image analysis. Increase in magnification at the imaging stage tends to alleviate this problem and also reduces the minimum drop size that the Quantimet can reliably measure. However, these improvements can be realized only by forfeiting depth of field [6].

A calibration procedure for determining the effective depth of field has been described by Chigier [40]. Particles of known size, deposited on a glass slide, are traversed across the region of focus of the camera and photographs are taken at regular intervals. Particle images are determined by measuring each particular diameter at two different light intensity levels. The average of these measurements is related to the particle diameter, and the difference between the diameters is a measure of the thickness of the blurred "halo" at the edge of the particle. By this means the effective depth of field is determined for each particular optical system and particle size.

High-speed photography can also be used to obtain information on drop velocities. If two light pulses are generated in rapid succession, a double image is obtained of a single drop on the photographic plate, from which the velocity of the drop can be determined by measuring the distance traveled by the drop and dividing it by the time interval between the two pulses. The direction of movement of the drop can also be directly determined from the photograph as an angle of flight with respect to the central axis of the spray. The method provides instantaneous measurements for individual drops, and from a series of such measurements time-average and space-average quantities can be determined as well as standard deviations. Measurement techniques based on this principle have been used successfully by many workers including De Corso and Kemeny [41], Mellor et al. [38], and Chigier [40].

High-magnification double-pulse laser photography has the capability for measuring simultaneously both drop sizes and velocities. It cannot normally be used in regions close to the atomizer where drops are traveling at speeds in excess of 100 m/s, but particles down to 5 μm in diameter have been photographed when moving at speeds of 30 m/s [1].

TV Image Scanning Spray Analyzer

The data-processing problems associated with photographic methods are alleviated to a great extent by the TV analyzer developed by the Parker-Hannifin Corporation [42]. This instrument has been used to study in detail the spray characteristics of large numbers of pressure-swirl and airblast atomizers. It uses a 0.5-μs flash to photograph the drops contained within a small frame of dimensions 1.5 × 2.0 mm, approximately 1 mm in depth. The resolution of the system is within 4 μm. The rate of photography is 15 frames/s. A complete test takes about 20 min and encompasses more than 14,000 drops. The instrument is considered precise in the SMD range from 80 to 200 μm. The term "precise" in this context is intended to imply a repeatability of ±6%.

Holography

Although holographic techniques have much in common with high-speed photography, they have the advantage of allowing a sizable region to be captured rather than the limited depth of field afforded by photographic methods. The application of holographic techniques to spray systems has been described by Chigier [2], Jones [6], MacLoughlin and Walsh [43], Murakami and Ishikawa [44], and Thompson [45].

With this method a sample volume of moving drops is illuminated with a coherent beam of light in the form of a short pulse. The measurement volume is a cylinder with length equal to the total width of the spray at the particular axial station and with diameter equal to that of the laser beam. As the duration of the laser pulse is extremely short (20 ns), the drops contained within the measurement volume are effectively "frozen." The resulting hologram provides a complete three-dimensional image of the spray in which drops as small as 15 μm are clearly visible. The hologram can then be illuminated with a coherent beam of light to produce a stationary image of all the drops at their correct relative locations in space. Thus, the holographic method is essentially a two-step imaging process that captures in permanent form the size and location of a moving system of drops and then produces a stationary three-dimensional image of all the drops contained within the sample volume.

One advantage of the holographic technique is that it produces a recorded image that can be studied later at leisure. Another asset is that, in principle, it requires no calibration. The accuracy of the method is fundamentally set by the wavelength of light; hence, in theory, about 2-μm resolution can be obtained. The main drawback of the method is that its application is limited to dilute sprays [6]. A typical laser holographic system is illustrated in Fig. 9.4.

A laser holographic system for the study of sprays has been developed at the

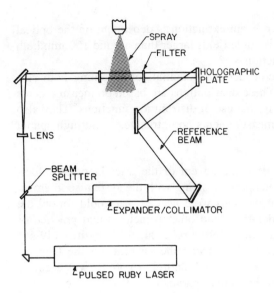

Figure 9.4 Schematic diagram of laser holographic system.

Marchwood Engineering Laboratories in England [6]. Holographic techniques are also being used for spray analysis at the United Technology Research Center in East Hartford, Connecticut [46].

At the Kyushu University in Fukuoka, Japan, Murakami and Ishikawa [44] have developed a technique employing two pulse lasers of different wavelengths so that two holograms are recorded on two different films separated by a suitable time interval. The displacements of moving particles (i.e., particle velocities) are measured on a superimposed picture of these two films.

Single-Particle Counters

As their name implies, single-particle counters measure individual particles or drops that pass through a focused laser beam. Because a single-particle counter can see only one drop at a time, the size of the *measurement* or *probe* volume sets the limitation on the drop number densities in which these counters will operate accurately. In practice, the measurement volume is controlled by the diameter of the focused beam, the f number of the receiver lens and its angle to the transmitted beam, and the aperture of the photodetector [3].

Although, in the past, single-particle counters were limited to fairly dilute sprays, the more recent use of large off-axis angles of light scatter detection has allowed these instruments to be used in relatively dense sprays [47]. They are now restricted to sprays with number densities similar to those to which the Fraunhofer diffraction method is restricted. It should be noted that single-particle counters are limited by high particle number densities and beam extinction, whereas diffraction instruments are limited by low number densities and by extinction in high number density environments. Since beam extinction is dependent on the optical path length and number density, in particle fields larger than around 100 mm both methods are limited by beam extinction.

Single-particle counters have considerable potential for obtaining both size and velocity distributions directly. These distributions are based on measurements of individual drops and do not require the use of distribution functions. They also have the capability to perform the measurements nonintrusively with high spatial resolution over a large size range.

Light-scattering interferometry. With the advent of the laser Doppler velocimeter (LDV), interest in simultaneous particle sizing with the system soon arose. This was due to the realization that particles were needed that could follow the flow to be measured. These considerations and other principles and practice of laser anemometry are outlined in the book by Durst et al. [48]. Farmer [49,50] investigated the possibility of particle sizing using the visibility of the Doppler burst signal. The analysis of Farmer, although not entirely rigorous, was accurate given the appropriate conditions on the optical parameters. Perhaps the greatest limitation of the technique was the requirement of using on-axis forward- or back-scattered light detection. This resulted in an excessively large sample volume and the detection of particles that passed outside the region of complete overlap of the two laser beams. In addition, the size range of the method was limited to a factor of 10 or less. Other efforts by Chigier and co-workers ([51], for example) sought to combine the measurement of the scattered light intensity with the LDV to obtain simultaneous particle size and velocity measurements. Because near-forward light scatter detection was used, this method was also limited to very dilute particle fields. It was not until Bachalo [47] derived the theoretical analysis for dual-beam light scattering at large off-axis angles that the method of light-scattering interferometry appeared to be a viable means for spray diagnostics.

Beginning with the approach of Farmer, the measurement of drop size and velocity is based on the observation of the light scattered by drops passing through the crossover region of two intersecting laser beams. In practice, a single laser beam is split into two coherent beams of equal intensity and parallel polarization, which are made to cross as illustrated in Fig. 9.5. Drops passing through the intersection of the two beams scatter light that produces information from which the drop velocity U is calculated as

$$U = \frac{\lambda f_D}{2 \sin(\theta/2)} \qquad (9.1)$$

where λ is the wavelength of laser light, f_D the Doppler frequency, and $\theta/2$ the laser beam intersection half-angle.

Size information is contained in the relative modulation (visibility) of the scattered signal. The term visibility may be understood by reference to Fig. 9.6. It was defined by Michelson as

$$\text{Visibility} = \frac{I_{max} - I_{min}}{I_{max} + I_{min}} \qquad (9.2)$$

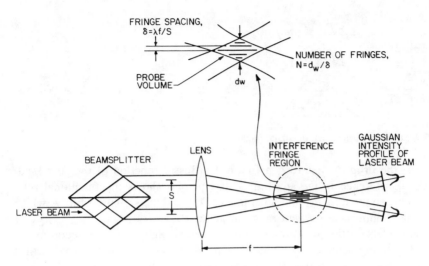

Figure 9.5 Drop sizing interferometer optics *(Mularz et al. [55])*.

where I_{max} and I_{min} are as shown in Fig. 9.6. The phenomena may be viewed in two ways. In Fig. 9.7 the fringe model is used to show how, as particles increases in size relative to the fringe spacing, they scatter light from more than one fringe at a time to reduce the signal modulation, or visibility. A more rigorous description considers the superposition of the light scattered from each beam separately which then interferes. In the near-forward direction, and for particles much larger than the light wavelength, this scattered light may be described by Fraunhofer diffraction theory or more accurately, for spherical particles of arbitrary size, by the Lorenz-Mie theory. The relative overlap of the forward-scattered light intensities then determines the relative visibility of the scattered interference fringe pattern. For example, larger particles scatter light that is more narrowly distributed in the forward direction, hence producing less intensity overlap for a given beam intersection angle. Based on the scalar diffraction theory and the derivation of Farmer, the relationship between drop size and visibility can be expressed as

$$\text{Visibility} = \frac{2J_1\,(\pi D/\delta)}{\pi D/\delta} \qquad (9.3)$$

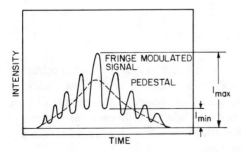

Figure 9.6 Doppler burst signal and pedestal components.

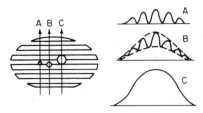

Figure 9.7 Effect of particle size on signal modulation.

where D is the drop diameter, J_1 a first-order Bessel function of the first kind, and δ the interference fringe spacing. This equation shows the functional relationship between the dimensionless parameter D/δ and the signal visibility, which allows a maximum measurement range of about 10:1.

In the implementation by Farmer using the on-axis light scatter detection mode, the visibility sizing method is suitable only for drops with diameters smaller than the fringe spacing. This forces the use of rather large beam diameters. Also, if the fringe spacing is increased to measure drop sizes around, say, 200 μm, the beam diameter must be approximately 10 times that diameter and the beam intersection angle must be made very small. For sprays containing a broad range of drop sizes, the excessively large sample volume will no longer leave an adequate probability that only one particle will reside in the sample volume at any one time, which leads to particle coincidence errors.

Large off-axis light scatter detection. The need to measure large drops on the order of 100 μm in diameter suggested a reevaluation of the light-scattering mechanism used. Bachalo [47] considered the light scattered by reflection and refraction as a more reliable means of sizing large spherical particles. These light-scattering components require the use of large off-axis detection, which, of course, is one of the advantageous sought for handling high number density environments. The sample volume is thus reduced in size by as much as two to three orders of magnitude. Bachalo's analysis showed how the phase shift of the light scattered by refraction or reflection could be related to the particle diameter. For example, for light scattered by refraction in the off-axis forward direction, the scattering angle θ and phase shift ϕ are given as

$$\theta = 2\tau - 2\tau' \tag{9.4}$$

$$\phi = \frac{2\pi D}{\lambda} (\sin \tau - m \sin \tau') \tag{9.5}$$

where τ is the incidence angle of a ray with a surface tangent, τ' is the refracted angle of the transmitted ray given by Snell's law, and m is the drop index of refraction. Note that τ and τ' are fixed by the receiver aperture, so the phase shift ϕ is directly proportional to the drop diameter D. With dual-beam light scattering, the phase shift manifests as an interference fringe pattern with a spatial frequency that is inversely proportional to the drop diameter.

To avoid the detection of light scattered by diffraction and to detect light scattered predominantly by the mechanism of refraction, an angle 30° off the forward optical axis and orthogonal to the plane of the two beams is used. For this optical arrangement, the required focused beam diameters and length of the sample volume are significantly reduced and more precisely defined by the off-axis detection. Using refracted or reflected light, good sensitivity to drops in the size range from 5 to 3000 μm can be obtained using optical configurations allowing discrete measurement steps. However, the dynamic size range for any one optical configuration is still limited to less than 10:1.

A drop sizing interferometer incorporating a laser velocimeter and an interferometer system with off-axis collection has been described by Bachalo et al. [52] and is available commercially through Spectron Development Labs (Costa Mesa, Calif.) The system is shown schematically in Fig. 9.8. It comprises two essential units, the transmitter and the receiver. The transmitter optics define the fringe spacing (determined by the intersection angle of the laser beams) and the position of the probe volume. The receiving optics collect the light scattered from drops moving through the probe volume. In the system shown in Fig. 9.8 the collection angle θ is 30° from the forward direction. A collecting lens is used to image the probe volume on the photomultiplier's aperture or pinhole, where the drop signals are registered. The electronic components consist of a visibility processor that is used to process the input signals, which consist of Doppler signals superimposed on Gaussian pedestals as illustrated in Fig. 9.6. The processor separates the Doppler component from the pedestal component, integrates the areas under the respective signals, and divides the result to produce the signal visibility measurement. The drop size is determined from the signal visibility, i.e., the magnitude of the ratio between the Doppler and the pedestal components, and the velocity is determined from the Doppler frequency.

The visibility technique can provide accurate drop size measurements only if

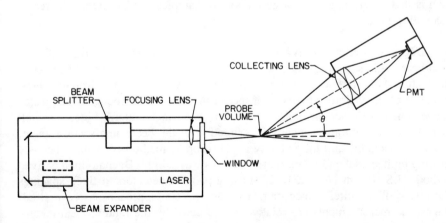

Figure 9.8 Schematic diagram of optical arrangement for drop sizing interferometer with off-axis collection.

the fringe visibility (or contrast) is known. In dense sprays, drops interacting with the transmitted beam upstream of the probe volume can randomly disrupt the conditions necessary for perfect fringe contrast at the probe volume. (These conditions include highly coherent beams, equal beam intensities, and complete beam overlap at the crossover.) The result is reduction in fringe visibility, which causes the measured drop size to exceed the actual drop size. It is also possible for the visibility to increase due to masking of the signal from out-of-focus drops by the pinhole or by the visibility processor's response to a signal with a high rise [53].

Spectron has now extended the capability of their drop sizing interferometer to include the amplitude of the scattered light in addition to the visibility [54]. The signal from each drop yields both parameters, and their cross-correlation can be used to select the right signals.

A correlation between the measured size of the drop and the amount of scattered light (Mie theory) is used to eliminate signals that indicate the wrong size. This *intensity validation* (IV) technique is based on the fact that drops that produce a certain visibility must have a given size, and consequently they must scatter light with a given intensity. In this manner, limits are fixed for every measured visibility [55].

Two of the main problems are solved by this technique. First, drops with erroneous visibility (due to beam blockage that causes a disturbed probe volume) will scatter light with an intensity lower than that pertaining to their apparent size. Second, drops crossing the probe volume at the point of less intensity (the tail of the Gaussian intensity profile) will have the right visibility but will scatter light with different intensity. The IV technique sets limits for the probe volume and rejects invalid signals by establishing limits on the intensity of the pedestal.

An advantage of this system is that it obviates the need for a direct calibration of intensity by an automated setting of the voltage to the photomultiplier. This added capability has resulted in a significant improvement in instrument accuracy, especially for measurements in dense sprays [3]. The disadvantage of this approach is that it rejects a very large number of samples, which greatly increases the sampling time.

Phase/Doppler particle analyzer. The theoretical description by Bachalo [47] of dual-beam light scattering showed that, at off-axis angles, the spatial frequency of the scattered interference fringe pattern is inversely related to the drop diameter. Limitations of the visibility method for measuring the spatial frequency suggested the development of other methods of measuring the fringe pattern. Since it is well known that the temporal frequency is a function of the beam intersection angle, the light wavelength, and the velocity of the drop, a method for simultaneously measuring both spatial and temporal frequencies was sought. Bachalo then derived a method (U.S. Patent No. 4,540,283) using pairs of detectors at fixed spacings in the image of the interference or fringe pattern. The approach has several advantages: the measurements are relatively unaffected by the random beam attenuations provided the signal-to-noise ratio is sufficient, the instrument response is linear over the entire working range, and it has a potentially large size dynamic range. The dynamic range is limited only by the detector response and signal-to-noise ratio. An example of the instrument response function for monodisperse drops is shown in Fig. 9.9.

The method has been developed into a phase/Doppler particle analyzer by Aerometrics, Inc. (Mountain View, Calif.). The instrument is similar to a conventional dual-beam laser Doppler velocimeter except that three detectors are used in the receiver. The optical arrangement is shown schematically in Fig. 9.10. A standard system consists of a 10-mW HeNe laser light source followed by a combination beam expander and diffraction grating to split the beam. The diffraction grating is rotated at precisely controlled rates to provide frequency shifting for measurements in complex sprays. The receiver consists of a 108-mm-diameter $f/5$ lens located at 30° to the plane of the transmitted beams. Three detectors are employed to eliminate measurement ambiguity, add redundant measurements to improve reliability, and provide high sensitivity over a large size range. Their juxtaposition is such that Doppler burst signals are produced by each detector but with a relative phase shift. This phase shift is linearly related to the drop size. Signals from the photodetectors are amplified and transferred to the signal processor. Data acquisition and transfer to computer memory requires 20 μs per drop. Histograms of drop size and velocity distributions can be displayed on a video monitor in essentially real time. At the completion of the data acquisition, the various mean sizes, mass flux, number density, and mean and rms velocity are determined.

The overall measurement size range of the instrument is from 0.5 to 3000 μm, provided the drops remain spherical. A size range of 105 may be covered at a single optical setting. The dynamic range is a factor of 35, which is limited by signal level and signal-to-noise considerations. Extensive testing of the method has been carried out, sometimes in high-density environments [56, 57].

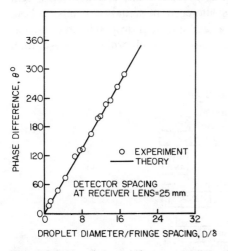

Figure 9.9 Plot of phase shift versus nondimensional drop size; comparison of theory and experiment *(Bachalo and Houser [56]).*

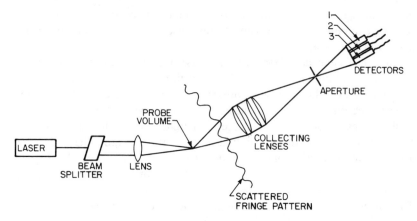

Figure 9.10 Schematic of the phase/Doppler particle analyzer.

Intensity deconvolution technique. In addition to interferometric techniques, single-particle counting can also be accomplished by measuring the *absolute intensity* of light scattered by particles traversing a single focused laser beam. The PCSV instrument manufactured by Insitec (San Ramon, Calif.) makes use of an intensity deconvolution technique to measure particle size, concentration, and velocity (Fig. 9.11). The principle of operation is explained in publications by Holve and co-workers [58–62]. Basically, this instrument measures the absolute intensity of light scattered by drops or solid particles passing through the sample volume of a single focused HeNe laser beam. The sample volume is defined by the intersection of the laser beam and the detector optics. A photomultiplier in the receiver measures the absolute intensity of light scattered by each particle passing through the sample volume. Since the sample volume has a nonuniform light intensity distribution, the intensity of the scattered light signals depends on both drop size and drop trajectory. The potential ambiguity in drop size and concentration measurements is resolved using a deconvolution algorithm. This solution

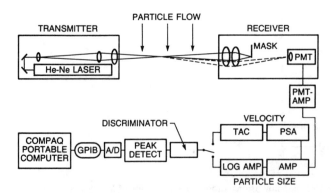

Figure 9.11 Schematic diagram of particle counter-sizer velocimeter (PSCV).

has two requirements: (1) the absolute scattered light intensity of a large number of drops must be measured (a process that typically requires less than 1 min) and (2) the sample volume intensity distribution must be known.

Knowing the sample volume intensity distribution, the count spectrum of scattered light intensities can be *deconvoluted* to yield the absolute drop concentration as a function of drop size, hence the name *intensity deconvolution* technique. Although the physics of the two measurement techniques is entirely different, the solution strategy of the intensity deconvolution technique is analogous to that of the Malvern particle sizer.

An important asset of the intensity deconvolution method is its ability to make measurements of drops, slurries, and solid particles. Since this measurement technique analyzes light scattered in the near-forward direction, the scatters may be spherical or irregular in shape. Use of an interference filter allows particle measurements to be made in combustion conditions. The overall measurement size range is 0.2 to 200 μm. The optical configuration of a typical instrument allows measurements from 0.3 to 100 μm. Particle velocities up to 200 m/s are determined using a transit timing technique [62]. Rapid signal processing (40–150 kHz) permits on-line data analysis.

Other miscellaneous techniques. Several other techniques have been proposed for simultaneous measurements of velocity and drop size in sprays. Work at Sheffield University by Yule et al. [51] has led to the development of a drop sizing technique based on laser velocimetry. Rather than measure signal modulation, they measure the peak mean value of the Doppler signal for drops larger than the fringe spacing. Their analysis shows a linear relationship between pedestal peak amplitude (Fig. 9.6) and drop diameter for drops between 30 and 300 μm in diameter. Good agreement was found between drop size distributions measured in sprays with their laser velocimeter and size distributions measured by collecting the drops on a slide and using an image analysis computer.

TSI Inc. (St. Paul, Minn.) has developed a visibility module for use in conjunction with a TSI laser anemometer system that provides simultaneous measurement of velocity, visibility, and pedestal amplitude. It operates in conjunction with a TSI LDV signal processor. The system has a large input Doppler frequency range from 1 kHz to 2 MHz with high data rates up to 50 kHz [2]. As the system is still relatively new, insufficient operational experience has been gained to allow performance comparisons to be made with the more established instruments.

Light-Scattering Technique

The optical properties of a medium are characterized by its refractive index, and as long as this is uniform, light will pass through the medium undeflected. Whenever there are discrete variations in the refractive index due to the presence of particles, part of the radiation will be scattered in all directions while the other part is transmitted unperturbed. In drop size analysis, provided the number of drops under observation is large enough to ensure that a representative sample is obtained, the properties of the scattered light can be used to indicate the drop size distribution. The formulation of a theory for the scattering properties of particles of arbitrary size and arbitrary refractive index occurring in polydispersions of finite optical depth was first derived by Dobbins et al. [63]. For a polydisperse system, the radiant intensity $I(\theta)$ scattered at a small angle θ from the forward direction due to an incident planar wave of irradiance E_0 can be written as

$$\frac{I(\theta)}{E_0} = \frac{D^2}{16}\left\{\alpha^2\left[\frac{2J_1(\alpha\theta)}{\alpha\theta}\right]^2 + \left[\frac{4m^2}{(m^2-1)(m+1)}\right]^2 + 1\right\} \qquad (9.6)$$

where E_0 = incident irradiance
  $I(\theta)$ = radiant intensity
  θ = scattering angle
  α = size number $\pi d/\lambda$, where λ is the wavelength of incident light
  m = refractive index
  J_1 = Bessel function of first kind of order unity

The three terms inside the braces in Eq. (9.6) represent, respectively, the Fraunhofer diffraction, the optical scattering due to refraction of the centrally transmitted ray, and the optical scattering due to incident rays. Equation (9.6) requires that (1) the incident radiation is planar and monochromatic, (2) the forward angle θ is small, (3) the particle size number α and phase shift $2\alpha(m-1)$ are large, and (4) the distance between particle and observer is large compared to D^2/λ.

If a polydispersion of particles is present, the integrated intensity of all particles is found by summing over all diameters. The expression for the intensity of scattering due to polydispersion is normalized by dividing by the intensity of diffractively scattered light in the forward direction, $\theta = 0$. It is then found that the second and third terms in Eq. (9.6) are small and can be ignored. The normalized integrated intensity of forward-scattered light, $I(\theta)$, due to a polydispersion of relatively large particles is given as

$$I(\theta) = \frac{\displaystyle\int_0^\infty \left[\frac{2J_1(\alpha\theta)/}{\alpha\theta}\right]^2 N_r(D)D^4\,dD}{\displaystyle\int_0^\infty N_r(D)D^4\,dD} \qquad (9.7)$$

where $N_r(D)$ is defined in such a way that the integral of $N_r(D)$ over a given diameter interval represents the probability of occurrence of particles within the specified interval.

Equation (9.7) represents a relationship between the angular distribution of scattered light and the particle size distribution. The basic problem then becomes one of evaluating the particle size distribution from an experimentally determined angular distribution of scattered light, $I(\theta)$. Dobbins et al. [63] and Roberts and Webb [64] have demonstrated that, for a number of drop size distribution functions with varying size parameters, a unique illumination profile (or scattered light distribution) of $I(\theta)$ is obtained when plotted against an abscissa of $\pi D_{32}\theta/\lambda$, as shown in Fig. 9.12. Thus, the experimental determination of $I(\theta)$ versus θ does permit an evaluation of D_{32} (SMD).

The original optical system employed by Dobbins et al. [63] is shown in Fig. 9.13. The advent of laser technology allowed various improvements to be made, and a more recent version, due to Lorenzetto [65], is shown schematically in Fig. 9.14. The light source is a 5-mW HeNe laser operating at wavelength of 6328 Å. The laser beam is spatially filtered to pass only the fundamental laser mode. An expanding telescope is used to produce a highly collimated beam, which is then "chopped" by a rotating perforated disk to reduce the stray light and increase the sensitivity of the system.

The parallel light beam is diffracted through the spray under investigation and focused by a receiver of 600 mm focal length onto a circular aperture 20 μm in diameter. The light then enters a photomultiplier tube that is located at the focal plane of the receiving lens and is arranged to traverse horizontally in a direction at right angles to the optical axis. The signals from both the photomultiplier tube

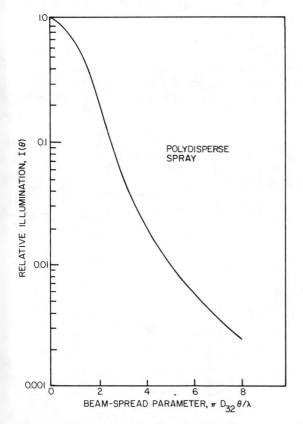

Figure 9.12 Mean theoretical illumination profile for mean drop size D_{32} with sizes distributed according to the upper-limit distribution function *(Dobbins et al. [63])*.

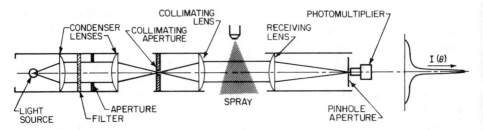

Figure 9.13 Schematic drawing of the light-scattering system used by Dobbins et al. [63].

and a position transducer are fed simultaneously into an X-Y plotter to obtain the light intensity profile. Typical plots for sprays produced by a prefilming type of airblast atomizer are given in Fig. 9.15, while Fig. 9.16 shows the normalized light intensity profiles. The SMD is determined by measuring the radial distance from the optical axis to a point on the curve at which the light intensity is one-tenth of its maximum value [64]. The SMD is then read off the curve drawn in Fig. 9.17, which is based on a relationship between the relative light intensity and the scattered angle, as determined by Roberts and Webb [64].

Lefebvre and co-workers have carried out an extensive series of measurements using this technique [66–70]. The good agreement found between their results and other independent measurements indicates the practical usefulness of the method. However, all methods have certain limitations, and this is no exception. One obvious drawback is a size limitation, which is evident in Fig. 9.17. The intensity

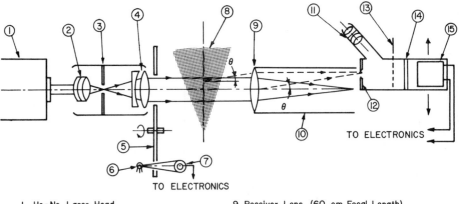

I He–Ne Laser Head	9 Receiver Lens (60 cm Focal Length)
2 Condensing Lens	IO Stray Light Shield
3 Spatial Filter (Aperture)	II Eyepiece
4 Telescope–Collimating Lens	I2 Pin–Hole Aperture
5 Rotating Disc (Chopper)	I3 Shutter
6 Small Lamp	I4 Neutral Density Filter
7 Photo Cell	I5 Traversing Photomultiplier
8 Spray Under Test	

Figure 9.14 Lorenzetto's improved light scattering system [65].

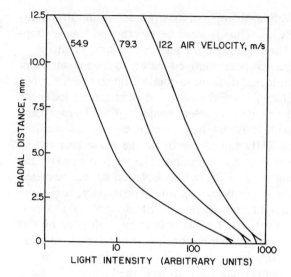

Figure 9.15 Data obtained with an airblast atomizer illustrating effect of air velocity on light intensity distributions *(Rizk and Lefebvre [71])*.

variation is weak for small particles, resulting in a lower size limit of about 10 μm in practice. For large particles the light intensity is concentrated near the axis, and it is difficult to obtain accuracy for particles with diameters larger than around 250 μm. Another problem with this method is that some ambiguity exists con-

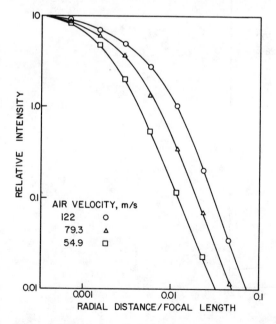

Figure 9.16 Variation of relative intensity with radial distance *(Rizk and Lefebvre [71])*.

cerning the maximum light intensity at the optical axis (I_0) due to the strong on-axis peak from the unscattered beam. Thus I_0 must be determined from extrapolation, as shown in Fig. 9.15. The resulting error proves to be rather small.

To date, the Dobbins technique has been confined mainly to the measurement of SMD. It has generally been considered difficult to obtain drop size distributions from measurements of light intensity, since the scattered light profile lacks the sensitivity needed to determine the shape parameters uniquely [63]. To overcome this limitation, Rizk and Lefebvre [71] have evolved a technique whereby the light intensity plots used to determine SMD can also provide the basic information needed to estimate the complete drop size distribution. The method employs the theory developed by Swithenbank et al. [72], which is based on the scattered energy distribution rather than the intensity distribution. Fortunately, it is quite straightforward to convert plots of light intensity into distributions of light energy. The light intensity at any given radial distance has only to be multiplied by this distance and then replotted. The energy distributions shown in Fig. 9.18 were obtained in this manner from the light intensity profiles plotted in Fig. 9.16.

In comparison with the Malvern particle sizer, as described below, the Dobbins technique has the advantages of simplicity and lower cost. It also has the advantage of recording the light intensity profile for inspection so that results that are clearly spurious can be discarded. However, it lacks the speed, convenience, and data-storing capacity of the Malvern instrument.

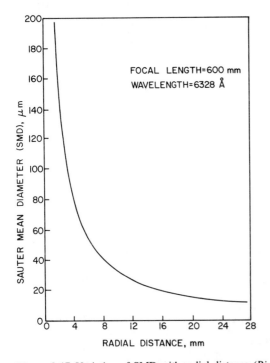

Figure 9.17 Variation of SMD with radial distance *(Rizk and Lefebvre [71]).*

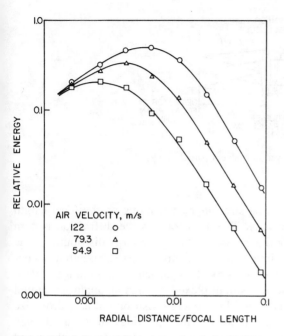

RADIAL DISTANCE/FOCAL LENGTH

Figure 9.18 Influence of atomizing air velocity on light energy distribution *(Rizk and Lefebvre [71]).*

Malvern Particle Sizer

Although many nonintrusive techniques have been developed for the measurement of drop size distribution, the most widely used is the Malvern particle sizer. For a rapid measurement of the global (ensemble) characteristics of a spray, this is currently one of the most effective, simple, and reliable methods that is commercially available. It is easy to use, and very little knowledge of its basic principles is required for operation. For applications where it is adequate to characterize a spray solely in terms of its SMD and an exponent in a size distribution equation, the Malvern system with its associated microcomputer yields computer printouts of the desired information. The instrument owes its initial development to Swithenbank et al. [72] but is now manufactured and distributed by Malvern Instruments Ltd. (Malvern, England, and Framingham, Mass.).

The technology employed is based on the Fraunhofer diffraction of a parallel beam of monochromatic light by a moving drop. The optical system is shown schematically in Fig. 9.19. When a parallel beam of light interacts with a drop, a diffraction pattern is formed in which some of the light is diffracted by an amount depending on the size of the drop. The diffraction pattern produced when the spray is monodisperse is of Fraunhofer form; i.e., it consists of a series of alternate light and dark concentric rings, the spacing of which depends on the drop diameter. However, for the more commonly encountered polydisperse sprays, a number of these Fraunhofer patterns are produced, each by a different group of drop sizes. This results in a series of overlapping diffraction rings, each of which

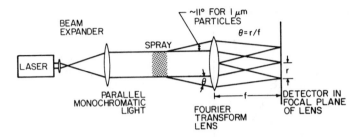

Figure 9.19 Optical arrangement employed in Malvern particle sizer.

is associated with a characteristic drop size range, depending on the focal length of the Fourier transform receiver lens. This lens focuses the diffraction pattern onto a multielement photodetector that measures the light energy distribution. The photodetector comprises 31 semicircular photosensitive rings surrounding a central circle. Each ring is therefore most sensitive to a particular small range of drop sizes. The output of the photodetector is multiplexed through an analog-to-digital converter. Interpretation of the measured light energy distribution as a drop size distribution is then carried out by a computer, which provides an instant display of the measured distribution. Thus, a major advantage of the system is the speed at which data can be both accumulated and analyzed, allowing complete (ensemble) characterization of an atomizer over its entire operating range within a few hours. Another important attribute is that the diffraction pattern generated by drops is independent of the position of the drop in the light beam. This allows measurements of size distribution to be made with the drops moving at any speed.

Experience has shown that good alignment of the Malvern optical system is essential for accurate measurement. The correct focal length of lenses should be chosen according to the drop size range to be measured. Each detector/lens combination employed gives approximately a 100-to-1 drop size range; a 63-mm lens gives a drop size range from 1.2 to 118 μm, a 100-mm lens from 2 to 197 μm, and a 300-mm lens from 5.8 to 564 μm. Other lenses are available having focal lengths of 600, 800, and 1000 mm for use with coarser sprays. It is also important to align the laser beam so that it is perpendicular to the spray axis at a suitable distance downstream from the atomizer.

Developments in computer software have produced a number of user-selectable features (Malvern Instruments Ltd.). Data sets may be stored in any one of 32 RAM memory blocks. Mass storage of data is available on magnetic tape or disks. The data may be analyzed in terms of a histogram with 15 size classes or presented in the format of normal, lognormal, Rosin-Rammler, or model-independent modes. When the Rosin-Rammler function is used, attention must be paid to the log error reading, which represents the closeness of the data fit to the indicated values of SMD and q (see Chapter 3). If the log error exceeds 4.5 it may be desirable to check the data against other drop size distribution models or use the model-independent mode.

Accuracy of Malvern Particle Sizer

In view of its widespread usage and unique capabilities, it is important for the user to fully appreciate the limitations on the Malvern instrument's accuracy and repeatability. For example, the Fraunhofer diffraction theory on which the instrument relies applies only to drops whose diameter is much larger than the wavelength of the incident light. This means that inaccuracies may be incurred when measuring drop sizes below 5 μm using a 300-mm receiver lens. It is also important to remember that the instrument is capable of measuring drop sizes only in a certain range. By a proper choice of lens, most practical sprays can be analyzed without difficulty. However, the indicated histogram of drop sizes should always exhibit a definite peak. Where no peak is evident, it may mean that a significant number of drops lie outside the range of measurement capability, so the instrument cannot interpret properly the observed scattering pattern.

Other problems that have arisen with the Malvern particle sizer include variations in detector sensitivity, multiple scattering, vignetting, and beam steering. How and to what extent these problems influence data accuracy are discussed below.

Variations in detector responsivity. One of the earliest doubts expressed about the accuracy of the Malvern system arose from a reported lack of consistency between results obtained at different laboratories in tests carried out on the same atomizer. The interpretation of such discrepancies between different Malvern instruments in their application to sprays that are ostensibly the same is complicated by the numerous possible reasons for the observed differences. These reasons include actual differences in the spray, even when using the same atomizer at nominally the same flow conditions, errors in the instrumentation used to measure nozzle pressures and flow rates, and failure to align the instrument properly. However, according to Dodge [73], the major cause of inconsistencies between different Malvern instruments is an inherent limitation on the accuracy of the photodiode detectors. These detectors are assumed to have a uniform responsivity to the incident light energy but, in fact, the responsivity of different detectors in the same set can vary appreciably, and these variations are not repeatable between different detector assemblies.

To overcome this problem, Dodge [73] devised a method that involves the separate calibration of each detector ring. Each detector is exposed to several different light levels to determine its response factor $F = C/R$, where C is a normalization constant that is different for each light level and R is the responsivity. The value of F determined from this procedure is then entered into the Malvern computer program to replace the existing response factor. The latter factor, termed LA in the Malvern 2200 user's manual, was determined using a procedure similar to that followed by Dodge, but the values actually employed for calibration purposes are the same for all instruments and were calculated by averaging the results obtained from a few detector assemblies. The values of F determined by the Malvern company range from 0.551 for the first (innermost) detector, through 0.848

for the fifteenth (central) detector, to 0.996 for the thirtieth (outermost) detector. The corresponding values of F as determined by Dodge [73] were 0.288, 0.746, and 0.996, respectively. Thus, significant improvements in accuracy can be achieved by using Dodge's calibration procedure.

It is the author's understanding that Malvern Instruments Ltd. now uses an undescribed procedure similar to that of Dodge for the calibration of current production particle analyzers.

Multiple scattering. When spray densities are high, light that is scattered by a drop may be scattered by a second drop before reaching the detector. Since the theory for laser diffraction-based instruments assumes scattering from a single drop, this multiple scattering introduces errors in the computed size distribution.

When measuring drop sizes in a spray, the Malvern instrument indicates the extinction or percentage of the light removed from the original direction. As Dodge [74] has emphasized, extinction (sometimes called obscuration) is of primary importance since it is directly dependent on the cross section of the drops present in the beam and is a major indicator of the extent to which the measurements are being affected by multiple scattering. The density of the spray is not a good parameter because extinction can vary over a wide range for a given density, depending on the size distribution and optical path length in the spray. The major consequence of multiple scattering is distortion of the diffraction pattern toward larger angles and a wider angle range. When the distorted data are processed, they cause the indicated drop sizes to be broader in distribution and smaller in average size than the actual distribution. The problem becomes significant when more than 50% of the incident light is scattered by the drops. However, measurements can still be performed in dense sprays where the extinction is as high as 95%, provided a correction is applied to the results.

Dodge [74] examined this problem experimentally and developed an empirical correction scheme. This was followed by more extensive theoretical efforts that led to new correction techniques that are more generally applicable, as discussed below.

Felton et al. [75] have developed a theoretical model of multiple light scattering whereby the light path is divided into a sequence of slices of equal extinction and the effect of multiple scattering on the diffraction pattern is evaluated. Predictions based on this model were compared with experimental results obtained using dense suspensions of glass beads, and excellent agreement was obtained. The influence of multiple scattering was found to depend on size distribution parameters, and a set of correction equations was derived for both Rosin-Rammler and lognormal distributions. For a Rosin-Rammler distribution, the relevant correction factors are

$$C_X = \frac{\bar{X}_0}{\bar{X}} \qquad (9.8)$$

and

$$C_q = \frac{q_0}{q} \qquad (9.9)$$

where X and q are the actual measured values with high extinction, and X_0 and q_0 are the corresponding calculated values for low extinction. The parameters X and q are defined in Chapter 3. They represent, respectively, the characteristic diameter and size distribution functions in the Rosin-Rammler expression, Eq. (3.14).

The following expressions were derived for C_X and C_q:

$$C_X = 1.0 + [0.036 + 0.4947(EX)^{8.997}]q_{\text{dens}}^{1.9 - 3.437(EX)} \qquad (9.10)$$

$$C_q = 1.0 + [0.035 + 0.1099(EX)^{8.65}]q_{\text{dens}}^{0.35 + 1.45(EX)} \qquad (9.11)$$

where EX is the extinction and q_{dens} is the Rosin-Rammler distribution parameter as measured for the dense spray.

The results of applying these correction factors to Dodge's [74] experimental data are shown in Fig. 9.20, in which SMD is plotted against light extinction. The corrected values are clearly much closer to the true values than the raw data, but it is also apparent that at the highest extinction levels some differences between theory and experiment still exist.

Hamadi and Swithenbank [76] recently extended the procedure of Felton et al. for the correction of multiple scattering to include any type of drop size distribution, including bi- and trimodal size distributions.

Vignetting. When taking measurements in sprays, it is necessary to limit the maximum distance between the spray and the receiving optics. This maximum distance depends on the focal length and diameter of the receiver lens. Exceeding this distance results in vignetting of the signal on the outer detector rings, thereby skewing the measured size distribution toward the larger diameters. Dodge [73] has discussed this problem and provided the following expression for calculating the maximum allowable distance x:

$$x = f\left(\frac{D_1 - D_b}{D_d}\right) \qquad (9.12)$$

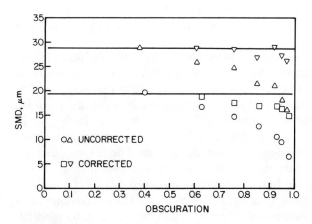

Figure 9.20 Dodge's results [74], with and without the correction of Felton et al. [75] for multiple scattering.

where f = lens focal length
  D_1 = lens diameter
  D_b = laser beam diameter
  D_d = detector diameter (twice the distance from center to outermost edge of detectors)

Dodge [73] has given specific results for the Malvern system using the three standard lenses. With the standard instrument it is difficult to exceed x with the 300-mm lens, but with the 100-mm lens the spray should be within 120 mm of the lens, and with the 63-mm lens it should be within 55 mm.

Wild and Swithenbank [77] have also studied the problem of vignetting and developed a theory that can be used to correct for vignetting errors. Their method has been validated by comparison with computational and practical vignetting experiments. It can be applied wherever the particle field is homogeneous and is of large extent or low concentration.

Beam steering. Problems can arise with light-scattering techniques in their application to sprays in a high-temperature environment. Even with no spray present, thermal gradients in the hot air refract the laser beam in a random high-frequency pattern (called *beam steering*), resulting in a spurious reading on the innermost detectors. When the spray is present, additional thermal and concentration gradients are created that result in the first few detectors receiving abnormally large signals. Unfortunately, the problem cannot be solved merely by subtracting out the background when no spray is present.

Dodge and Cerwin [78] have suggested that, as the 15 detector pairs of the Malvern analyzer form an overdetermined data set, if only the two or three innermost detectors are involved, their contribution can be ignored by appropriate changes in the computer software. However, if the spray contains many large drops that scatter light onto the inner detector channels, it is not possible to ignore these channels and still arrive at a reasonable drop size distribution. In practice, a subjective judgment must be made on how many channels contain spurious data. Usually this is fairly straightforward if the drops in the spray are small enough so that the peak detector signal is located well outside the region of the disturbed channels.

Dodge and Cerwin [78] recommend three checks for verifying that the corrected light distribution is a reasonable one. First, the corrected light distribution should show a smooth transition between the outer channels that were used to compute the drop size distribution and the inner channels in which the recorded data have been replaced by the new computed light pattern that best fits the overall pattern. Second, there should be at least four or five good detector signals to the left of the peak of the light distribution. When the perturbed detector signals approach the peak of the light distribution, this technique becomes invalid. Third, the log error as indicated by the Malvern should not exceed 5.0.

The errors and uncertainties associated with this correction technique must obviously increase with the number of detector signals that have to be ignored. Best results are obtained with finely atomized sprays for which the light intensity signal reaches a maximum in the outer half of the detector channels. Examples illustrating the use of this correction technique have been provided by Dodge and Cerwin [78].

As a footnote to the subject of potential errors and inconsistencies with the Malvern instrument, Jackson and Samuelson [53] have drawn attention to significant differences in the data depending on whether Rosin-Rammler or model-independent distribution is used to process the data. They clearly favor the model-independent mode, but the study carried out by Dodge et al. [79] indicates no particular advantage for this approach.

Intensity Ratio Method

When a light beam is passed through a spray the drops scatter light in all directions. For drops larger than around 2 μm, the relative intensity of the light scattered in the forward direction can be calculated as a function of angle by using diffraction theory.

This smooth variation of scattered intensity with angle in the forward direction led Hodkinson [80] to suggest a simple means of particle sizing based on the ratio of measured light intensity at two different angles. An instrument based on this idea was developed and used by Gravatt [81]. The method has the advantage of being fairly insensitive to refractive index, but its drop size range is limited to below 5 μm. Larger drops call for accurate measurements at very small angles, where interference with the direct unscattered beam becomes significant [81, 82].

The single intensity ratio method as outlined above is prone to serious error if the drop size distribution is broad. Hirleman and Wittig [83] proposed a multiple ratio system based on different wavelengths and/or different pairs of scattering angles to minimize these errors. The multiple ratio system actually used by Hirleman [84] has four angles, but clearly the system becomes less practical as the number of angles increases.

A ratioing processor has been developed by Spectron [85] and is shown schematically in Fig. 9.21. An annular mask pair is incorporated in the collection optics, so light scattered at two angles (5° and 2.5°) in the near-forward direction is observed. The ratio of the two measurements is a function of particle size, the lower limit on size being around 0.3 μm.

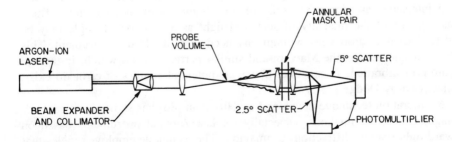

Figure 9.21 Intensity ratio system for particle sizing *(Wuerer et al. [85]).*

Calibration Techniques

Most of the optical methods employed in drop size measurement require calibration of some kind. The calibration techniques in common use have been reviewed by Hirleman [86].

Highly monodisperse polystyrene microspheres (available from Duke Scientific Co., Palo Atlo, Calif.) sprayed from a dilute liquid suspension and subsequently dried can be used for the calibration of particle sizing instruments up to a maximum particle diameter of around 5 μm. For larger particle sizes, polystyrene or glass spheres must be used in liquid suspensions. As glass spheres have broader size distributions than polystyrene, sieve or sedimentation techniques are used to produce well-defined size groups. Glass microspheres can also be attached to microscope slides or rotating glass disks and placed in the instrument sample volume [51]. Problems that arise with these techniques include gravitational settling and wall deposition, which are size-dependent phenomena, the need for a sample cell and flow system, and interference with the scattering process by the glass slide used to support the calibration spheres [86].

Vibrating-orifice techniques are sometimes used to generate a stream or cloud of nearly monodisperse drops in sizes ranging from 10 to 150 μm in diameter [87]. Spinning disk generators have also been used to generate relatively small drops. These mechanical techniques are useful for calibrating single-particle optical counters, but they cannot generate sprays containing the broad spectrum of drop sizes needed for the calibration of ensemble scattering instruments [86].

The fact that in the near-forward direction the scattering properties of drops, pinholes, and opaque disks of the same diameter are equivalent allows precisely formed circular pinholes to be used for the calibration of drop sizing instruments that operate as single-particle counters [58]. This approach also has limited utility for ensemble scattering instruments.

All the calibration methods described above have limited applicability to the calibration and performance verification of diffraction-based drop sizers. In one sense it might be argued that such instruments do not require calibration because the diffraction angles and intensities of scattered light are defined rigorously in the Mie scattering theory. If the scattering measurements are restricted to small angles (forward diffraction) and the particle diameters are much larger than the wavelength of the radiation, the particle properties are unimportant and the scattering intensities are dependent only on the diameters of the scatterers. Thus, measurements of the intensity of scattered light as a function of angle may be used to determine drop sizes without any need to calibrate the instrument. This is the principle used in the Malvern and similar particle sizers, which, in theory, require no calibration. In practice, however, potential sources of error do exist, as discussed by Dodge [73, 74].

A calibration technique developed by Hirleman [86] has some significant advantages over other methods, especially for laser/optical instruments based on forward light scatter (diffraction) or imaging. The technique employs reticles that provide a two-dimensional sample array of typically 10,000 circular disks of chrome film photoetched on a glass substrate. These disks (simulated drops) are randomly positioned in a circular sample area 8 mm in diameter. Typically, 23 discrete diameters with one to several thousand replications of each discrete size are used to approximate a specific particle size distribution. Reticles can be used in series to simulate multimodal distributions, and the number of potential array configurations appears to be virtually unlimited. Measurement accuracy for drops larger than 10 μm is around ±1 μm [86].

The manufacturer of the reticles (Laser Electro-Optics) supplies a predicted light-scattering intensity distribution relative to the detector sizes of the Malvern instrument. This can be entered into the Malvern computer as ideal measured data and the corresponding values computed for X and q in the Rosin-Rammler distribution parameter [Eq. (3.14)].

CONCLUDING REMARKS

It is clear that no single drop sizing method is completely satisfactory. Many useful techniques have been developed, each with its own advantages and limitations.

Mechanical methods have the virtues of simplicity and low cost. Their main problem lies in the extraction and collection of representative spray samples.

Of the various electrical methods discussed, the hot-wire technique appears to be the most useful. However, the device can operate only at relatively low velocities (<10 m/s) due to the risk of drop breakup on impact with the wire. There is also some question about the measurement of liquids that could leave a residue on the wire.

Imaging systems allow the spray to be "seen," but a major difficulty is the determination of the size of the viewing volume to be assigned to given drop sizes. For ensemble measurements the light diffraction methods of Dobbins and Swithenbank have much to commend them. The latter, in the form of the Malvern particle sizer, is now in widespread use as a general-purpose tool for spray analysis. Of the remaining optical methods discussed, the advanced laser light scatter interferometry techniques show considerable promise for obtaining simultaneous drop size measurements and drop velocity data in the dense sprays produced by most practical atomizers.

Dodge [88] has recently compared the performance of 17 drop size measuring instruments including six different types. Ten instruments were compared on one type of spray and eleven on another, with some instruments being used on both types. The results indicated systematic differences in instrument performance beyond what could be attributed to problems in spray reproducibility. Differences in average drop size were as large as a factor of five. However, good to excellent agreement was obtained between observations for some types of instrument.

REFERENCES

1. Chigier, N. A., Instrumentation Techniques for Studying Heterogeneous Combustion, *Prog. Energy Combust. Sci.*, Vol. 3, 1977, pp. 175–189.

2. Chigier, N. A., Drop Size and Velocity Instrumentation, *Prog. Energy Combust. Sci.*, Vol. 9, 1983, pp. 155–177.

3. Bachalo, W. D., Droplet Analysis Techniques: Their Selection and Applications, *Liquid Particle Size Measurement Techniques*, ASTM STP 848, J. M. Tishkoff, R. D. Ingebo, and J. B. Kennedy, eds., American Society for Testing and Materials, 1984, pp. 5–21.

4. Ferrenberg, A. J., Liquid Rocket Injector Atomization Research, *Liquid Particle Size Measurement Techniques*, ASTM STP 848, J. M. Tishkoff, R. D. Ingebo, and J. B. Kennedy, eds., American Society for Testing and Materials, 1984, pp. 82–97.

5. Hirleman, E. D., Particle Sizing by Optical, Nonimaging Techniques, *Liquid Particle Size Measurement Techniques*, ASTM STP 848, J. M. Tishkoff, R. D. Ingebo, and J. B. Kennedy, eds., American Society for Testing and Materials, 1984, pp. 35–60.

6. Jones, A. R., A Review of Drop Size Mesurement—The Application of Techniques to Dense Fuel Sprays, *Prog. Energy Combust. Sci.*, Vol. 3, 1977, pp. 225–234.

7. Wittig, S., Aigner, M., Sakbani, Kh., and Sattelmayer, Th., Optical Measurements of Droplet Size Distributions: Special Considerations in the Parameter Definition for Fuel Atomizers, paper presented at AGARD meeting on Combustion Problems in Turbine Engines, Cesme, Turkey, October 1983.

8. Chin, J. S., Nickolaus, D., and Lefebvre, A. H., Influence of Downstream Distance on the Spray Characteristics of Pressure-Swirl Atomizers, *ASME J. Eng. Gas Turbines Power*, Vol. 106, No. 1, 1986, p. 219.

9. Tate, R. W., Some Problems Associated with the Accurate Representation Droplet Size Distributions, *Proceedings of the 2nd International Conference on Liquid Atomization and Spray Systems*, Madison, Wis., June 1982, pp. 341–351.

10. Lewis, H. C., Edwards, D. G., Goglia, M. J., Rice, R. I., and Smith, L. W., Atomization of Liquids in High Velocity Gas Streams, *Ind. Eng. Chem.*, Vol. 40, No. 1, 1948, pp. 67–74.

11. Bowen, I. G., and Davies, G. P., Report ICT 28, Shell Research Ltd., London, 1951.

12. Chin, J. S., Durrett, R., and Lefebvre, A. H., The Interdependence of Spray Characteristics and Evaporation History of Fuel Sprays, *ASME J. Eng. Gas Turbines Power*, Vol. 106, 1984, pp. 639–644.

13. Crosby, E. J., Atomization Considerations in Spray Processing, *Proceedings of 1st International Conference on Liquid Atomization and Spray Systems*, Tokyo, August 1978, pp. 434–448.

14. Pilcher, J. M., Miesse, C. C., and Putnam, A. A., Wright Air Development Technical Report WADCTR 56-344, Chapter 4, 1957.

15. May, K. R., The Measurement of Airborne Droplets by the Magnesium Oxide Method, *J. Sci. Instrum.*, Vol. 27, 1950, pp. 128–130.

16. Elkotb, M. M., Rafat, N. M., and Hanna, M. A., The Influence of Swirl Atomizer Geometry on the Atomization Performance, *Proceedings of the 1st International Conference on Liquid Atomization and Spray Systems*, Tokyo, 1978, pp. 109–115.

17. Kim, K. Y., and Marshall, W. R., Drop-Size Distributions from Pneumatic Atomizers, *J. Am. Inst. Chem. Eng.*, Vol. 17, No. 3, 1971, pp. 575–584.

18. Rao, K. V. L., Liquid Nitrogen Cooled Sampling Probe for the Measurement of Spray Drop Size Distribution in Moving Liquid-Air Sprays, *Proceedings of the 1st International Conference on Liquid Atomization and Spray Systems*, Tokyo, 1978, pp. 293–300.

19. Hausser, F., and Strobl, G. M., Method of Catching Drops on a Surface and Defining the Drop Size Distributions by Curves, *Z. Tech. Phys.*, Vol. 5, No. 4, 1924, pp. 154–157.

20. Karasawa, T., and Kurabayashi, T., Coalescence of Droplets and Failure of Droplets to Impact the Sampler in the Immersion Sampling Technique, *Proceedings of the 2nd International Conference on Liquid Atomization and Spray Systems*, Madison, Wis., 1982, pp. 285–291.

21. Rupe, J. H., A Technique for the Investigation of Spray Characteristics of Constant Flow Nozzles, *Third Symposium on Combustion, Flame, and Explosion Phenomena*, Williams & Wilkins, Baltimore, 1949, pp. 680–694.

22. Spray Droplet Technology, brochure published by Delavan Corporation, West Des Moines, Iowa, 1982.

23. Joyce, J. R., The Atomization of Liquid Fuels for Combustion, *J. Inst. Fuel*, Vol. 22, No. 124, 1949, pp. 150–156.

24. Longwell, J. P., Fuel Oil Atomization, D.Sc. thesis, Massachusetts Institute of Technology, 1943.
25. Taylor, E. H., and Harmon, D. B., Jr., Measuring Drop Sizes in Sprays, *Ind. Eng. Chem.*, Vol. 46, No. 7, 1954, pp. 1455–1457.
26. Choudhury, A. P. R., Lamb, G. G., and Stevens, W. F., A New Technique for Drop-Size Distribution Determination, *Trans. Indian Inst. Chem. Eng.*, Vol. 10, 1957, pp. 21–24.
27. Nelson, P. A., and Stevens, W. F., Size Distribution of Droplets from Centrifugal Spray Nozzles, *J. Am. Inst. Chem. Eng.*, Vol. 7, No. 1, 1961, pp. 80–86.
28. Street, P. J., and Danaford, V. E. J., A Technique for Determining Drop Size Distributions Using Liquid Nitrogen, *J. Inst. Pet. London,* Vol. 54, No. 536, 1968, pp. 241–242.
29. Kurabayashi, T., Karasawa, T., and Hayano, K., Liquid Nitrogen Freezing Method for Measuring Spray Droplet Sizes, *Proceedings of 1st International Conference on Liquid Atomization and Spray Systems,* Tokyo, 1978, pp. 285–292.
30. May, K. R., The Cascade Impactor; an Instrument for Sampling Coarse Aerosols, *J. Sci. Instrum.,* Vol. 22, 1945, pp. 187–195.
31. Wicks, M., and Dukler, A. E., *Proceedings of ASME Heat Transfer Conference,* Vol. V, Chicago, 1966, p. 39.
32. Gardiner, J. A., *Instrum. Pract.,* Vol. 18, 1964, p. 353.
33. Mahler, D. S., and Magnus, D. E., Hot-Wire Technique for Droplet Measurements, *Liquid Particle Size Measurement Techniques,* ASTM STP 848, J. M. Tishkoff, R. D. Ingebo, and J. B. Kennedy, eds., American Society for Testing and Materials, 1984, pp. 153–165.
34. Dombrowski, N., and Fraser, R. P., A Photographic Investigation into the Disintegration of Liquid Sheets, *Philos. Trans. R. Soc. London Ser. A,* Vol. 247, No. 924, 1954, pp. 101–130.
35. Dombrowski, N., and Johns, W. R., The Aerodynamic Instability and Disintegration of Viscous Liquid Sheets, *Chem. Eng. Sci.,* Vol. 18, 1963, pp. 203–214.
36. Fraser, R. P., Dombrowski, N., and Routley, J. H., The Production of Uniform Liquid Sheets from Spinning Cups; The Filming by Spinning Cups; The Atomization of a Liquid Sheet by an Impinging Air Stream, *Chem. Eng. Sci.,* Vol. 18, 1963, pp. 315–321, 323–337, 339–353.
37. Mullinger, P. J., and Chigier, N. A., The Design and Performance of Internal Mixing Multi-Jet Twin-Fluid Atomizers, *J. Inst. Fuel,* Vol. 47, 1974, pp. 251–261.
38. Mellor, R., Chigier, N. A., and Beer, J. M., Pressure Jet Spray in Airstreams, ASME Paper 70-GT-101, ASME Gas Turbine Conference, Brussels, 1970.
39. Chigier, N. A., McCreath, C. G., and Makepeace, R. W., Dynamics of Droplets in Burning and Isothermal Kerosine Sprays, *Combust. Flame,* Vol. 23, 1974, pp. 11–16.
40. Chigier, N. A., The Atomization and Burning of Liquid Fuel Sprays, *Prog. Energy Combust. Sci.,* Vol. 2, 1976, pp. 97–114.
41. De Corso, S. M., and Kemeny, G. A., Effect of Ambient and Fuel Pressure on Nozzle Spray Angle, *ASME Trans.,* Vol. 79, No. 3, 1957, pp. 607–615.
42. Simmons, H. C., and Lapera, D. L., A High-Speed Spray Analyzer for Gas Turbine Fuel Nozzles, paper presented at ASME Gas Turbine Conference, Session 2b, Cleveland, March 1969.
43. MacLoughlin, P. F., and Walsh, J. J., A Holographic Study of Interacting Liquid Sprays, *Proceedings of 1st International Conference on Liquid Atomization and Spray Systems,* Tokyo, 1978, pp. 325–332.
44. Murakami, T., and Ishikawa, M., Laser Holographic Study on Atomization Processes, *Proceedings of 1st International Conference on Liquid Atomization and Spray Systems,* Tokyo, 1978, pp. 317–324.
45. Thompson, B. J., Droplet Characteristics with Conventional and Holographic Imaging Techniques, *Liquid Particle Size Measurement Techniques,* ASTM STP 848, J. M. Tishkoff, R. D. Ingebo, and J. B. Kennedy, eds., American Society for Testing and Materials, 1984, pp. 111–122.
46. McVey, J. B., Kennedy, J. B., and Owen, F. K., Diagnostic Techniques for Measurements in Burning Sprays, paper presented at Meeting of Western States Section of the Combustion Institute, October 1976.
47. Bachalo, W. D., Method for Measuring the Size and Velocity of Spheres by Dual-Beam Light Scatter Interferometry, *Appl. Opt.,* Vol. 19, No. 3, 1980, pp. 363–370.

48. Durst, F., Melling, A., and Whitelaw, J. H., *Principles and Practice of Laser-Doppler Anemometry,* Academic Press, New York, 1976.

49. Farmer, W. M., The Interferometric Observation of Dynamic Particle Size, Velocity, and Number Density, Ph.D. thesis, University of Tennessee, 1973.

50. Farmer, W. M., Sample Space for Particle Size and Velocity Measuring Interferometers, *Appl. Opt.,* Vol. 15, 1976, pp. 1984–1989.

51. Yule, A., Chigier, N., Atakan, S., and Ungut, A., Particle Size and Velocity Measurement by Laser Anemometry, AIAA Paper 77-214, 15th Aerospace Sciences Meeting, Los Angeles, 1977.

52. Bachalo, W. D., Hess, C. F., and Hartwell, C. A., An Instrument for Spray Droplet Size and Velocity Measurements, ASME Winter Annual Meeting, Paper No. 79-WA/GT-13, 1979.

53. Jackson, T. A., and Samuelson, G. S., Spatially Resolved Droplet Size Measurements, ASME Paper 85-GT-38, 1985.

54. Hess, C. F., A Technique Combining the Visibility of a Doppler Signal with the Peak Intensity of the Pedestal to Measure the Size and Velocity of Droplets in a Spray, paper presented at the AIAA 22nd Aerospace Sciences Meeting, Reno, Nev., January 9–12, 1984.

55. Mularz, E. J., Bosque, M. A., and Humenik, F. M., Detailed Fuel Spray Analysis Techniques, NASA Technical Memorandum 83476, 1983.

56. Bachalo, W. D., and Houser, M. J., Phase Doppler Spray Analyzer for Simultaneous Measurements of Drop Size and Velocity Distributions, *Opt. Eng.,* Vol. 23, No. 5, 1984, pp. 583–590.

57. Bachalo, W. D., and Houser, M. J., Spray Drop Size and Velocity Measurements Using the Phase/Doppler Particle Analyzer, *Proceedings of the 3rd International Conference on Liquid Atomization and Spray Systems,* London, 1985, pp. VC/2/1–12.

58. Holve, D. J., and Self, S. A., Optical Particle Sizing Counter for In Situ Measurements, Parts I and II, *Appl. Opt.,* Vol. 18, No. 10, 1979, pp. 1632–1645.

59. Holve, D. J., In Situ Optical Particle Sizing Technique, *J. Energy,* Vol. 4, No. 4, 1980, pp. 176–182.

60. Holve, D. J., and Annen, K., Optical Particle Counting and Sizing Using Intensity Deconvolution, *Opt. Eng.,* Vol. 23, No. 5, 1984, pp. 591–603.

61. Holve, D. J., and Davis, G. W., Sample Volume and Alignment Analysis for an Optical Particle Counter Sizer, and Other Applications, *Appl. Opt.,* Vol. 24, No. 7, 1985, pp. 998–1005.

62. Holve, D. J., Transit Timing Velocimetry (TTV) for Two Phase Reacting Flows, *Combust. Flame,* Vol. 48, 1982, pp. 105–108.

63. Dobbins, R. A., Crocco, L., and Glassman, I., Measurement of Mean Particle Sizes of Sprays from Diffractively Scattered Light, *AIAA J.,* Vol. 1, No. 8, 1963, pp. 1882–1886.

64. Roberts, J. M., and Webb, M. J., Measurement of Droplet Size for Wide Range Particle Distribution, *AIAA J.,* Vol. 2, No. 3, 1964, pp. 583–585.

65. Lorenzetto, G. E., Influence of Liquid Properties on Plain Jet Atomization, Ph.D. thesis, School of Mechanical Engineering, Cranfield Institute of Technology, 1976.

66. Lorenzetto, G. E., and Lefebvre, A. H., Measurements of Drop Size on a Plain Jet Airblast Atomizer, *AIAA J.,* Vol. 15, No. 7, 1977, pp. 1006–1010.

67. Rizk, N. K., and Lefebvre, A. H., Influence of Airblast Atomizer Design Features on Mean Drop Size, *AIAA J.,* Vol. 21, No. 8, 1983, pp. 1139–1142.

68. Rizkalla, A., and Lefebvre, A. H., The Influence of Air and Liquid Properties on Air Blast Atomization, *ASME J. Fluids Eng.,* Vol. 97, No. 3, 1975, pp. 316–320.

69. Rizk, N. K., and Lefebvre, A. H., Spray Characteristics of Plain-Jet Airblast Atomizers, *Trans. ASME J. Eng. Gas Turbines Power,* Vol. 106, July 1984, pp. 639–644.

70. El-Shanawany, M. S. M. R., and Lefebvre, A. H., Airblast Atomization: The Effect of Linear Scale on Mean Drop Size, *J. Energy,* Vol. 4, No. 4, 1980, pp. 184–189.

71. Rizk, N. K., and Lefebvre, A. H., Measurement of Drop-Size Distribution by a Light-Scattering Technique, *Liquid Particle Size Measurement Techniques,* ASTM STP 848, J. M. Tishkoff, R. D. Ingebo, and J. B. Kennedy, eds., American Society for Testing and Materials, 1984, pp. 61–71.

72. Swithenbank, J., Beer, J. M., Abbott, D., and McCreath, C. G., A Laser Diagnostic Technique

for the Measurement of Droplet and Particle Size Distribution, Paper 76-69, 14th Aerospace Sciences Meeting, Washington, D.C., January 26–28, 1976, American Institute of Aeronautics and Astronautics.

73. Dodge, L. G., Calibration of Malvern Particle Sizer, *Appl. Opt.*, Vol. 23, 1984, pp. 2415–2419.

74. Dodge, L. G., Change of Calibration of Diffraction Based Particle Sizes in Dense Sprays, *Opt. Eng.*, Vol. 23, No. 5, 1984, pp. 626–630.

75. Felton, P. G., Hamidi, A. A., and Aigal, A. K., Measurement of Drop Size Distribution in Dense Sprays by Laser Diffraction, *Proceedings of the 3rd International Conference on Liquid Atomization and Spray Systems*, London, 1985, pp. IVA/4/1–11.

76. Hamadi, A. A., and Swithenbank, J., Treatment of Multiple Scattering of Light in Laser Diffraction Measurement Techniques in Dense Sprays and Particle Fields, *J. Inst. Energy*, Vol. 59, 1986, pp. 101–105.

77. Wild, P. N., and Swithenbank, J., Beam Stop and Vignetting Effects in Particle Size Measurements by Laser Diffraction, *Appl. Opt.*, Vol. 25, No. 19, 1986, pp. 3520–3526.

78. Dodge, L. G., and Cerwin, S. A., Extending the Applicability of Diffraction-Based Drop Sizing Instruments, *Liquid Particle Size Measurement Techniques*, ASTM STP 848, J. M. Tishkoff, R. D. Ingebo, and J. B. Kennedy, eds., American Society for Testing and Materials, 1984, pp. 72–81.

79. Dodge, L. G., Rhodes, D. J., and Reitz, R. D., Comparison of Drop-Size Measurement Techniques in Fuel Sprays: Malvern Laser-Diffraction and Aerometrics Phase Doppler, Spring Meeting of Central States Section of the Combustion Institute, Cleveland, May 1986.

80. Hodkinson, J., Particle Sizing by Means of the Forward Scattering Lobe, *Appl. Opt.*, Vol. 5, 1966, pp. 839–844.

81. Gravatt, C., Real Time Measurement of the Size Distribution of Particulate Matter by a Light Scattering Method, *J. Air Pollut. Control Assoc.*, Vol. 23, No. 12, 1973, pp. 1035–1038.

82. Stevenson, W. H., Optical Measurement of Drop Size in Liquid Sprays, Gas Turbine Combustion Short Course Notes, School of Mechanical Engineering, Purdue University, West Lafayette, Ind., 1977.

83. Hirleman, E. D., and Wittig, S. L. K., Uncertainties in Particle Size Distributions Measured with Ratio-Type Single Particle Counters, Conference on Laser and Electro-Optical Systems, San Diego, Calif., May 25–27, 1976.

84. Hirleman, E. D., Optical Techniques for Particulate Characterization in Combustion Environments, Ph.D. thesis, School of Mechanical Engineering, Purdue University, 1977.

85. Wuerer, J. E., Oeding, R. G., Poon, C. C., and Hess, C. F. (Spectron Development Labs), The Application of Nonintrusive Optical Methods to Physical Measurements in Combustion, American Institute of Aeronautics and Astronautics 20th Aerospace Sciences Meeting, Paper No. AIAA-82-0236, January 1982.

86. Hirleman, E. D., On-Line Calibration Technique for Laser Diffraction Droplet Sizing Instruments, ASME Paper 83-GT-232, 1983.

87. Berglund, R. N., and Liu, B. Y. H., Generation of Monodisperse Aerosol Standards, *Environ. Sci. Technol.*, Vol. 7, 1973, pp. 147–153.

88. Dodge, L. G., Comparison of Performance of Drop-Sizing Instruments, *Appl. Opt.*, Vol. 26, No. 7, 1987, pp. 1328–1341.

AUTHOR INDEX